T0317593

MATERIALS SCIENCE OF CONCRETE:

The Sidney Diamond Symposium

Other Volumes in the *Materials Science of Concrete* series:

Materials Science of Concrete V
Edited by Jan Skalny and Sidney Mindess
© 1998, ISBN 1-57498-027-0

Materials Science of Concrete IV
Edited by Jan Skalny and Sidney Mindess
© 1995, ISBN 0-944904-75-0

Materials Science of Concrete III
Edited by Jan Skalny
© 1992, ISBN 0-944904-55-6

Materials Science of Concrete II
Edited by Jan Skalny and Sidney Mindess
© 1991, ISBN 0-944904-37-8

Materials Science of Concrete I
Edited by Jan Skalny
© 1989, ISBN 0-944904-01-7

For information on ordering titles published by The American Ceramic Society, or to request a publications catalog, please contact our Customer Service Department at

Phone:	614-794-5890
Fax:	614-794-5892
E-mail:	customersrvc@acers.org

Visit our on-line book catalog at <www.acers.org>.

special volume

MATERIALS
SCIENCE OF
CONCRETE:
The Sidney
Diamond
Symposium

Edited By
Menashi Cohen
Sidney Mindess
Jan Skalny

Published by
The American Ceramic Society
735 Ceramic Place
Westerville, OH 43081

Proceedings of the Sidney Diamond Symposium held in Honolulu, Hawaii, August 30–September 3, 1998.

Library of Congress Cataloging-in-Publication Data

A CIP record for this book is available from the Library of Congress.

For information on ordering titles published by The American Ceramic Society, or to request a publications catalog, please call 614-794-5890

1 2 3 4–01 00 99 98

ISSN 1042-1122
ISBN 1-57498-072-6

Contents

Mechanical Properties, Testing, and Fiber Reinforcement

Rheology

Durability

Admixtures

Special Cements and Concretes

Foreword

Many of the friends and colleagues of Dr. Sidney Diamond gathered in Honolulu from August 30 to September 3, 1998, to attend a symposium in his honor. Fortunately, Sid was not retiring; he was merely slowing down a bit and redirecting some of his energies. However, after more than 35 years as a Professor of Civil Engineering at Purdue University, during which time he was a mentor and sometimes a father-figure both to his own graduate students and to many other younger cement scientists, it seemed an appropriate time to recognize his monumental contributions to the field.

During his career, Sid carried out research in many areas, including the fundamentals of cement chemistry, the structure of the hydration products, alkali-silica reactions interfacial phenomena, admixtures, cracking and fracture, and the use of the SEM in cement and concrete research. Indeed, the wide variety of topics covered in this symposium volume is a reflection of Sid's broad interests. In each of these areas, he provided a special insight, and those of us who were fortunate enough to work with him directly learned a tremendous amount not only about cementitious materials, but also about the methodology and integrity of scientific research.

When Sid began his academic career, concrete technology was taught primarily as a collection of more or less unrelated experimental "facts" and empirical "laws". It was through the pioneering efforts of Sid and a few of his contemporaries that the chemical and physical principles underlying concrete behavior began to form the basis for the education of the succeeding generations of students. To help spread this message, Sid was always generous with his time and with access to his laboratories, to all of those who wanted to benefit from his knowledge and experience.

This is not the place to enumerate all of the awards and honors that have been bestowed upon Sid over the years, or to try to summarize his several hundred technical papers. It is sufficient to say that Sid has really left his mark on the ways in which we now study concrete, and that his work has profoundly influenced all of us who continue to work in this area.

The papers collected together in this volume are, in a sense, our way of thanking Sid for all that he has done for us, both individually and collectively; we are truly grateful.

We would like to thank the sponsors and donors who helped to make this symposium possible. And, in particular, we would like to express our appreciation to Harriet Diamond, who was always there to provide a solid anchor to Sid's work and was always willing to open her home to the countless visitors to Purdue. Without her support, much of what we are celebrating here could not have occurred.

The editors would like to thank Catherine Ralston for her work in putting together this proceedings volume and for handling the details involved in organizing the conference.

Menashi Cohen
Sidney Mindess
Jan Skalny
August 1998

Sponsors and Donors

We would like to acknowledge and thank the following sponsors/donors of the Sidney Diamond Symposium:

American Concrete Institute

American Society for Testing and Materials

Canada Centre for Mineral and Energy Technology

Denki Kagaku Kogyo Ltd.*

Holnam Inc.*

Japan Cement Association

Japan Concrete Institute

Joint Transportation Research Program—Purdue University and Indiana Department of Transportation

Master Builders Inc.*

NSF Center for Science and Technology of Advanced Cement-Based Materials*

Portland Cement Association*

Precast/Prestressed Concrete Institute

Schlumberger Cambridge Research*

The Ceramic Society of Japan

Toagosei Co., Ltd.*

U.S. Army Engineer Waterways Experiment Station

U.S. Department of Transportation/Federal Highway Administration*

*Financial contribution provided

Biographical Notes of Professor Sidney Diamond

Sidney Diamond was born Nov. 10, 1929, in New York City. He attended local schools and was graduated from the Bronx High School of Science in 1946. He received his B.S. Degree from Syracuse University in 1950 in Forestry, and a Master's degree in Forest Soils from Duke University in 1951. After serving in the U.S. Army Corps of Engineers Research and Development Laboratories from 1951 to 1953, he worked as a research engineer in the laboratories of the Materials Division of the U.S. Bureau of Public Roads (now Federal Highway Administration) for a number of years. He obtained his Ph.D. from Purdue University in 1963 and joined the Purdue faculty as an Associate Professor in 1965. Since 1969 he has served as Professor of Engineering Materials.

In 1953 he was married to Harriet Urish. They have two daughters, Florence and Julia, and are immensely proud of their four grandchildren.

Professor Diamond has published more than 250 technical papers, mostly in the field of physical chemical and microstructural aspects of cement and concrete. He is a founding member of the Cements Division of the American Ceramic Society and is a Fellow of that Society and of the

American Concrete Institute. He is a long-time member of the Board of Editors of Cement and Concrete Research and has served on various national and international scientific boards and committees.

Dr. Diamond has lectured extensively in universities and technical institutes around the world and has been a visiting professor at a number of institutions including the Universities of Aberdeen, British Columbia, and Toronto; the Technical University of Denmark; the Imperial College of Science and Technology; and NIST.

Dr. Diamond has also maintained an active consulting practice. His clients have included a number of Fortune 500 companies, some nationally-known law firms, and various consulting companies and firms active in the cement and concrete industries.

MICROSTRUCTURE, POROSITY,

PERMEABILITY

CONCRETE POROSITY REVISITED

Sidney Diamond
Purdue University
West Lafayette, IN 47907, U.S.A.

ABSTRACT

Concrete is an intrinsically porous material, and the nature and extent of the various kinds of pores present strongly influence its properties. In this paper the writer attempts a synthesis of the results of various investigations carried out in his laboratory on (a) the microstructural characteristics of the pores found in concrete and their spatial distribution within the paste, especially with respect to the widely held but not necessarily correct concepts of the interfacial transition zone, (b) the proper quantification of the distribution of pore sizes in concrete, and (c) the nature and character of the solution occupying the pores in concretes.

INTRODUCTION

It is well understood that, unlike ceramics and metals, concrete is intrinsically a porous material, and the pore system plays a major role in determining the behavior of the material. The pore system may incorporate pore spaces in the aggregate as well as in the cement paste. Additionally, concrete inevitably incorporates air voids which, properly speaking, are neither paste nor aggregate but constitute a third phase which may be of considerable importance.

The nature, sizes, and spatial distributions of pores in the paste are well known to vary with the proportioning of the concrete (especially the water:cement ratio) with the hydration history, and sometimes with exposure to external solutions. Paste pore structure may be influenced by the use of superplasticizers, and by incorporation of various supplementary cementing components, especially silica fumes and fly ashes. It is usually considered that the paste pore structure is influenced by the nearby presence of aggregates; the existence of an interfacial transition zone (ITZ) of significantly higher porosity and greater pore interconnectivity around aggregate and sand grains is commonly assumed.

Unless concrete is artificially dried, most or all of the paste pores are occupied by pore solutions, usually containing significant concentrations of dissolved substances derived from the cement, from admixtures, and sometimes from external sources. Pore solutions may also penetrate some or all of the pores in the aggregate; they may also invade air voids, often leaving behind evidence of their presence in the form of calcium hydroxide, ettringite, and other precipitates.

The writer and his colleagues have devoted considerable effort over the years to elucidating some of the characteristics and details of pore systems in concrete. It seemed to him that this might be an appropriate occasion to revisit some of the findings produced in these investigations, and especially to document certain characteristics that are not commonly understood or appreciated.

HOW POROUS ARE CONCRETES?

Obviously, concretes vary in their porosity depending on the mix design, the porosity of the aggregate, the amount of air entrained or entrapped, and especially the porosity of the cement paste (including the interfacial zone). Variations in aggregate porosity and in the air void content depend on decisions made by the concrete producer in his selection of materials, and on the details of the mix design selected for a particular concrete. The paste porosity in concrete is, of course, a function of the water:cement ratio and of the degree of hydration.

Many years ago Winslow and Diamond (1) carried out measurements of total pore volume of cement pastes of various water:cement ratios and degrees of hydration, for comparison with mercury intrusion results. To do this saturated pastes were weighed in air and in water; the solid volume is given by the bouyant loss (difference between weights in air and in water). The bulk volumes of the same samples were measured in a mercury intrusion penetrometer after evacuation but before applying pressure to intrude the pores. The porosity measurement is provided by the difference between bulk volume and solid volume. The results, as expressed in cm^3/g of oven dry weight, are provided in Table I.

Table 1. Total Porosities Measured in Hydrated Cement Pastes (cm^3/g)

Age, days	w:c 0.40	w:c 0.60
1	0.32	0.41
7	0.26	0.36
60	0.22	0.32

It is evident that paste pore volume decreases significantly with hydration, and that for a reasonably mature concrete the water:cement ratio exercises a controlling influence, facts well known in concrete technology.

POROSITY IN CONCRETE AS REVEALED BY MICROSTRUCTURAL INVESTIGATION

Estimation of the content of pores in the paste of a particular concrete provides helpful but necessarily incomplete information. Details of the pore structure, especially as conveyed by visual examination at appropriate magnifications, are necessary to properly understand the pore structure. An attempt at a visual exposition of some of the features of the pore system in a 'generic' concrete is provided below, using backscatter SEM micrographs.

Fig. 1 shows a backscatter SEM of a 3-day old w:c 0.50 lab-mixed and cured concrete, taken at a magnification low enough to see the character of the concrete as a composite material. The aggregate in this concrete was a crushed limestone of 12.5 mm maximum size. The sand was crushed from the same material, and its size distribution falls approximately midway in the envelope used to delineate acceptable size distributions in the ACI C-33 concrete specification. In other words, this is a 'practical' concrete, although laboratory mixed, and the image is selected to focus attention on the first-order features of ordinary, conventional concrete.

Much of the area of Fig. 1 is occupied by sand-sized aggregate grains, which are separated to varying extents by bands of cement paste. One might note that the limestone is porous to some degree, with isolated black pores visible in many, but not all, of the individual sand grains. Some of these pores are seen to exist as elongated channels. At the magnification used, only relatively large pores can be seen in the aggregate; finer pores may also be present.

This aggregate was prepared from a homogeneous material. Many concrete aggregates, especially gravels, are intrinsically heterogeneous. Thus, aggregate grains with pores of various size ranges (and with varying degrees of interconnectedness) may be expected in many real concretes. Comparatively little attention has been paid to this aspect of concrete porosity in the literature. The subject deserves serious study.

The next largest obvious feature in Fig. 1 is the presence of perhaps a dozen empty air voids, ranging in size from perhaps 10 μm to about 100 μm in this particular figure, and up to half a millimeter in many concretes. No air entraining agent was used in this concrete; accordingly the air void content is small, perhaps 2% by total volume. Air voids are much more prominent in air-entrained concrete and may constitute ca. 6% of the total volume or about 15% of the total space between the aggregate grains. Air voids tend to be isolated from each other, although at high air contents there is some clustering. They usually develop a lining of calcium hydroxide that partly isolates them from paste pore solutions; linings in the larger air voids are visible even at the low magnification of Fig. 1. However, the isolation of air voids from the pore solution is imperfect, and deposits of calcium hydroxide, ettringite, and other substances into many air voids attest to the penetration of solution into air voids in many field concretes.

Finally in Fig. 1, the composite character of the paste itself is obvious. The bright grains are residual unhydrated cement particles; their prominence in Fig. 1 indicates that this is a young concrete. As hydration proceeds the residual grains grow progressively smaller and fewer in number. The ill-defined gray mass between the bright cement grains is, of course, the hydrated cement paste. At this age and water:cement ratio it contains a relatively large proportion of pores, although they are difficult to distinguish at the magnification employed.

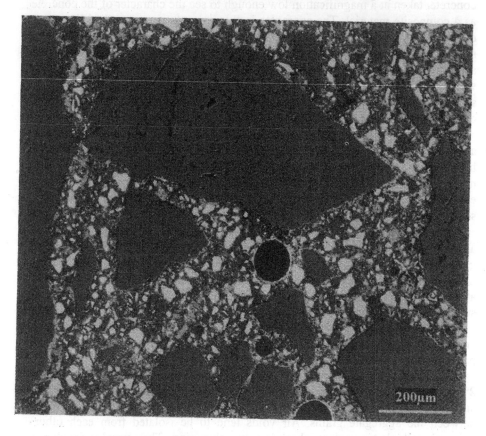

Fig. 1. Low magnification SEM micrograph of 3 day-old w:c 0.50 concrete.

Fig. 2 is presented at a significantly higher magnification, so as to show details of at least the larger paste pores. The figure was taken from a cement paste specimen rather than from concrete, but the paste is of the same age and water:cement ratio as the concrete of Fig. 1. At this magnification the bright residual cement grains are seen to be surrounded or partly surrounded by a thin shell of dense C-S-H gel. The cement grains are obviously not porous; the surrounding dense hydration product is porous, but the pores are too fine too see at this magnification. The dense hydration product ('inner product' or 'phenograin C-S-H') is obviously different from the visibly porous 'groundmass' C-S-H, deposited in what was originally water-filled space between the cement grains.

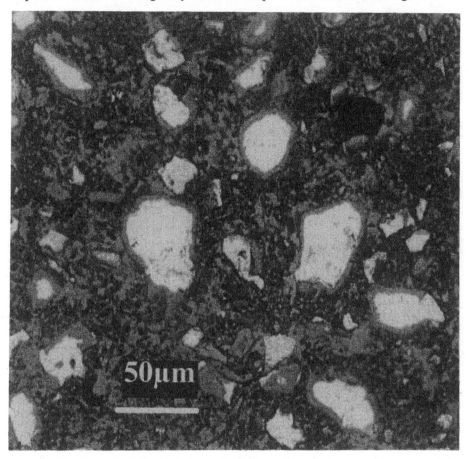

Fig. 2. Higher magnification backscatter SEM micrograph of 3-day old w:c 0.50 cement paste.

The pores within the groundmass mostly correspond to the classical definition of capillary pores, and range in size up to ca. 10 μm. Naturally as hydration proceeds the content and sizes of these pores tend to diminish, and the visual distinction between the not-visibly porous shells of phenograin C-S-H and the obviously porous groundmass C-S-H tends to become less pronounced.

Fig. 2 also contains a single large air void.

Fig. 3, at a still higher magnification, was obtained from the same concrete as Fig. 1, and represents the paste adjacent to the middle of the long sand grain at the left side of that figure.

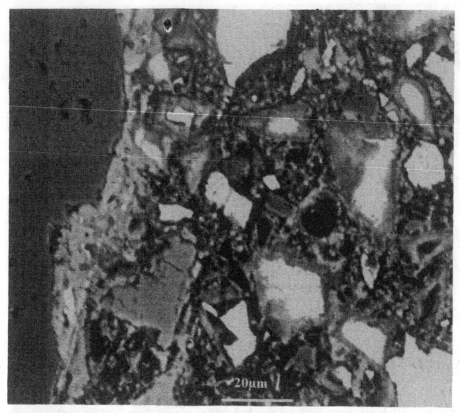

Fig. 3. Still higher magnification SEM of a region of Fig. 1.

Close examination indicates that not all of the pores within the paste are direct remnants of original water-filled space. A number of them, such as the pore within the triangular grain at the extreme right, represent space hollowed out from the interior of certain hydrating cement grains, in the variant mode of hydration known as 'Hadley grain' or 'hollow shell hydration grain' formation. The pores within some Hadley grains may be annular spaces surrounding interior cores of unhydrated clinker material, rather than completely hollowed out spaces.

In our experience, pores derived from Hadley grains are <u>always</u> present in concrete or cement paste, although they are not always prominent and easy to recognize. They do show up particularly well in mature low water:cement ratio concrete where the sizes of the capillary pores are much reduced but many Hadley grain spaces remain in the 5 - 10 μm range. The distinction between Hadley grain spaces and ordinary capillary pores may be important; there is reason to believe that Hadley grain spaces pores are less well interconnected with the rest of the paste pore system than the capillary pores per se.

For many years the pores with hydrated cement paste have been traditionally classified into "capillary pores" and "gel pores", the latter thought to be of almost molecular sizes, i.e. in the 1 - 3 nm size range. Gel pore space is thought to be associated with (and form the boundaries of) the high internal surface measured for hardened cement paste by various techniques. Water (or pore solution) in gel pores is presumably evaporable by low temperature drying. However there is no clear distinction between gel pore water and more firmly bound water constituting part of the C-S-H gel structure, and the nature and content of gel porosity in concrete remains difficult to assess. Gel pores are certainly not visible using backscatter SEM or optical microscopic techniques, although high-magnification TEM might provide some information. Gel pores presumably play little part in fluid transmission or ion diffusion through concrete.

POROSITY NEAR AGGREGATE SURFACES – THE INTERFACIAL TRANSITION ZONE

For many years it has been believed that cement paste in the immediate vicinity of sand or coarse aggregate particles is somehow different from 'bulk' cement paste far removed from aggregates. The zone over which these differences can be observed is now universally designated as the interfacial transition zone or ITZ. Many treatments and various detailed models have been constructed to assess and quantify the various effects expected from the differences between the ITZ paste and the non – ITZ or bulk paste in concrete.

The special characteristics of the ITZ are considered to be the result of a so-called 'wall effect', i.e. a partial exclusion of ground cement grains from the areas adjacent to the aggregates and sand grains. The partial exclusion of cement grains leaves such areas locally richer in water and poorer in cement; accordingly, the cement paste is thought to develop a greater local content of capillary pores. In consequence these zones are generally considered to be zones of weakness, making the concrete weaker than it otherwise would be.

ITZ paste is considered to be locally more permeable than bulk cement paste in the same concrete. If the gradation of the concrete sand is fine enough so that a high population of sand grains exists, the individual ITZs surrounding adjacent sand grains are thought to overlap significantly. Under such conditions the entire pore system is thought to 'percolate', i.e. the concrete is thought to be significantly more permeable than it otherwise would be.

Over the last several years at Purdue University we have carried out detailed microstructural examinations of ITZs and corresponding bulk pastes in concrete using image analysis and other means. A summary of our findings was recently published in the Proceedings of the just-concluded RILEM International Conference on The Interfacial Transition Zone in Cementitious Composites (2).

Our results confirm the existence of the wall effect, and the extent of the partial exclusion of cement grains from the vicinity of the aggregates. Narrow bands of paste immediately surrounding the aggregate grains have typically only one-third or one-fourth the content of unhydrated cement grains than bands of paste 50 μm or more from the nearest aggregate. The effect is statistically very strong, and there is an approximately linear increase in content of unhydrated cement with distance starting from the aggregate and extending for about 50 μm into the surrounding cement paste. Examples of the trend are plotted in Fig. 4, for young

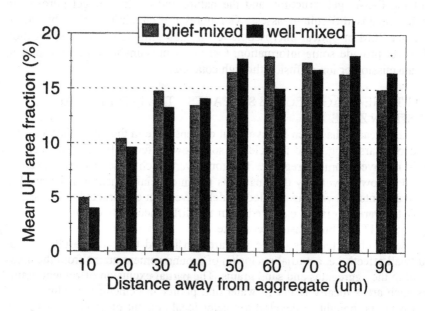

Fig. 4. Illustration of the ITZ wall effect: variation in content of unhydrated cement grains with distance from the aggregate, as recorded in separate 10 μm – wide strips.

(3 day old) briefly-mixed and very well-mixed quartzite aggregate concretes, respectively. The degree of mixing obviously has little influence.

It should be kept in mind that these are averaged data, broadly representative of a given concrete. Examination of Fig. 1, and indeed of almost any concrete micrograph reveals the presence of some unhydrated cement particles adjacent to sand grains and others are found even in narrow channels where sand grains approach each other closely. The wall exclusion effect is statistically pronounced but is not complete.

However, the significance of the ITZ in the present context is not the extent of exclusion of cement particles, but rather its effect on local pore systems adjacent to aggregates. We found such an effect, but only a limited one, in all of the concretes examined. Previous results reported by Scrivener and various co-workers (for example, (3)) suggested that very high pore contents, as much as 25 to 30% of the paste, were to be found in the zone of the ITZ nearest the aggregate. This was very much higher than the 7 to 10% of pores found by these authors in the bulk paste of the same concretes. Our results are in quite good agreement with respect to the pore content found in bulk paste, but show comparatively little extra pore space near the aggregate surfaces. Indeed, we find so little extra pore space in the ITZ that its importance seems to be marginal.

Fig. 5 provides typical examples of our results showing the pattern of % of pore space with distance from the nearest aggregate for young (3 day old) and mature (100 day old) well-mixed quartzite aggregate concretes. The percentage of detectable pores is higher near the interface than in the bulk paste, but even in the innermost zone it is only of the order of 10% or 13%, depending on age.

In reading the description accompanying Fig. 5, it may have occurred to the reader to wonder why the writer has bothered to specify that the concretes in question were "well-mixed". The reason is that we had included a comparison of the extent of mixing in our studies. In one extreme the concretes were mixed only 30 seconds, albeit in a very effective mixer. In contrast the "well-mixed" concretes were mixed for 10 minutes, and then lightly consolidated by vibration. Unexpectedly, we found that the degree of mixing had a much larger effect on the percentage of pores detected than location within or away from the ITZ. The brief mixing generated a cement paste structure that was much more visibly porous throughout the paste than thoroughly mixed concrete of the same components.

Fig. 6 shows data for contents of detectable pores in briefly mixed and well mixed 3-day old quartzite aggregate concretes. The detectable pores in briefly-mixed concretes was as much as 17% in bulk pastes. There was some excess pore space within the ITZ, but the value reached only to about 21% near the actual interface. Thus the relative increase of pore content near interface compared to the bulk paste was even smaller for briefly-mixed concrete than for well-mixed concrete.

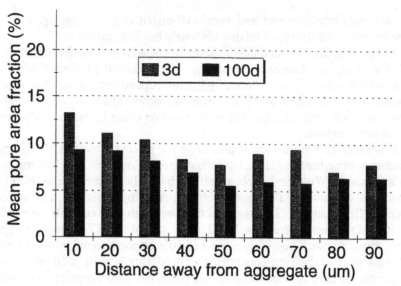

Fig. 5. Variation in area % pores. with distance from the aggregate, as recorded in separate 10 μm – wide strips in well-mixed 3-day old and in 100-day old quartzite aggregate concrete.

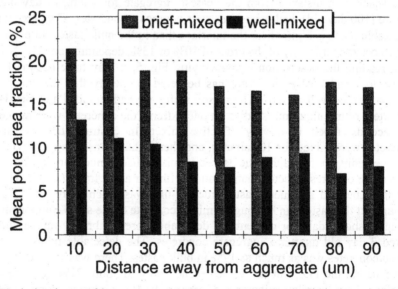

Fig. 6. Variation in area % pores vs. with distance from the aggregate, as recorded in separate 10 μm – wide strips in well-mixed and brief-mixed 3-day old quartzite aggregate concrete.

Our results thus indicate that the ITZ in conventional concrete, although statistically deficient in unhydrated cement grains, is not a zone of much higher porosity than the bulk paste. It appears that in some areas within the ITZ, some of what would otherwise be excess pore space is filled with deposits of calcium hydroxide, mostly nucleating on the aggregate surface itself. More generally, it appears that groundmass C-S-H gel, deposited from solution, tends to accumulate in what otherwise would be empty space in the ITZ. Indeed, our data indicate that the contents of undifferentiated C-S-H (determined after subtracting unhydrated cement, pores, and calcium hydroxide) is uniform throughout a given concrete, and shows no variation at all with distance from the aggregate.

Our findings as reported in (2) also raise a important point with respect to the local spatial distribution of porosity within the cement paste in concrete. Our data indicate that cement paste porosity is distributed in a highly non-uniform fashion. Patches of locally porous material alternate in highly irregular fashion with patches of less porous material. This is true of the bulk paste; it is also true within any particular shell within the conventional ITZ. An illustration of the effect is provided in Fig. 7, which records the pore contents of adjacent sampling units taken end-to-end in a strip next to and running completely around a representative sand grain. The average area % of detectable pores for the full innermost 10 μm ITZ strip is a little over 8%. However, as seen in Fig. 7, the area % detected within any particular sampling unit may be as low as 4% or as high as 13%.

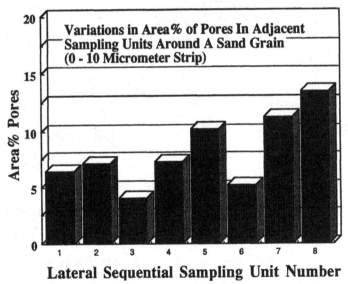

Fig. 7. Variation in area % pores in successive lateral sampling units comprising the 0 – 10 μm strip adjacent to and running entirely around a sand grain.

It should be noted that each sampling unit is about 250 μm long and 10 μm wide and contains about 12,500 pixels; this is more than enough that statistical error in evaluating its pore content is not a problem.

Even within an individual 2500 square μm sampling unit, porosity is not uniformly distributed; there are patches of small collections of pores here and there. Fig. 8 is a reasonably representative sampling unit of an innermost strip next to a sand grain. The backscattered information has been subjected to pixel-by-pixel binary segmentation, with the pixels representing pore space turned black and those representing solid matter turned white. In some sampling units there are local clusters of pores in linear strings at the actual interface with the sand grain, the pores signaling local failure of the paste to fill in completely against the aggregate.

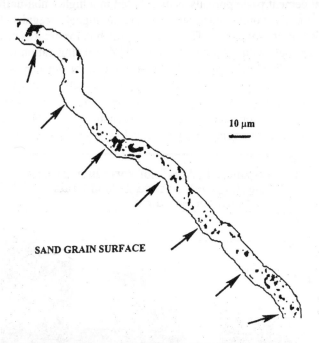

Fig. 8. Segmented image showing the distribution of pore pixels in an individual 0 – 10 μm wide sampling unit next to a sand grain.

The general picture that emerges from these considerations is that the occurrence of statistically excess porosity can be demonstrated within the ITZ, but the magnitude of the statistical excess is small, and the local porosity is distributed irregularly. Under these circumstances the magnitude of the effect of the ITZ on concrete properties might be much smaller than usually anticipated.

The results of careful studies carried out recently by Prasada Rangaraju at Purdue University (4) suggest that this may very well be the case. Working under the supervision of Professor Jan Olek, Rangaraju prepared several suites of concretes similar in all respects except for the grain size distribution of the sands. The sand grain size distribution was deliberately varied to produce concretes of average inter-aggregate distances varying from only about 60 μm at one extreme to as much as 150 μm at the other. The paste in the concrete with closely-spaced aggregate would be expected to be nearly all influenced by ITZ effects; that in the concrete with widely-spaced aggregate much less so. Rangaraju measured compressive and tensile strengths and rapid chloride permeability, among other concrete properties. He found essentially no quantifiable effect of inter-aggregate spacing on any of the concrete properties measured.

SIZE DISTRIBUTION OF PORES IN CONCRETE

For about 30 years the standard method of attempting to determine pore size distributions in cementitious materials has been mercury intrusion porosimetry, and papers reporting MIP size distribution measurements continue to appear on an almost monthly basis. Actually most determinations are carried out on cement paste or mortar. It is extremely difficult to carry out determinations on concrete, because of the small specimen size required by MIP instrumentation. It not usually possible to secure small concrete samples that have proportions of paste and aggregate that is representative of the whole.

Unfortunately, the pore size distributions obtained with MIP are grossly in error for cementitious materials, and underestimate the true range of sizes of pores present by several orders of magnitude. The MIP method assumes unrestricted access of mercury to all pores, such that the pores can be successively filled with mercury in decreasing order of size by the effects of successively increasing pressures. In fact, access by intruding mercury to the pore systems of cement pastes and concretes is severely restricted; access to nearly the entire interior pore assemblage does not take place until a very high 'breakthrough pressure' is reached. As soon as this breakthrough pressure is reached, most of the interior space is flooded with mercury. The result is an intrusion vs. pressure curve that is not properly translatable into an intrusion vs. pore diameter curve through the Washburn equation, as is commonly assumed.

Several years ago Diamond and Leeman (5) reported the results of a series of experiments in which MIP size distributions of a variety of cement pastes were compared with size distributions determined by an image analysis system from backscatter SEM micrographs of the same pastes. The results of the two methods were wildly divergent, as is illustrated below.

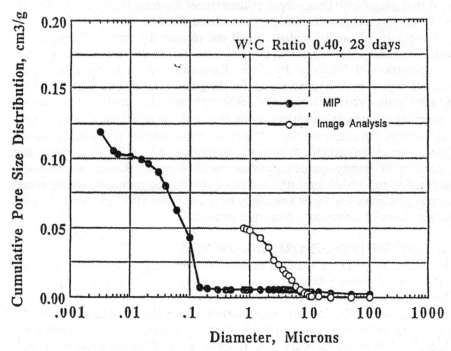

Fig. 9. Comparison of MIP and image analysis pore size distributions for a 28-day old w:c 0.40 cement paste.

Fig. 9 shows MIP and image analysis cumulative pore size distributions for a 28-day old w:c 0.40 cement paste, the image analysis results being limited to sizes above 0.8 μm by the limitations of the system used.

The MIP results indicate that appreciable porosity exists only in sizes below about 0.15 μm, in common with the usual MIP results for such pastes. Nearly all the pore space tallied is in the narrow range between the 0.15 μm value and about 0.01 μm. In contrast, the image analysis results show pore space in sizes up to about 10 μm; a great deal of pore space is tallied between 10 μm and 1 μm. Visual examination of the SEM micrographs, similar to Fig. 2, shows clear evidence of the extent of these 1 – 10 μm sized pores.

The cement paste of Fig. 9 was vacuum mixed to avoid entraining air voids. In a critical experiment, identical pastes were prepared with a heavy dose of air entraining agent to entrain a significant quantity of air voids. SEM examination confirmed the physical presence of these voids in sizes up to several hundred μm, as expected. The air-entrained pastes were then also examined by the MIP and image analysis procedures.

Fig. 10 compares the image analysis pore size distributions for the air-entrained and vacuum-mixed pastes. The extra porosity induced by the incorporation of the air voids shows up clearly in the size range between 10 μm and 300 μm, a range

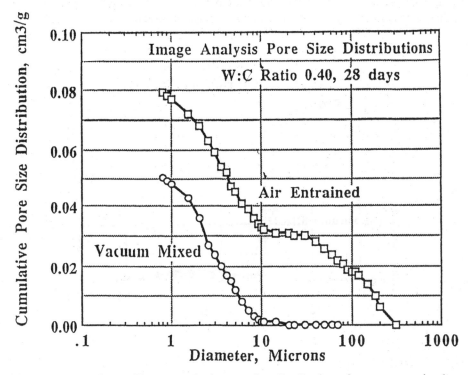

Fig. 10. Comparison of image analysis pore size distributions for vacuum-mixed and for air-entrained 28-day old w:c 0.40 cement pastes.

of sizes that is appropriate air voids. In contrast, as seen in Fig. 11, the MIP results do not indicate the presence of any air voids in this size range. No extra pore space is detected in the air entrained pastes at any intrusion pressure above that of the breakthrough point. Instead, the extra volume of pore space contributed by the entrained air voids is tallied in the same 0.15 − 0.01 μm range as are the capillary pores. This is obviously an absurd result.

Recently Willis et al (6) studied the intrusion of molten Wood's metal under successively higher pressures into w:c cement pastes,. After freezing the metal at each stage, they used image analysis to examine the sizes of the pores intruded by the Wood's metal, and the content and sizes of unintruded pores. They found that general intrusion into the main mass of the paste sample did not take place until a threshold pressure of approximately 1,500 psi was reached; this was about the same value of threshold pressure required to generate the breakthrough intrusion in MIP. At pressures less than this, large voids in the sample that did not happen to be connected by open cracks remained unintruded. Intrusion with Wood's metal was sensibly complete at an intrusion pressure of 5,000 psi. Image analysis of the pores intruded by Wood's metal after intrusion pressures in excess of the

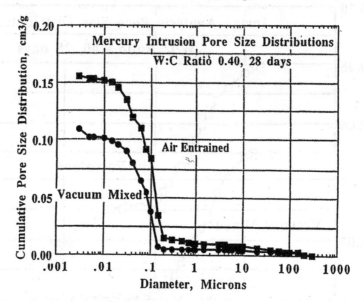

Fig. 11. Comparison of MIP pore size distributions for vacuum-mixed and for airentrained 28-day old w:c 0.40 cement pastes.

breakthrough pressure showed a distribution of sizes ranging in diameter from about 9 μm down, i.e. the same size range found by visual examination of micrographs and by image analysis of unintruded samples.

The deficiencies of MIP pore size distributions are often noted in the literature and more or less casually dismissed as resulting from the "ink-bottle" effect. Indeed, this is the description offered by Willis et al. (6). In point of fact the problem is much more fundamental than that. The ink-bottle effect refers to the failure of an *individual* pore to fill with mercury until the pressure necessary to intrude its entrance diameter is reached. The problem with MIP for cementitious materials is the failure of the mercury to obtain access *to essentially the whole interior* of the sample until a general breakthrough pressure is reached; the entrance diameters of individual pores are not really at issue.

Unfortunately, image analysis methods only provide part of the solution to the problem of obtaining proper size distribution information. The size distributions obtained are correct as far as they go, but the minimum pore size that can be tallied must be at least several times the size of a single pixel of the image. Pixel size depends on magnification and on the resolution at which the image is acquired. Useful magnifications much larger than about 2,000x are not readily obtained with current-generation backscatter detectors. Images can readily be acquired at higher resolutions than the commonly used 512 x 512 or 1024 x 1024 pixels, but the mathematical overhead in image processing rapidly goes up with resolution. Accordingly, extension of the range of image analysis pore size distributions down to ca 0.01 μm does not seem currently feasible.

Despite this lower bound limitation, comparisons of reliable pore size distributions obtained for cement pastes and concretes by image analysis can be of significant interest. For example, Fig. 12 presents the results of such comparisons

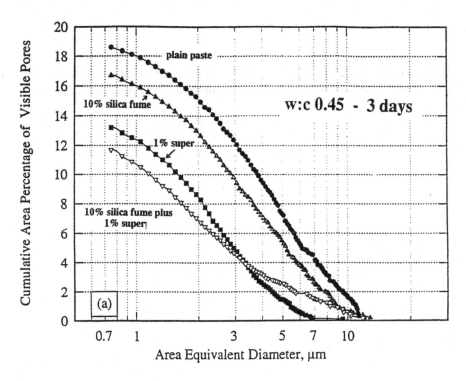

Fig. 12. Image analysis pore size distributions for 3-day old w:c 0.5 pastes: plain, 10% silica fume replacement, 1% superplasticizer, and a combined treatment of 1% superplasticizer and 10% silica fume.

for several mature w:c 0.45 pastes obtained several years ago in the writer's laboratory (7). The comparisons involve plain paste, paste with 10% silica fume replacement, superplasticized paste, and paste with a combined superplasticizer – silica fume treatment. It is seen that while the silica fume treatment refines the pore structure slightly, use of the superplasticizer, alone or in combination with silica fume, has a much greater effect in reducing both maximum size of pores detected and percentage of coarse pores. The superplasticizer and superplasticizer-silica fume distributions are not identical, the combined treatment producing a distribution with substantially higher contents of coarser pores and substantially lower contents of finer pores in the size range covered.

THE CHEMISTRY OF CONCRETE PORE SOLUTIONS

In extensive perusals of the nature and structure of the pore system of concretes such as constitute the bulk of this paper, it is easy to lose sight of the fact that the pore system is in fact not a dry geometric construct, but a internal storage system for pore solutions derived originally from excess mix water, and modified by various chemical processes inherent in cement hydration. Indeed, the pore system is a venue for ongoing chemical reactions that may strongly affect the service life and functioning of the concrete. Freezing damage, alkali aggregate problems, sulfate attack, and steel corrosion problems all take place through the media of concrete pore solutions.

As stated earlier, the pores in concretes are never completely dry unless the concrete is artificially heated. Under conditions of cold weather exposure the pore solutions may be partially frozen, but much of the solution remains unfrozen down to exremely low temperatures. In dry climates, surface concrete layers dry out but the interior portions of concretes retain considerable pore solution; indeed the undersides of slabs on ground may remain wet even in dry climates like that of Arizona.

Studies of concentrations of dissolved substances in concrete pore solutions have been extensively pursued for many years. Many laboratories have constructed pore solution expression devices patterned after that of Barneyback and Diamond (8) and have obtained useful results with them.

When first mixed, concrete mix water quickly acquires appreciable concentrations of dissolved alkali, calcium, sulfate, and hydroxide ions from the cement, and sometimes additional contributions from any soluble admixtures that may be used. Only trace amounts of silicon, aluminum and iron species are found. As a general rule the formation of ettringite rapidly depletes the sulfate content, and the precipitating sulfate ions are replaced by an equivalent concentration of OH⁻ ions. The increased OH⁻ ion concentration reduces the concentration of calcium ions to negligible amounts. As a result of these interactions, usually by the end of one day the pore solution consists of a high concentrations of potassium and sodium hydroxide and very little else unless soluble admixtures have been used. To the degree that the concrete is impermeable or otherwise isolated from its surroundings, such solution compositions may be maintained indefinitely.

These solutions are extremely alkaline; pH values of the order of 13.5 to as much as 14.0 are often found, depending on the alkali content of the Portland cement used. In most modern cements, potassium contents far exceed sodium contents, and this is reflected in the relative proportions of the two alkali hydroxides found in solution.

Admixtures may or may not remain in solution for long periods of time. Organic admixtures, particularly superplasticizers, tend to be rapidly removed from solution by sorption into newly-forming hydration products. For superplasticizers, this can result in premature slump loss if the dosage is low. Interestingly, it has been found that sodium-neutralized superplasticizers leave their ionizable sodium behind in solution when the organic portion of the admixture is sorbed; the sodium remains behind as the hydroxide, and thus increases the alkalinity of the pore solution (9). When calcium chloride is used as an admixture, depending on the dosage and on the C_3A content of the cement, significant amounts of chloride can remain in solution indefinitely (10), thus almost guaranteeing prospective steel corrosion problems.

Some admixtures are designed to protect concrete against various long-term environmentally-induced problems. Proper functioning of such admixtures depends on retaining a significant proportion of the added substance in solution indefinitely. Examples of such admixtures include calcium nitrite used as a steel corrosion inhibitor, and various lithium salts used to inhibit the effects of alkali aggregate reactions. Expression and analysis of pore solutions over time can provide useful information concerning the dosages needed to insure long-term retention of these protective substances in the pore solution.

Supplementary cementing components such as silica fume tend to react with dissolved alkali hydroxide, (as well as with solid calcium hydroxide), thus reducing the pH of the pore solution. Such responses lower the risk of long-term alkali aggregate reactions, and indeed supplementary components based on dehydroxylated kaolinite clay have been marketed expressly for that purpose. With a heavy dosage, the pore solution pH can approach the 12.4 value below which calcium hydroxide becomes unstable, and the remaining crystalline calcium hydroxide may dissolve, further reducing the risk of alkali silica reactions.

Occasionally it is found that concrete that has been partly dried (in the presence of atmospheric carbon dioxide) and then rewetted shows lower than expected concentrations of alkali hydroxide; the drying has resulted in partial 'fixation' of the alkali hydroxide. This effect has been studied in detail in laboratory investigations (11) and may be a factor in limiting near-surface alkali aggregate attack on concrete members.

Recently the writer and various colleagues have had occasion to study the details of sulfate attack on particularly porous high water:cement ratio concretes exposed to sulfate-bearing ground waters. In these concretes appearance of crystalline sodium sulfates (thenardite and mirabilite) as deposits on the upper surfaces of concrete slabs on grade attest to the penetration of the ground-water derived substances entirely through the concretes. Pore solutions in such concretes are drastically altered from the normal expectation. They are found to contain very little potassium, and much lower OH⁻ ion concentrations than usual. Calcium hydroxide is found to have been dissolved out of portions of the concrete. In a particular concrete in one site that has undergone exposure to sulfate-bearing ground water for more than 20 years, the pore solution is so altered as to be completely unrecognizable as a concrete pore solution, and exhibits a pH value close to 7.

CONCLUDING REMARKS

Concrete research has provided the writer with a fascinating field of activity for many years, much of it concerned with microstructure and pore structure. He hopes that the informal synthesis provided in this paper may be useful to those similarly addicted to the study of the endless complications exhibited by this "simple" material.

The writer is greatly appreciative of the honor done to him in the organization and dedication of this Symposium. He is grateful to his colleague, Professor Menashi Cohen for conceiving the idea and for organizing and administering the symposium; to his many friends and colleagues who have assisted in the organizing effort; and to all those who have provided technical contributions. He is particularly indebted to Dr. Jan Skalny and Professor Sidney Mindess for their editorial contributions and for suggesting that the proceedings be incorporated into the well-known series on the Science of Cement and Concrete

ACKNOWLEDGMENTS

Much of the research that formed the basis for this synthesis was supported by the U.S. National Science Foundation as part of the program of the Science and Technology Center for Advanced Cement Based Materials (ACBM). The writer is grateful for this continued support. He is pleased to acknowledge the contributions to this synthesis made by his present and former Purdue University colleagues, Professors W. L. Dolch, M.D. Cohen, Jan Olek, and D. N. Winslow, by Mrs. Janet Lovell and Mr. Jingdong Huang, and by many former graduate students.

REFERENCES

[1] D. N. Winslow and S. Diamond, J. Materials 5, 564 (1970).

[2] S. Diamond, Proc. RILEM Intern.Conf. on the Interfacial Transition Zone in Cementitious Composites, Haifa, 1998, A. Bentur, ed. (1998).

[3] K L .Scrivener and E. M. Gartner, Matls. Res. Soc. Proc. 114, 77 (1988).

[4] P. Rangaraju, Ph.D. thesis, School of Civil Engineering, Purdue University (1997).

[5] S. Diamond and M. E. Leeman, Matls. Res. Soc. Proc. 370, 217 (1994).

[6] K. L. Willis, A. B. Abell, and D. A. Lange, accepted for publication in Cem. Concr. Res. 28 (1998).

[7] Y. Wang, Ph. D. Thesis, School of Civil Engineering, (1995).

[8] R. L. Barneyback, Jr. and S. Diamond, Cem. Concr. Res. 11, 279 (1981).

[9] K. Matsukawa and S. Diamond, Ceramic Transactions 16, S. Mindess, ed., 41 (1991).

[10] S. Diamond, Cem., Concr., and Aggregates, 97 Winter (1986).

[11] D. Constantiner and S. Diamond, in "Mechanisms of Chemical Degradation of Cement Based Systems, R. L. Scrivener and J. F. Young, eds., E & F. N. Spon, London (1997).

INFLUENCE OF CURING CONDITION ON THE MICROSTRUCTURE FORMATION IN HIGH-STRENGTH CONCRETE

Shunsuke HANEHARA and Hiroshi UCHIKAWA
Chichibu Onoda Cement Corporation,
2-4-2, Ohsaku,Sakura, Chiba, Japan 285-0802

ABSTRACT

Microstructure of high-strength concrete prepared at low water cement ratio is quite different from that of ordinary concrete. Both steam and autoclave curing applied to high strength concrete also have a large influence on the structure formation of concrete. Heat curing at 250-400°C is applied to manufacture RPC - reactive powder concrete - which is ultra-high strength and ductile properties. In order to study the influence of water cement ratio, curing condition and addition of silica fume on the hydration of cement and microstructure formation, and to discuss the relationship between microstructure and compressive strength, the high strength mortars prepared at W/C=0.25 with S/C=1 cured at 20°C, 250°C, 80°C-steam and 180°C-10atm were characterized and stress-strain properties and clack propagation of high strength mortar were evaluated.By the steam curing, the strength of SFC was remarkably increased through the acceleration of cement hydration and pozzolanic reaction. The strength of mortar after steam-curing was ,however, hardly increased though it was additionally autoclave-cured and oven-cured. The linearity of stress-strain curve of mortar under the loading of compressive strength was much improved by temperature-increasing curing. The static modulus of elasticity of the SFC mortar was lower than of NPC mortar especially in temperature-increasing curing. The reason for these phenomena were discussed in relation to the microstructure-related characters including porosity, pore size distribution, crack of mortar, and kind, crystallinity, quantity and rearrangement of produced hydrates.

INTRODUCTION

Temperature-increasing curing including steam and autoclave curings is widely used for quickly developing the strength of concrete in the manufacturing process of the concrete products. Reactive powder concrete (RPC) which has recently been developed[1] for producing ultra-high strength concrete product is cured at a temperature from 250 to 400°C under normal pressure. Some papers concerning the pore structures of hydrate and hardened mortar formed during the curing have been reported[2, 3]. In those papers, however, the mechanical properties including strength and modulus of elasticity and microstructures of the hardened mortar are hardly described.

This paper traces the effects of the mixture proportion of mortar, curing conditions and addition of admixture on the hydration reaction of cement and the composition and formation of microstructure of cement paste, correlates those characters with the compressive strength and stress-strain characteristics of hardened mortar. Based on those data, the relationship between the microstructure of high-strength mortar and its properties were discussed.

SAMPLE AND EXPERIMENTAL METHODS
Preparation of samples and test specimens

Normal portland cement (NPC), silica fume cement (SFC) containing 10% of silica fume and Ogasa sand were used to prepare the test specimens of mortar with the dimensions of 4 x 4 x 16 cm at W/C ratio of 0.25 and S/C of 1.0. Character of these materials are listed in Table 1. Two percent (2.0%) of a naphthalene sulfonic acid-based high-performance water-reducing agent was added to the mixture for keeping the fluidity of fresh mortar. Cylindrical test specimens 10 cm high and 5 cm in diameter were prepared from the same mixture as above for measuring the compressive strength, static modulus of elasticity and strain.

Table 1 Character of cement, silica fume and sand

	Chemical composition (%)								Specific surface Area
	SiO_2	Al_2O_3	Fe_2O_3	CaO	MgO	SO_3	Na_2O	K_2O	
Normal portland cement (NPC)	21.6	5.3	3.3	64.9	1.1	2.3	0.45	0.35	$3,290cm^2/g^*$
Silica fume	96.0	0.0	0.4	0.8	-	-	0.21	0.45	$18.3 \, m^2/g^{**}$

*; Blaine specific surface area   **; BET surface area

	Paticle size distribution (Residue ,%)						Specific gravity (g/cm^3)
	5mm	2.5mm	1.2mm	0.6mm	0.3mm	0.15mm	
Sand (Ogasa)	0.1	11.2	43.5	65.3	79.2	91.9	2.55

The mortar was released from the mold after cured at 20°C for one day in wet air (briefly "1-day standard curing at 20°C"), then steam-cured at 80°C for three days (briefly "steam curing") and heat-cured at 250°C for one day in a dryer (briefly "oven curing") or cured at 180°C under 10 atm for 12 hours in an autoclave (briefly "autoclave curing"). A test specimen was also prepared by water curing at 20°C for 27 days (briefly "28-day standard curing at 20°C") after releasing the mortar from the mold for comparison.

Experimental methods
Physical property tests

The strength of hardened mortar was measured according to JIS R-5201. The strain of it was measured by sticking two strain gauges 30 mm long in the center of the side of cylindrical test specimen.

Characterization of hardened mortar

The sample was prepared by cutting cubes with approximately 5 mm of each side out of the hardened mortar with a diamond cutter and immediately dipped into acetone to terminate the hydration and D-dried. The sample for electron probe microanalysis (EPMA) was prepared by grinding the surface of one of them. The samples for the other measurements were prepared by pulverizing them and adjusting the particle size to the specific value. The particle size distribution of the sample for the specific surface area with a scanning BET apparatus was adjusted to 1.0 to 0.5 mm. The hydrate was identified with a rotating cathode type high-power powder X-ray diffractometer. A differential scanning calorimeter was used for determining $Ca(OH)_2$. The pore structure was determined with a mercury-intrusion porosimeter and the microstructure was observed and the composition of it was determined with EPMA. The combined water was measured by the ignition loss test and the insoluble matter was determined according to JIS R-5202.

EXPERIMENTAL RESULTS AND DISCUSSIONS
Physical properties of hardened mortar
Compressive and bending strengths

The compressive and bending strengths of each sample of hardened mortar are illustrated in Figure 1. The strengths of SFC at the age within 28 days prepared by the standard curing were equal to or lower than those of NPC, especially the bending strength was conspicuous. Maybe this is because the strength-increasing effect by the increase of the production of C-S-H caused by the pozzolanic reaction does not exceed the strength-decreasing effect by the dilution with silica fume.

Although the compressive and bending strengths developed by the steam curing were remarkably improved, those strengths were not so much improved by the oven- and autoclave-curings. Judging from the analytical result of $Ca(OH)_2$ shown in Figure 5, the reason why the strength developed of SFC were higher than those of NPC is considered that the content of calcium silicate hydrate in SFC was increased by the accelerated pozzolanic reaction during the steam curing.

The bending strength of the hardened mortar after the oven curing is lower than that just after the steam curing. Figure 2 reveals that the size of the test specimen is approximately 5% shrinked by the oven curing, thereby generating a lot of fine cracks on the surface of it. Perhaps the bending strength was decreased by the

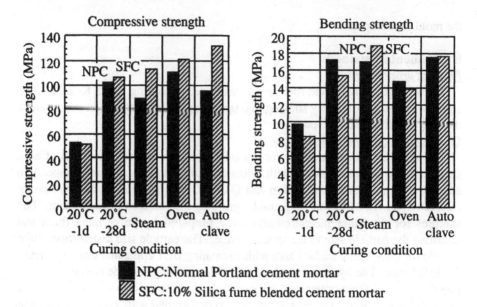

Figure 1 Strength of hardened mortar cured at various conditons

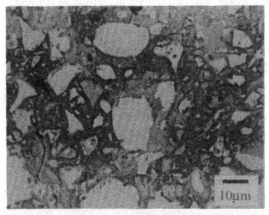

Figure 2 Cracks generated on the surface of oven
cured NPC mortar

generation of those cracks on the surface of it.

The ratio of bending strength/compressive strength of SFC was smaller than that
of NPC. This suggests that the production of C-S-H or tobermorite by the
pozzolanic reaction largely affects the improvement of the compressive strength.

Stress-strain characteristics

Figure 3 illustrates stress-strain curves of hardened mortar under the loading of compressive strength. The figure reveals that the stress-strain curves of hardened mortar prepared by the steam, oven and autoclave curings are almost linear and hardly depend upon the type of cement. The stress-strain curve of a hardened body of uniform material is generally almost linear, while that of uneven composite including concrete and mortar is convex[4]. The reason is considered to be as follows. Concrete and mortar have structurally weak transition zones[5] on the interface between the aggregate and the cement paste, and the strengths of both materials are different from each other. Stress is, therefore, concentrated into the interface between the aggregate and the cement paste, thereby generating fine cracks and fluctuating the modulus of elasticity [6, 7, 8]. Since the W/C ratio of the hardened mortar prepared in this study was 0.25, the transition zone hardly existed in it and the structure of hardened mortar [9] was uniformalized by the temperature-increasing curing. This may be a reason why the stress-strain curve is almost linear.

To the relationship in concrete between the modulus of elasticity and the compressive strength, the following formula applies[10] :

$$E = 3,320F^{1/2} + 6,900 (MPa) \quad (21 < F < 83 MPa) \qquad (1)$$

where E = static modulus of elasticity and F = compressive strength

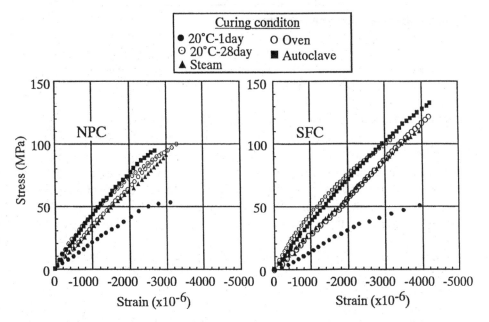

Figure 3 Stress-strain curve of hardened mortar cured at various condition

The static modulus of elasticity of concrete in a range of the compressive strength from 21 to 83MPa is determined according to Formula (1). Substituting the values of E and F for a formula E = F/ ε, the strain of concrete(ε) was 950 to 2,200 μm. Determining the strain of the hardened mortar from the limit values of the straight line region of the curve drawn in Figure 3, the strain of NPC and SFC mortars were 2,800 to 3,200 m and 3,200 to 4,600 μm, respectively, which were larger than that of concrete.

Meanwhile, the static modulus of elasticity can be determined from the gradient of a straight line connecting the straight line region of the stress-strain curve to the origin. Thus determined values of the static modulus of elasticity are listed in Table 2. The modulus of elasticity of SFC mortar was lower the that of NPC mortar, especially conspicuously lower in the temperate-increasing curing. Maybe this is mainly because the contents of highly distortion resistant well-crystallized large-sized crystals including Ca(OH)$_2$ and gehlenite hydrate are low.

Table 2   Static modulus of elasticity of hardened mortar (GPa)

Curing conditon	20℃-1d	20℃-28d	Steam	Oven	Autoclave
NPC	20	40	34	35	40
SFC	15	38	28	29	32

Composition of hardened mortar
Hydration reaction products

Powder X-ray diffraction profiles of various types of hardened mortar are illustrated in Figure 4. All of the samples contained ettringite (AFt) and Ca(OH)2 and the intensity of diffracted x-ray of AFt had a tendency to slightly decrease in the temperature-increasing curing. That of Ca(OH)$_2$ had also a tendency to slightly decrease and a peak considered to be gehlenite hydrate (C$_2$ASH$_8$) was observed at 2 θkα of 17.5°. This decreasing tendency of the intensity of diffracted x-ray for Ca(OH)$_2$ was conspicuous for SFC mortar. This tendency agrees with the determination result of Ca(OH)$_2$ shown in Figure 5. The decrease in the intensity of diffracted x-ray of Ca(OH)$_2$ in the temperature-increasing curing of NPC mortar may be caused by the reaction with the aggregate as mentioned later. That of SFC mortar is considered to be mainly caused by the consumption in the pozzolanic reaction with silica fume.

The existence of C-S-H, xonotlite and tobermorite could not be determined by X-ray diffractometry (XRD). Calcium aluminate sulfate produced by the curing at room temperature is mainly AFm at high W/C ratio, but it is AFt in early age and converted to AFm in later age at low W/C ratio[11]. The result of this study indicates that the disappearance of AFt and the conversion to AFm at low W/C ratio are

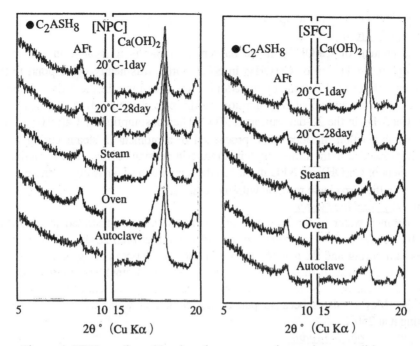

Figure 4 XRD profice of hardened mortar cured at various conditions

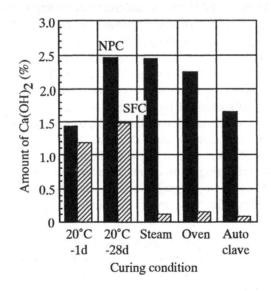

Figure 5 Amount of Ca(OH)$_2$ in hardened mortara cured at various condition

hardly accelerated by the temperature-increasing curing.

Content of Ca(OH)₂

The content of $Ca(OH)_2$ in the hardened mortar is illustrated in Figure 5. Although the reduction of $Ca(OH)_2$ in SFC mortar was nearly correspond to the reduction of portland cement by the addition of silica fume in the 1-day curing at 20°C, there was additional consumption by the pozzolanic reaction of silica fume with $Ca(OH)_2$ in the 28-day curing at 20°C. The consumption of $Ca(OH)_2$ by the pozzolanic reaction of silica fume proceeded in a moment in steam curing but it hardly proceeded in the following temperature-increasing curing. The reduction of the content of $Ca(OH)_2$ in NPC mortar in the temperature-increasing curing may be caused by the pozzolanic reaction with the aggregate.

Content of combined water

Combined water in the hardened mortar measured are illustrated in Figure 6. The figure reveals that the contents of bound water after both 28-day curing at 20°C and steam curing and the reaction rates of hydration in both curings are almost the same as each other. Although the content of combined water after the steam curing was hardly changed even by the autoclave curing, it was reduced by the oven curing. Maybe this is because part of the combined water was evaporated by heating it at 250°C.

Content of insoluble residue

The apparent rates of reaction of aggregate (fine sand) and silica fume with $Ca(OH)_2$ calculated from insoluble residue measured according to Formula (2) are illustrated in Figure 7. Assuming that silica fume selectively reacts with $Ca(OH)_2$ in SFC, the rate of reaction was calculated. The apparent rate of reaction of silica fume was sharply increased by the steam curing and as if reaction were proceeded by the following oven and autoclave curings. On the contrary, the analytical result of the content of $Ca(OH)_2$ mentioned before indicated that the content of $Ca(OH)_2$ in SFC mortar was hardly changed by the oven and autoclave curings after the steam curing. It is, therefore, considered that the apparent increase of the rate of reaction of silica fume does not correspond to the proceeding of the pozzolanic reaction but to the dissolution of part of silica fume in high-temperature water existing in the atmosphere during the oven and autoclave curings.

The reaction of NPC with fine aggregate showed the same tendency as that with silica fume. The rate of reaction was, however, as low as 3% or less.

$$R=(1-N/M) \times 100 \qquad (2)$$

where R = rate of reaction (%), M = content of aggregate or silica fume in mortar (mg), N = content of aggregate or silica fume in insoluble residue (mg). N was determined by separately measuring the content of insoluble residue in each component.

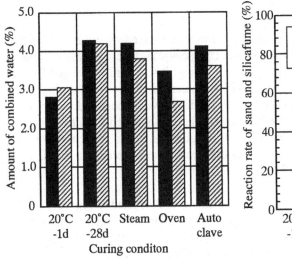

Figure 6 Amount of combined water of hardened mortar cured at various condition

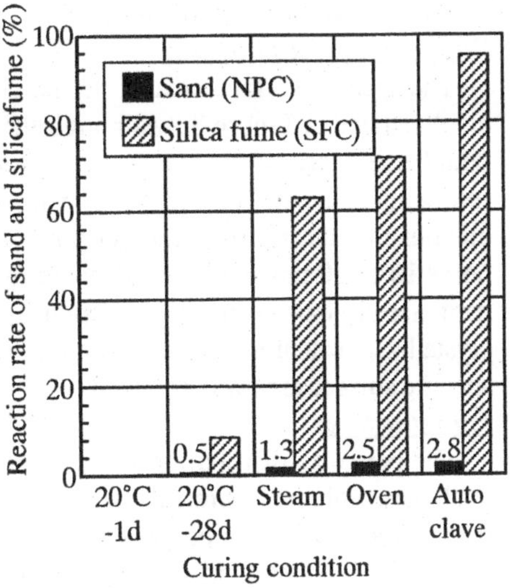

Figure 7 Reaction rate of sand and silica fume of hardened mortar cured at vairous condition

Table 3 Chemical composition of C-S-H in hardened mortar measured by EPMA

Cement	Curing condition	Chemical composition (%)									molar ratio	
		SiO₂	Al₂O₃	Fe₂O₃	CaO	MgO	SO₃	Na₂O	K₂O	Total	C/S	C/(S+A)
NPC	20°C-1day	26.4	2.97	1.18	55.3	0.35	2.26	0.19	0.23	88.9	2.25	2.11
	20°C-28day	23.9	3.17	1.58	45.7	0.51	3.34	0.24	0.18	78.6	2.05	1.90
	Steam	25.5	2.83	1.22	45.5	0.32	2.67	0.07	0.05	78.1	1.91	1.79
	Oven	26.2	2.21	1.11	46.0	0.54	3.45	0.20	0.17	79.9	1.88	1.79
	Autoclave	26.9	3.08	1.87	46.7	0.32	2.67	0.07	0.05	81.7	1.86	1.74
SFC	20°C-1day	21.0	3.12	2.40	39.5	0.35	5.20	0.06	0.13	71.8	2.02	1.85
	20°C-28day	31.2	3.28	2.18	43.7	0.33	3.01	0.32	0.23	84.3	1.50	1.41
	Steam	31.2	2.63	0.71	39.9	0.19	2.67	0.31	0.17	77.8	1.37	1.30
	Oven	33.7	2.38	0.64	43.2	0.25	2.44	0.11	0.20	82.9	1.37	1.32
	Autoclave	33.4	2.31	0.75	43.5	0.26	1.72	0.11	0.29	82.4	1.40	1.34

Composition of calcium silicate hydrate

The result obtained by the point analysis of EPMA for the composition of calcium silicate hydrate in the hardened mortar is listed in Table 3. The Ca/Si ratio in calcium silicate hydrate produced was lowered with the proceeding of the hydration, especially the lowering of the Ca/Si ratio in SFC mortar was conspicuous. Although the Ca/Si ratio in calcium silicate hydrate was sharply lowered by the steam curing, it was hardly changed by the following oven and autoclave curings.

Structure of hardened mortar

BET's specific surface area

The specific surface area of the hardened mortar measured by the BET method are illustrated in Figure 8. Although the BET's specific surface area was increased with the advance of the age and by the steam curing, it was hardly changed by the following autoclave curing but it was sharply decreased by the oven curing. As mentioned later, most of the gel pores once produced were converted to capillary pores 10 to 50 nm in diameter by the oven curing. It is inferred from this that the hardened structure was broken when C-S-H or tobermorite was converted to well-crystallized calcium silicate hydrate including xonotlite, whereupon the BET's specific surface area was reduced.

Pore structure

The pore size distribution in the hardened mortar is illustrated in Figure 9. The total pore volume decreased and the pore size distribution shifted to the small diameter side with the progress of curing except the oven curing. The total pore volume in the steam curing was larger than that in the 28-day curing at 20°C but the pore size distribution in steam curing was shifted to the small diameter side. The total pore volume was not reduced even by the autoclave curing after the steam curing, but it was increased by the increase of the pores 6 to 50 nm in diameter. Although the pore diameter in the SFC mortar was reduced more than that in the NPC mortar by the temperature-increasing curing, the total pore volume was increased. The pores 10 to 50 nm in diameter was more remarkably increased and total pore volume was increased by the oven curing than that after the steam curing. Maybe this is because the hardened structure was changed by the formation of new pores by dehydration and the changes of size, shape, crystallinity and structure of the hydration products by the temperature increase. The decrease of the pores 3 to 6 nm in diameter and the increase of the pores 10 to 50 nm in diameter agree with the measurements of the BET's specific surface area mentioned before.

The changes of the pore volume by the curing after the 1-day curing at 20°C and steam curing used as the basis are illustrated in Figure 10.

The pores 10 to 100 nm in diameter after the 1-day curing as the basis at 20°C were decreased by the 28-day curing at 20°C and the pores 30 nm in diameter showed the maximum decrease. The pores 10 to 100 nm in diameter were also largely decreased by successive steam curing similar to that in the case of 28-day curing at 20°C and the pores 3 to 10 nm in diameter were increased showing maximum at 4 nm in diameter. This suggests that the calcium silicate hydrate is increased and the produced quantities of calcium silicate hydrate by the steam curing in the NPC and SFC mortars are nearly equal to each other.

The gel pores approximately 3 nm in diameter after the steam curing as the basis were decreased by the oven curing and the capillary pores with 60 and 30 nm in

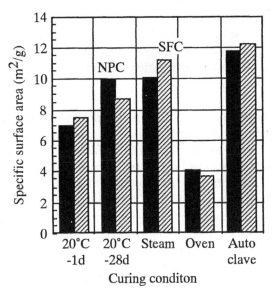

Figure 8 BET supecific surface area of hardened mortar cured at various conditions

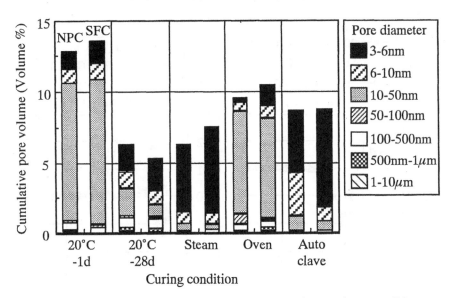

Figure 9 Pore volume of hardened mortar cured at various conditions

diameter as the central figure in the NPC and SFC mortars, respectively, were increased. Meanwhile, the volume of small diameter pores in the NPC mortar was remarkably increased by the autoclave curing and that of 6 nm in diameter showed the maximum increase. There was little difference in the SFC mortar before and after autoclave curing.

Microstructure

The back scattered electron images of the polished surface of hardened mortar are illustrated in Figure 11. The thickness of the inner-hydrate layers around unhydrated cement particles was increased with the advance of the ages of both NPC and SFC mortars in the standard curing and the gaps between the unhydrated cement particles were filled with the outer- hydrate. Thus the hardened structure was being densified. Large-sized crystals of $Ca(OH)_2$ were observed in the vicinity of the interfaces between cement paste and aggregate in NPC mortar. The densest structure was produced by the 28-day curing at 20°C compared with other curings. Spherical and ellipsoidal pores of one to several microns were formed in the hardened structure by the steam curing and the volume of them were increased by the oven and autoclave curings. Pores between the interfaces of aggregate and the cement particles considered to be caused by dehydration were slightly increased by the oven curing.

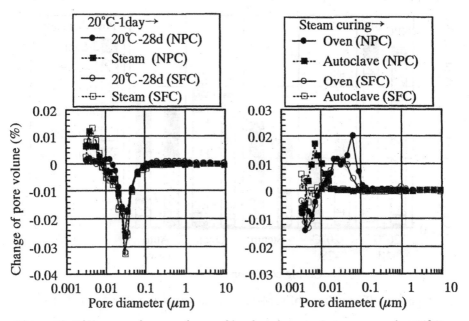

Figure 10 Difference of pore volume of hardened mortar between cured at 20°C-1day and under steam, added curing under various conditions

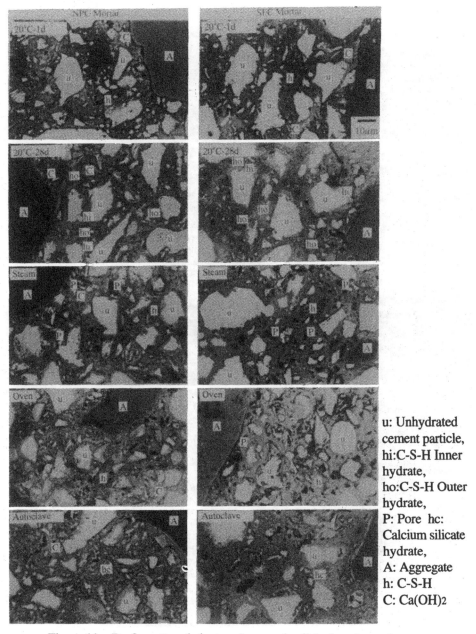

u: Unhydrated
cement particle,
hi:C-S-H Inner
hydrate,
ho:C-S-H Outer
hydrate,
P: Pore hc:
Calcium silicate
hydrate,
A: Aggregate
h: C-S-H
C: Ca(OH)2

Figure 11   Back-scattered electron image of polished surface of
hardened mortar cured at various conditions

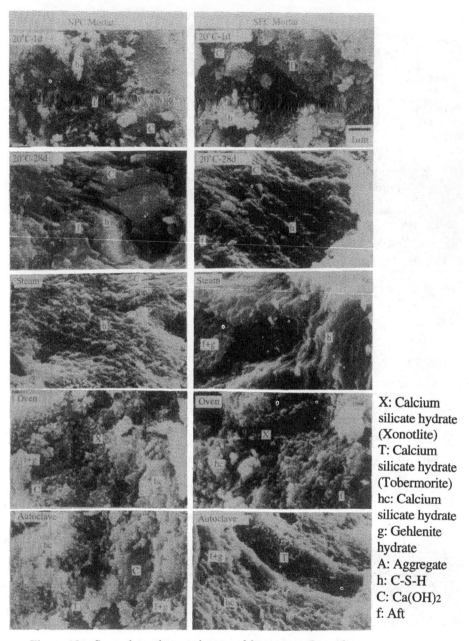

X: Calcium silicate hydrate (Xonotlite)
T: Calcium silicate hydrate (Tobermorite)
hc: Calcium silicate hydrate
g: Gehlenite hydrate
A: Aggregate
h: C-S-H
C: Ca(OH)2
f: Aft

Figure 12   Secondary electron image of fracture surface of hardened mortar cured at various conditions

The secondary electron images of a fracture surface of hardened mortar are illustrated in Figure 12. In the standard curing, the densified hardened structure was observed at the age of 28 days. A lot of large-sized platy crystals of $Ca(OH)_2$, platy crystals of AFt and C-S-H filling the gaps between them were observed in the NPC mortar, while a few small-sized crystals of $Ca(OH)_2$ and a lot of C-S-H were observed in the SFC mortar. In the steam curing, aggregated structures of calcium silicate hydrate considered to be grown from the C-S-H crystals were observed in the NPC mortar. The hardened structure in the SFC mortar was not so much different from that produced by the 28-day curing at 20°C. The hardened structure produced by the oven curing was more porous than that produced by the steam curing. This suggests that the hardened structure produced by the steam curing was changed by dehydration of it. Aggregated structures of well-crystallized calcium silicate hydrate which may be xonotlite and those of small-sized platy crystals which may be gehlenite hydrate were also observed in that structure. Although structures similar to the structures produced by the oven curing were observed in the autoclave curing, the size of hydrate was generally smaller and the morphology of crystalline calcium silicate hydrate, which seems to be tobermorite, was slightly different from those produced by the oven curing.

Relationship in hardened mortar between hardened mortar and physical properties

Strength and static modulus of elasticity

The relationship in the hardened mortar between the strength and static modulus of elasticity and a curve representing Formula (1) are illustrated in Figure 13. Estrangement of the plots from the curve were large in the mortar after the 1-day curing at 20°C and in the SFC mortar after the steam, oven and autoclave curings. This is because the maturity of hardened structure produced by the 1-day curing at 20°C is low and porous and the hydrate is not sufficiently crystallized. The structure of SFC mortar produced by the temperature-increasing curing may also be easily deformable by the stress because the consumption of $Ca(OH)_2$ crystal playing a role of the skeleton of the hydrated structure and the production of low-stiffness low-C/S ratio calcium silicate hydrate and of small particle hydrates.

Relationship between pore volume and compressive strength

The relationship in the hardened mortar between the total pore volume and the compressive strength is illustrated in Figure 14. The compressive strength of the mortar produced by the temperature-increasing curing was higher than that produced by the standard curing when pore volume was same and the difference of compressive strength was increased with the increase of the curing temperature. It is known that the compressive strength depends upon the content of crystalline materials in the hardened mortar as well as the porosity[12]. The result of this study agrees well with the precious work.

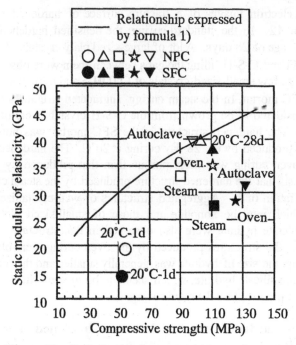

Figure 13   Relationship between compressive strength
and static modulus of elasticity

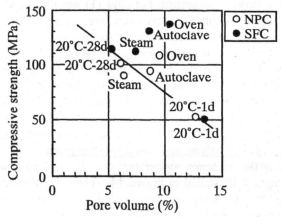

Figure 14   Relationship between compressive
strength and total pore volume

Relationship between microstructure and fracture

Applying compressive strength to a uneven composite materials containing pores including mortar and concrete, stress is concentrated into the points with peculiar shape around the pores in it and weak materials, thereby generating and propagating cracks resulting in fracture of the composite. Even applying a load smaller than the fracture stress to that, the hardened structure is changed in such a manner as peeling and breaking of hydrate and formation of large pores by linking small pores to each other when the applied stress exceeds the 60% of fracture stress[13]. The propagating velocity of cracks depends upon the denseness of hardened structure in such a manner that the propagating velocity of cracks is lowered with the increase of denseness of hardened mortar, so the growth to large cracks is retarded[13].

SUMMARY AND CONCLUSIONS

Normal portland cement mortar and silica fume cement containing 10% of silica fume mortar prepared at the W/C ratio of 0.25 and the S/C ratio of 1.0 were cured under various conditions. And the effects of the curing conditions and the addition of mineral admixture on the hydration reaction of cement and the formation of micro-structure of hardened mortar, the relationships in the hardened mortar of the composition and structure with the physical properties and the relationship between the microstructure and the fracture process were investigated.

The conclusions obtained in the study are summarized as follows:

1)The strength of SFC mortar cured under standard condition was lower than that of NPC mortar within the age of 28 days. Maybe this is because the dilution effect by the addition of silica fume exceeds the effect of the increasing production of C-S-H by the pozzilanic reaction.

2)The hydration reaction of cement was accelerated by the steam curing. The strength of SFC mortar was sharply improved by the steam curing through the increase of the production of calcium silicate hydrate with the proceeding of pozzolanic reaction, exceeding the strength of NPC mortar.

3)The strength of mortar after the steam curing was hardly increased though it was further autoclave-cured. Even if it was additionally oven-cured, the bending strength was sharply lowered though the compressive strength was slightly improved. The strength of mortar is generally negatively affected by the increase of the volume of pores, the increase of pore diameter and the generation of fine cracks, while it is positively affected by the increase of content of hydrate and the desirable degree of crystallinity of hydrate. It is considered that the compressive strength of oven-cured mortar is affected largely by the latter more than the former, while the bending strength is affected by the former, especially by the generation of fine cracks.

4)The volume of pores was increased and the pore size distribution shifted to the large diameter side in the oven-cured mortar. This may be brought by the generation of fine crack caused by the rearrangement of hardened structure by the conversion of tobermorite and others to well-crystallized calcium silicate hydrate including xonotlite by dehydration accompanied with the increase of curing temperature.

5)The strength of mortar was higher by the temperate-increasing curing than by the standard curing when the total pore volume is same. This suggests that the compressive strength in largely affected by the content of crystalline materials in the mortar, especially by the degree of crystallinity of calcium silicate hydrate as well as the porosity.

6)The stress-strain curve of the mortar under the loading of compressive strength was made more linear by the temperature-increasing curing. Maybe this is because the homogeneity of the hardened structure is improved by the temperature-increasing curing, whereupon the concentration of stress is alleviated.

7)The static modulus of elasticity determined from the gradient of the stress-strain curve of the SFC mortar was lower than that of the NPC mortar. This tendency was conspicuous in the temperature-increasing curing. This is mainly because highly distortion-resistant, well-crystallized large-sized $Ca(OH)_2$ crystals are remarkably reduced and the content of well-crystallized gehlenite hydrate crystal is low.

REFERENCES

1. Richard, P. and Cheyrezy, M.H. (1994) Reactive Powder Concrete with High Ductility and 200-800MPa Compressive Strength. Concrete Technology, SP144-24, 507-517.
2. Dugat, J., Roux, N. and Bernier, G. (1996) Mechanical Properties of Reactive Powder Concretes, Material and Structure. 233-240.
3. Kalousek, G.L. (1966) Autoclave Curing of Concrete in Soviet Union and United State, J. of ACI, 817-832.
4. Swamy, R.N. (1971) Fracture Phenomena of Hardened Paste, Mortar and Concrete, Proceeding of the International Conference on Mechanical Behavior of Materials, Kyoto Japan, 132-142.
5. Uchikawa, H. (1988) Similarities and Discrepancies of Hardened Cement Paste, Mortar and Concrete from the Standpoints of Composition and Structure, Proceedings of Engineering Foundation Conference 'Advances in Cement Manufacture and Use' Potosi,Missouri, 1, 271-294.
6. Tanigawa, Y. (1976) Model Analysis of Fracture and Failure of Concrete as a Composite Material, Cem.and Concr. Res., 679-690
7. Hsu, T.T.C., Slate, F.O., Struman, G.M. and Winter, G. (1963) Microcracking of Plane Concrete and the Shape of the Stress-Strain Curve, J. of ACI, 209-223.
8. Shah, S.P. and Winter, G. (1966) Inelastic Behavior and Fracture of Concrete, J. of ACI, 925-930.
9. Uchikawa,H. (1997) Role and Regulation of the Transition Zone in Realizing

High Performance Concrete, Proceedings of Third CANMET/ACI International Conference on 'Advances in Concrete Technology', Auckland, New Zealand,109-129

10. ACI Committee 363, (1984) State-of-the-Art Report on High-Strength Concrete, J. of ACI, 364-411.

11. Uchikawa, H. (1993) Characterization and Material Design of High-Strength Concrete with Superior Workability, Proceeding of Cement Technology Symposium, PAC RIM Ceramic Meeting, Honolulu, Hawai, Ceramic Transactions, 40, 143-186

12. Crennan,J.M., El-Hemaly,S.A.S. and Taylor,H.F.W. (1977) Autoclaved Lime-Quartz Materials I. Some Factors Influencing Strength, Cem.and Concr. Res., 493-502

13. Uchikawa, H. Hanehara, S. and Hirao, H. (1977) Influence of Microstructural Change under Stress on the Strength-Related Properties of Hardened Cement Mortar and Paste, Advanced Cement Based Materials, 6 [3/4],87-98

PERMEABILITY OF CONCRETE -- RELATIONSHIPS TO ITS MIX PROPORTION, MICROSTRUCTURE AND MICROCRACKS

Kejin Wang, Takeru Igusa, and Surendra P. Shah
The NSF Center for Advanced Cement-Based Materials
Northwestern University
2145 Sheridan Road
Evanston, IL 60208

ABSTRACT

Compressive strength and water-to-cement ratio of concrete have been long and widely employed in international specifications and guidelines for structural design, mix design, and quality control during constructions, which often misleads engineer to think that the higher the concrete strength, the better the concrete performance. Actually, most failures of concrete structures are not due to concrete strength but its durability problems. It is essential to have a better understanding of the relationship between mix proportion, strength, permeability, and durability of concrete. This paper presents an extensive study relating cement paste and concrete permeability to its mix proportion, microstructure and microcracks. Based on a comprehensive concrete mix design, with possible large coverage of variation in volume fraction of aggregate, selected grading and maximum size of aggregates as well as w/c ratios, the relationships between concrete permeability, and compressive strength were studied. The effects of dilution, tortuosity, the interfacial transition zone (ITZ), and percolation of aggregate on concrete permeability were explored. A feedback controlled splitting technique was introduced to generate width-controlled cracks in permeability specimens. The permeability of the cracked concrete was evaluated. The research indicates that three criteria, strength, permeability, and cracking resistance, need to be considered together in concrete mix design to achieve a durable concrete.

INTRODUCTION

Concrete is the most extensively used construction material in the world. In the United States, about 500 million tons of concrete is produced every year. With increasing growth in human population, industrialization, and urbanization, application of concrete from buildings, roads, and dams has been rapidly extended to underground mass transit facilities, airport facility, waste water treatment systems, offshore oil platforms and marine structures. More and more concrete is subjected to aggressive environmental conditions. Durable concrete is increasingly requested not only for new constructions but also for old structure rehabilitation.

Investigation and research have shown that many durability problems, such as corrosion of reinforcing steel, spalling due to freezing-thawing cycles, cracking due to alkali-silica reaction and sulfate attack, are associated to permeability of concrete [1]. Permeability of concrete is closely related to its microstructure. Microstructure of concrete is predetermined by mix proportioning and manufacturing process, including casting and curing, and it changes with time, loading and environmental conditions. Although a great deal of work has been done on concrete permeability, the relationships between concrete mix proportion, microstructure and permeability is still not fully understood. Furthermore, most previous work has focused on the permeability of non-cracked concrete. Actually, most concrete structures in practice unavoidably have cracks due to loading and weathering. The cracks may connect flow paths, increase concrete permeability, and accelerate deterioration. However, limited research has been done quantitatively shown the effect of cracks on permeability.

Currently, an extensive study of concrete permeability, cracking and mix design criteria is undergoing at the NSF Center of Advanced Cement-Based Materials (ACBM). The research program is illustrated in Figure 1.

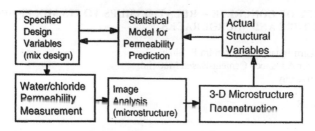

Figure 1: Research Program

A permeability model is being developed based on the comprehensive study of the statistically designed concrete mixes, with possible large range of variation in volume fraction of aggregate, selected grading and maximum size of aggregates as well as w/c ratios. Both water permeability and rapid chloride permeability tests are performed on cracked and non-cracked, normal strength and high strength concrete. The relationships and interactions between the batch design parameters, actual microstructure parameters, and concrete permeability are investigated. The actual microstructure of concrete is studied by image analysis plus a three-dimensional reconstruction modeling. The obtained actual microstructure parameters include the volume fraction of coarse aggregate, total surface area of the aggregate, the ratio of large and small aggregate, the radius and centroid location of aggregate particles, and the distance between the centroids. The relationship between concrete permeability and compressive strength is also studied.

The objectives of the present study are to (a) quantify the effects of mix design variables (such as w/c ratio, maximum aggregate size, grading, and volume fraction), microstructure factors (such as dilution, tortuosity, interfacial transition zone, percolation), and microcracks on concrete permeability; (b) explicate the relationship and interactions between concrete mix design, microstructure, cracking, and permeability; and (c) eventually provide rational mix design criteria for durable concrete. This paper presents some preliminary results of the study.

CONCRETE MICROSTRUCTURE AND ITS EFFECT ON PERMEABILITY

1. Permeability of Cement Paste

On a mezzo-scale, concrete consists of cement paste and aggregate. Cement paste contains a network of pores with a wide range of size, including air voids, capillary pores, and gel pores, and allows water and ion penetration. Permeability of cement paste is greatly related to its pores and pore structure, not only their amount and size but also the connectivity. The pores and pore structure of cement paste mainly depends on water/cement ratio and degree of hydration. At a given degree of hydration, the cement paste with a high w/c ratio generally contains large amount, large size and interconnected pores and has high permeability. A great deal of work has been done relating microstructure of cement paste to its conductivity and permeability. Since both conductivity and permeability depend critically on pore structure of concrete, conductivity is often used to represent permeability. By incorporating three relationships from (a) Powers-Brownyard's model, (b) Archie's model, and (c) Katz-Thompson's model, a nomogram (Figure 2) to predict permeability of cement paste can be developed.

Powers-Brownyard's model introduces the relationship between water/cement ratio and porosity [2]. The appropriate equations are:

$$\phi_{cap} = 1-(1+1.3\alpha)/(1+3.2(w/c)), \text{ and } \phi_{gel} = 0.21\alpha/(0.313+(w/c))$$

Here, ϕ_{cap} is the porosity of capillary pores; ϕ_{gel}, the porosity of gel pores; α, degree of cement hydration; and w/c, water-cement ratio.

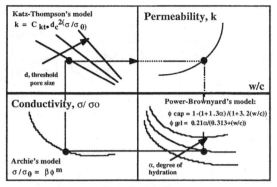

Figure 2: Permeability Nomogram

Archie's model presents the relationship between porosity and tortuosity [3]. The appropriate equation is: $\sigma/\sigma_0 = \beta\phi^m$. Here, σ is conductivity of a cement-based material, and σ_0 is conductivity of its pore solution. β indicaties the tortuosity, or interconnectivity, of the material. ϕ is the total porosity of cement paste ($\phi = \phi_{cap} + \phi_{gel} + \phi_{air}$). m, exponent, is about 1.3~1.7 for cement-based materials. In addition to Archie's model, there are also some other equations relating conductivity to porosity [4].

Katz-Thompson's model describes the relationship between tortuosity and permeability [5]. The appropriate equation is: $k = C_{kt} \cdot d_c^2(\sigma/\sigma_0)$. Here, k is permeability; C_{kt}, a constant, about 1/226, depending on material pore structure; and d_c, the threshold pore size.

Using this nomogram, one can predict permeability of a cement-based material from the knowledge of the w/c ratio and degree of cement hydration.

2. Permeability of Concrete

Concrete is composed by adding aggregate into a cement paste matrix. According to the discussion above, the permeability of a cement paste with a given w/c ratio and a certain degree of hydration can be considered as a constant. Therefore, the addition of aggregate, a relatively impermeable material, into a cement paste appears to reduce the permeability of the material. However, the consequence is not that simple. Adding aggregate into cement matrix makes the matrix become a three-phase composite material, including aggregate, cement paste, and the interfacial transition zone (ITZ) between aggregate and cement paste. Concrete has a much more complex microstructure than cement paste not only because of the characteristics of the third phase, ITZ, but also because of a very complicated geometric arrangement of grains, surfaces, interfaces and defects, involving phase volume fractions, orientations, sizes, shapes, distribution and connectivity of phases. Effect of concrete microstructure on permeability is generally considered as a result of four factors: dilution, tortuosity, interfacial transition zone (ITZ), and percolation (Figure 3).

Dilution effect results from the relatively impermeable aggregate particles that cut the flow paths and reduce the permeable area in a cross-section of concrete. The degree of dilution depends on the permeability and volume fraction of aggregate. Assuming that aggregate is impermeable, the dilution effect on concrete conductivity for a parallel ion flow with no tortuosity is:

$$\sigma = \sigma_p(1-VF) \quad \dots\dots\dots\dots\dots\dots\dots\dots\dots\dots\dots(1)$$

Where σ is conductivity of concrete, σ_p is conductivity of cement paste, and VF is volume fraction of aggregate.

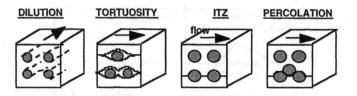

Figure 3: Effects of Concrete Mesostructure on Permeability

Tortuosity effect results from the impermeable aggregate that force a flow to go around the aggregate particles; thus increasing the length of flow paths and reducing the flow rate. As shown in Figure 3, the degree of tortuosity depends on aggregate size and distance between aggregate particles. Tortuosity effect can be expressed by Maxwell's equation [6]:

$$\sigma = \sigma_p(1 + m \cdot VF) \quad\text{...(2)}$$

Where, $m = -1.5$ for no interaction between aggregate particles; otherwise, $-1.5 < m < -1.0$. For the dilute limit, where the VF is around 30% and aggregate is spherical, Maxwell analytically derived the value $m = -1.5$. The fact that $m < -1.0$ is due to the tortuosity effect, as compared with equation (1). At higher levels of VF, aggregate interaction occurs. Although the total tortuosity of the ion flow increases, the tortuosity per unit VF of aggregate decreases resulting in increased concrete conductivity or $-1.5 < m$.

ITZ originates from a layer of bleeding water accumulated beneath large aggregate particles. This would account for a higher w/c ratio closer to the aggregate particle than that in cement matrix [7]. Existing as a thin shell (typically about 10 to 50 μm thick), ITZ generally features the weakest phase in normal strength concrete (NSC), with orientated, large crystals and high porosity, and greatly affect concrete properties. The amount of ITZ in concrete is greatly related to the volume fraction of aggregate. The thickness and weakness of ITZ primarily depends on the shape, size, and orientation of aggregate particles and the w/c ratio of surrounding cement matrix. In the dilute limit, an analytical expression for m, which included the ITZ effect, was obtained. Garboczi, et. al. [6] have expressed m as a function of α, aggregate size; σ_p, conductivity of cement paste; h_{ITZ}, thickness of the ITZ; and σ_{ITZ}, conductivity of the ITZ:

$$m = f(\alpha, \sigma_p, h_{ITZ}, \sigma_{ITZ}) \quad\text{................................(3)}$$

Detailed study on impact of the ITZ on concrete permeability has reported in reference [8].

Percolation in concrete generally indicates that a flow can go through a path connecting a number of ITZ's. The degree of percolation depends on volume and size of aggregate. It has been reported [9] that percolation occurs in mortars at a sand volume fraction of about 0.5. Considering percolation, Mclachlan, et. al. have proposed the following effective medium model for the conductivity of a composite medium:

$$\sigma = \sigma_p(1 - \frac{VF}{VF_c})^{m \cdot VF_c} \quad\text{..........................(4)}$$

Where, VF_c is the volume fraction at percolation. If the grading of aggregate is also considered, equation (4) can be re-written into

$$\sigma = \sigma_p \cdot \prod_i (1 - \frac{VF_i}{VF_c})^{m_i \cdot VF_c} \quad\text{..................(5)}$$

Where, VF_i is the volume fraction of aggregate with size i.

Based on the discussion above, we notice that dilution and tortuosity reduce concrete permeability and ITZ and percolation increase permeability. It is not clear which factor has dominant effect on concrete permeability, or what is the combined effect. However, all the four factors are influenced by volume fraction of aggregate. Therefore, volume fraction of aggregate may be one of important factor controlling permeability of concrete. Our initial research results have supported this preliminary inference.

The preliminary results from our permeability study demonstrate that:

(1) Unlike compressive strength, concrete permeability is more significantly influenced by volume fraction of aggregate than by w/c ratio. Figures 4 and 5 show that the measured concrete compressive strength has a linear relation with w/c ratio, regardless of volume fraction and the maximum size of aggregate. In contrast, the rapid chloride permeability of the corresponding concrete specimens does not relate to w/c ratio. As a result, some design codes attempt to use strength or w/c ratio as a sole concrete mix design criterion for durability control may be not appropriate.

(2) There is no direct relationship between compressive strength and chloride permeability. From Figure 6, it can be observed that a concrete having compressive strength of 41.5 MPa may have chloride permeability in terms of total charges ranging from as low as 3500 coulombs to as high as 9000 coulombs. This once again illustrates that using concrete strength as a sole concrete mix criterion can not ensure to produce a durable concrete.

(3) Considering concrete mix design variables and location of specimens, rapid chloride permeability value of concrete can be expressed by the following equation based on a linear regression:

$$log \text{ (total charge, coulombs)} = 9.0 + 0.079(top) - 0.13(max) + 0.435(w/c) - 1.55(vf)$$

and $R^2 = 0.86$ ....(6)

Here, *top* and *max* are indication variables. *top* = 1.0 for the top specimens from a 4"x8" concrete cylinder where a total of two 2"-specimens were taken from the middle of the cylinder, and *top* = 0 for the bottom specimens. *max* = 1.0 for MSA=19 mm and *max* = 0 for MSA=9.5 mm, *w/c* = water/cement ratio, (*vf*) = volume fraction of aggregate. *R* = standard deviation. Equation (6) and Figure 7 show that volume fraction of aggregate has a greater influence on concrete permeability than other variables. The aggregate grading used the present study does not appear in the equation because it seems insensitive to the chloride permeability value.

It is expected that the ongoing image analysis will provide us a quantitative result on concrete tortuosity. Continuous study is to relate the experimental results to modeling.

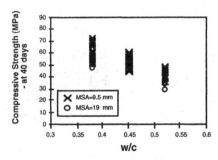

Figure 4: Effect of w/c on Strength

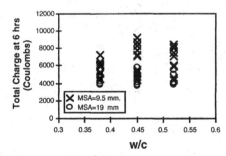

Figure 5: Effect of w/c on Permeability

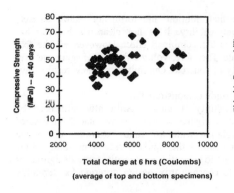

Figure 6: Relationship Between Strength and Permeability

Figure 7: Effect of Coarse Aggregate Volume Fraction on Permeability

MICROCRACKS IN CONCRETE AND THEIR EFFECT ON PERMEABILITY

Generally, a well proportioned concrete mix, not only having a low ratio of water to cementitious material but also correctly mixed, placed, compacted and cured, is known to be durable. However, concrete is weak in tension. When exposed to environment, concrete is more and less subjected to tension due to improper loading and shrinkage, and concrete almost unavoidably cracks even before some significant physical and chemical deterioration occurs. Cracks can be seen almost everywhere a concrete structure exists: sidewalks, highways, building walls and bridge decks. Cracks in concrete not only reduces strength and stiffness but also interconnect flow paths, increase permeability, allow more water and chemical ions to get into the concrete; thus facilitate concrete deterioration and finally result in destructive deterioration of the concrete structure.

In the concrete design and construction practice, crack widths are generally controlled by proper use of reinforcement. To prevent considerable shrinkage cracking, ACI Building Code requires a minimum amount of reinforcing steel for concrete beams, slabs, walls, etc.. It is well known that the steel reinforcement does not eliminate shrinkage cracking; but simply transforms wide cracks into fine cracks or microcracks. These microcracks may serve as potential pathways for a water flow.

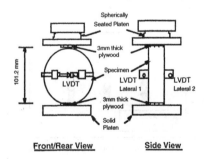

Figure 8: Feedback Controlled Splitting Test Setup

Darcy's Law

$$\frac{dq}{dt} = K \frac{\Delta H A}{L \mu}$$

$\frac{dq}{dt}$ = rate of flow

K = permeability coeffecient
ΔH = water head
A = area of specimen
μ = fluid viscosity
L = thickness of specimen

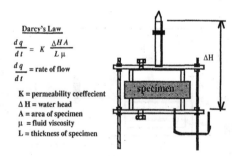

Figure 9: Water Permeability Test Setup

However, due to difficulties in generating desirable crack patterns in concrete specimens and in finding an appropriate method for the permeability measurement, limited research has been done to evaluate permeability of cracked concrete. In the present research, a new technique, feedback controlled splitting tests, is introduced to generate width-controlled cracks in concrete specimens, and then studies the water permeability of the cracked concrete [10].

The feedback controlled splitting test is a Brazilian splitting test controlled by a closed-loop feedback system, as shown in Figure 8. A cylindrical specimen (100 mm in diameter and 25 mm in thickness) was loaded diametrically under Brazilian test configuration. A LVDT was fixed on each side of the specimen, perpendicular to the loading direction, so as to monitor the crack opening displacement (COD). The average displacement of the two LVDTs was digitally calculated and used as the feedback control. The specimens were loaded to a given crack width under the feedback controlled condition; and then, unloaded under force control. The time, force, stroke, and crack opening displacement from each LVDT as well as their average (used as the feedback signal) were recorded during tests.

Water permeability tests were performed on the cracked specimens. The test setup is shown in Figure 9. Permeability coefficient was calculated from the cumulative water flow, based on Darcy's law for a falling water head.

Figure 10 shows a typical result from a feedback controlled splitting test. This specimen was unloaded when the measured crack width reached 350 microns. The remaining cracking in the specimen after unloading was about 180 microns. In the figure, it is noticed that the peak load generally occurred at an average transverse displacement of about 20 microns. If a specimen was unloaded before reaching or even at the peak load, about 80% of the displacement could be recovered, and the remaining crack opening displacement would be very small, less than a few microns. As a result, the crack may show little effect on concrete permeability.

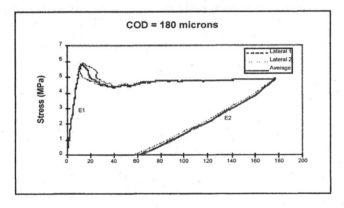

Figure 10: A Typical Result From Feedback Controlled Splitting Test

Figure 11 presents the relationship between water permeability and crack openings. It was observed that when specimens are loaded to crack opening displacements less than 50 microns, crack openings had little effect on concrete water permeability. When the crack opening displacement increased from 50 microns to about 200 microns, water permeability increased rapidly. With crack opening displacements larger than 200 microns, the rate of increase of water permeability becomes steady. Considering concrete deterioration due to cracking, ACI Building Code (ACI 318-89 Commentary, Section 10.6) limits crack widths by limiting the distribution of flexural reinforcement in reinforced concrete design. The Code limitations are based on crack widths of 0.016 in. (400 μm) for interior exposure and 0.013 in. (330 μm) for exterior exposure, which are arbitrary numerical values based on past experience. The present study provides engineers with insight into modifying the current design criteria for a durable concrete and in predicting service life of a concrete structure.

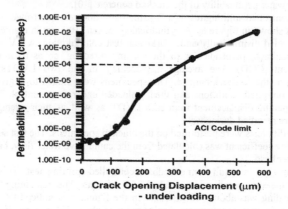

Figure 11: Relationship Between Water Permeability and Crack Widths

It is clear that in addition to crack width, the length, depth and number of cracks also influence concrete permeability. The current research in the ACBM Center on this topic is focusing on the effects of crack length and depth, connectivity and tortuosity, shape and orientation as, well as crack/specimen surface area ratio on concrete permeability. Both water permeability and chloride permeability are being studied.

CONCRETE MIX DESIGN CRITERIA AND PROCEDURE

More and more engineers have realized that a high quality concrete includes both appropriate strength and high durability. However, most currently used codes and specifications still use w/c ratio or compressive strength for concrete quality control. For instance, ACI Building Code (ACI 318-89 Commentary, Chapter 4 -- Durability Requirements) indicates that "Maximum water-cementitious materials ratios of 0.40 to 0.50 that may be required for concrete exposed to freezing and thawing, sulfate soils or waters, or for preventing corrosion of reinforcement will typically be equivalent to requiring an $f'c$ of 5000 psi (34.5 MPa) to 4000 psi (27.5 MPa), respectively." This often misleads engineers to think that durability is directly related to the w/c ratio and strength of concrete.

As discussed in the previous sections, concrete compressive strength is consistent with the water-cementitious material ratio, but not consistent with permeability, which plays the most important role in concrete durability. As shown in Figure 5, concrete having a w/c ratio of 0.38 may have chloride permeability value of 8000 coulombs, while a concrete have a w/c ratio of 0.52 may have chloride permeability value of 4000 coulombs. A concrete having compressive strength of 41.5 MPa may have chloride permeability value in terms of a total charge ranging from 3500 coulombs to 9000 coulombs. That is, designing a stronger concrete cannot ensure a lower permeable concrete. Using strength as a single concrete mix design criterion also misleads that the higher the concrete strength, the better the concrete performance. A study at ACBM indicates that high strength concrete (HSC), with a high cement content, may have a higher risk to drying-shrinkage cracking than normal strength concrete (NSC) [11]. These results imply that HSC does not necessarily has low permeability and a low permeability, high strength concrete is not necessarily durable. With an increasing number of concrete structures exposed to aggressive environmental conditions and increasing strength and durability requested by the modern concrete structures, it seems urgent to establish reliable criteria for concrete mix design.

Based on the results from researches at ACBM and a literature view, Shah and Wang [12] has proposed that to achieve a durable concrete, three criteria may need to be concerned in concrete mix design. The three criteria are strength, permeability, and cracking resistance. The strength criterion is to ensure that concrete is able to resist designed stress without failure. The permeability criterion is to ensure that concrete has a limited flow penetration rate and will not be vulnerable to water/ion-related attacks during a designed period of service life. The crack resistance criterion is to ensure that concrete has a minimum capability to resist the cracking due to environmental conditions, such as thermal and drying shrinkage.

Considering all the three criteria, an improved mix design procedure is proposed in Figure 12. To simplify the mix design procedure, a single crucial factor is considered in the initial mix design to meet the strength and permeability requirement. According to the present research, concrete strength (f'_c) and permeability (k_c) can be simply expressed as a function of water/cement ratio (w/c) and volume fraction of coarse aggregate $(VF)_c$, respectively. As we know, actually, both concrete strength and permeability are not simply related to a single variable. Therefore, it is necessary to have a checking procedure at the end of the mix design process and to ensure the designed concrete satisfies all three mix design criteria. A similar checking step has been used in structural design and should not be ignored in concrete mix design. The press need is to develop appropriate models for the checking.

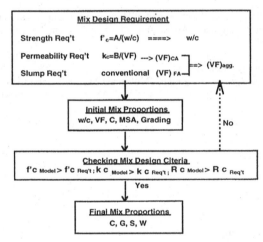

Figure 12: A Proposed Mix Design Procedure For Durable Concrete

The statistical models for predicting concrete strength and permeability for a given mix proportion are being developed at the ACBM Center, such as equation (6) for permeability. However, little work has been done relating cracking resistance to concrete mix proportion. Based on fracture mechanics, the resistance of concrete to crack propagation can by represented by $R = \beta \phi (a-a_0)^d$. Here, R is the resistance of concrete to crack propagation; a, crack length; F and ϕ are known functions; and β and d are material parameter determined from concrete fracture testing [11]. In addition, the creep effect should also be considered in concrete cracking resistance. Further research is needed on this subject.

CONCLUDING REMARKS

Based on the preliminary results from an extensive ongoing research program at the NSF Center of Advanced Cement-Based Materials (ACBM), the present paper explicates the relationships and interactions between concrete microstructure, permeability, cracking, and durability. The models for predicting permeability of cement paste and concrete are introduced. The effect of crack characteristics on concrete permeability is discussed. A statistic model relating concrete permeability to its microstructure, involving phase volume fractions, orientations, sizes, shapes, distribution and connectivity of phases, is being developed. The research indicates that three criteria, strength, permeability, and cracking resistance, need to be considered together in concrete mix design to achieve a durable concrete.

ACKNOWLEDGMENT

The authors gratefully acknowledge the support of the National Institute of Statistical Sciences and the NSF Center for Science and Technology of Advanced Cement Based Materials. The work summarized here has been conducted by several ACBM researchers.

Reference

[1] P. K. Mehta, "Concrete Technology at the Crossroads -- Problems and Opportunities", Concrete Technology -- Past, present, and Future, ACI SP-144, pp. 1-30, 1994

[2] H. F. W. Taylor, *Cement Chemistry*, Academic Press, San Diego, CA, 1990

[3] G. E. Archie, "The Electrical Resistivity Log as an Aid in Determining Some Reservoir Characteristics," AIME Trans., vol. 146, pp. 54, 1942

[4] B. J. Christensen, R. T. Coverdale, R. A. Olson, S. J. Ford, E. J. Garboczi, H.M. Jennings, and T.O. Mason, "Impedance Spectroscopy of Hydrating Cement-Based Materials: Measurement, Interpretation, and Application", J. of Am. Ceramic Society, vol. 77, no. 11, pp. 2789-2804, 1994

[5] A. J. Katz and A. H. Thompson, "Quantitative Prediction of Permeability in Porous Rock," Phys. Rev. B: Condens. Matter, vol. 34, no. 11, pp. 8179-81, 1986

[6] D. Bentz, E. Garboczi, and E. Landgren, "Multiscale Microstructural Modeling of Concrete Diffusivity: Identification of Significant Variables," ASTM Cement, Concrete and Aggregate, in press.

[7] P. Kumar Mehta and Paulo J.M. Monteiro, Concrete - Structure, Properties, and Materials, Second Edition, Prentice Hall, pp. 37, 1993

[8] Sanjay S. Jaiswal, Jeff Picka et. al., "Impact of the Interfacial Transition Zone on the Chloride permeability of Concrete", (to be published)

[9] D. Winslow, et. al., "Percolation and Pore Structure in Mortars and Concrete," Cement and Concrete Research, 24, 25-27, 1994

[10] K. Wang, D.C. Jansen, S.P. Shah, and A.F. Karr, "Permeability Study of Cracked Concrete" vol. 27, No. 3, pp. 381-393, 1997

[11] W. Yang, K. Wang, and S.P. Shah, "Prediction of Concrete Cracking Under Coupled Shrinkage and Creep Conditions", Proc. of the 4th Material Engineering Conference, ASCE, pp. 564-573, Nov. 10-14, 1996, Washington D.C.

[12] S. Shah and K. Wang, "Microstructure, Microcracking, Permeability, and Mix Design Criteria of Concrete" The fifth Int'l conf. on Structural Failure, Durability and Retrofitting, pp. 260-272, Singapore, November 27-28, 1997

CEMENT CLINKER

AND HYDRATION

TRACE ELEMENTS IN CLINKER AND THEIR USE AS "FINGERPRINTS" TO FACILITATE THEIR QUALITATIVE IDENTIFICATION

Ferenc D. Tamás*, **Arezki Tagnit-Hamou**** and **Josef Tritthart*****

*University of Veszprém, Dept. of Silicate and Materials Engineering,
H-8201 VESZPRÉM, P.O.B. 158, Hungary

**University of Sherbrooke, Dept. of Civil Engineering
SHERBROOKE, Québec, Canada J1K 2R1

***Technical University of Graz, TVFA
A-8010 GRAZ, Stremayrgasse 11, Austria

ABSTRACT
Some Hungarian and Austrian clinker and cement samples have been analysed (by ICP/ES - Inductively Coupled Plasma Emission Spectroscopy) for Sr, Ba, Mn, Mg, Ti and Zr content, and analytical data processed by statistical "pattern recognition" methods: dendrograms were constructed and the significance value (Fisher's weight) of the above elements computed.. The recognition is quite good in the case of ordinary Portland clinkers, worse in the case of sulfate resistant clinkers. It is hoped that a larger number of samples, and including other elements into statistical processing would improve results.

Introduction

The problem: how to determine the *origin of cement* in concrete. This is a challenging problem, both scientifically and practically (even in the case of criminal investigation).

The problem can be solved by the analytical determination of certain elements contained in cements, and the statistical processing ("Pattern Recognition") of analytical data. These elements act as „chemical fingerprints", by which the identification of the manufacturer can be made. As an extension, isotopic distributions can help too, but this is beyond the scope of this paper.

Selection of "Fingerprint" Elements

Not all elements are good for identification. Several aspects should be taken into consideration, when selecting the „fingerprint" (dactylogrammatically relevant) elements. Dactylogrammatically relevant elements

1. should not be the main elements, as cements manufactured all over the world contain the same main elements, and even their ratio is very similar, depending only on cement type, and not cement origin;
2. should come from the main raw materials (limestone, marl, clay, etc.) and not from the fuel, from refractories, or from grinding media wear;
3. should be thermally stable, as volatile ones evaporate in hot zones and condense in cooler places, thus their amount may vary by kiln conditions;
4. should be present in measurable quantities;
5. should not form soluble ions at pH > 10, as in the highly basic environment of concrete these ions can migrate.

The above considerations exclude several elements:

1. Main elements: Ca, Si, Al, Fe, S, alkalies. 2. Fuel may contain V (fuel oil), Zn, Pb (waste tyres), etc. Older kilns are frequently lined with magnesite-chrome refractories, thus excluding Cr. Grinding media contain steel alloy elements, thus excluding W, Mn and Cr. 3. Alkali-, Tl- , Sn-compounds (esp. chlorides) are easily volatilized. The two last items exclude rare earths, noble metals, as well as Cr, Mn, U.

What remains after that? It looks probable that best "fingerprint" elements are Sr, Ba, Mn, Mg, Ti, Zr, and possible one or two others.

Sr and Ba are always present in limestones; their quantity and ratio is fairly constant, depending on their geological origin, thus they are dactylogrammatically reliable. Mn, although mentioned above (points 2 and 5), can still be used: to investigate the effect of grinding media wear, pure quartz has been ground in a ball mill with steel media for extended time (3 days), but its Mn content was less than 20 ppm; besides the presence of manganates or permanganates have never been reported.. The Mn content of cement is usually higher by 1 or 2 orders of magnitude. Goguel and StJohn have used the Ca-Sr-Mn plot for identification [1].

Our experience [2] has shown that the Sr-Ba-Mn 3-element approach (3EA) is not fully adequate, especially in the case of sulfate resistant clinkers, having a high iron content. For the time being, three more elements (Mg, Ti and Zr) are added. According to recent experiences, this 6-element approach (6EA) gives a possibility of a more satisfactory identification.

The knowledge of the concentration of the "fingerprint" elements can be used, in itself, for a rough estimation only. For a more satisfactory resolution, analytical data should be statistically processed, using "pattern recognition" methods.

Principles of Pattern Recognition

Pattern recognition methods use measurement data (in our case, analytical data) simultaneously, rather than sequentially, to make significance testing (in our case determine the source of cement). The set of measurements which is used to characterize the sample is called *pattern*. Sometimes even two variables are sufficient to make recognition (e.g. to distinguish between „own" and „foreign" cements); in that case the pattern can be represented graphically by a point, where the two co-ordinates of the point are measurement values and the line from the origin to the point is called the *pattern vector*. In the case of more complicated systems two variables are not sufficient any more; in that case the vector should be drawn in an *n*-dimensional space; or the data processed mathematically to permit a two-dimensional approach. The basis of pattern recognition is that pattern vectors for similar samples lie close together in the pattern space, forming *clusters*. In two dimensions this clustering can be detected easily by the human eye, while in more complicated systems, mathematical methods are needed to detect clustering

There are essentially two different approaches to pattern recognition, depending on whether the *classes* into which a sample may fall are known, or not. In our case the clinker (or cement) may come from one of several sources (the cement factories), and each source forms a class. The role of pattern recognition in this case of *supervised learning* is to identify the class, and hence the source of the sample. In case of *unsupervised learning* the classes are not known, and the purpose of pattern recognition is to see whether the patterns obtained fall into natural groups. Although in our case supervised learning would be adequate, still, to find out the potentialities of clinker characterisation, unsupervised learning has been frequently used and is presented in the experimental part of the paper.

Several methods of pattern recognition exist; in this study, however, only three of them will be used and discussed in detail: the Karhunen-Loeve transformation, the calculation of Fisher's weights, and also dendrograms are constructed.

The KL transformation

One of the most frequently used methods is the Karhunen-Loeve (KL for short) transformation, a display method in which n-dimensional pattern vectors appear in m dimensions, with negligible loss of information. (Obviously $n > m$, in most cases $m = 2$). In KL-transformation an A(k,n) data matrix is constructed of n characteristics of k samples, and from this a co-variance matrix can be constructed:

$$C_{i,j} = \sum_{k=1}^{m} (x_{k,i} - \overline{x_i})(y_{k,j} - \overline{y_j}) \quad \dots [1]$$

where C_{ij} means the j-th element of the i-th line of the co-variance matrix, and $\overline{x}$ and $\overline{y}$ are averages of the x and y variables.

After that the eigenvalues λ_k and eigenvectors V_k are calculated:

$$C . V_k = \lambda_k . V_k \quad \dots [2]$$

Obviously, this data reduction means some loss in information. This loss can be exactly calculated by various statistical methods (omitted here); according to practical experience, this loss is usually less than 5-8%.

Dendrogram construction

In case of unsupervised learning, a *dendrogram* is usually constructed. The construction of a dendrogram can be seen in Figs. 1. and 2. A pattern space showing 5 points is shown in Fig. 1. open circles show the original points. Then the pairs of points which are closest are grouped together and replaced by a new point halfway between them (full circles). This procedure is repeated until all the points are grouped together (this requires in Fig. 1. four stages). The successive stages of this grouping can be shown on the dendrogram in Fig. 2. The vertical axis shows the similarity, S_{ij} between the points i and j, as calculated from

$$S_{i,j} = 1 - d_{i,j}/d_{max} \quad \dots [3]$$

where d_{ij} is the distance between points i and j and d_{max} is the maximum distance between any two points. Obviously the similarity takes values from 1 (coincident points) to 0 (for the two points farthest apart).

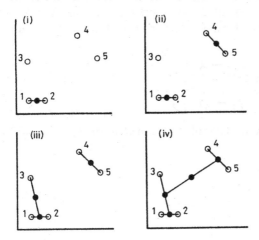

Fig. 1. Successive steps of dendrogram construction.

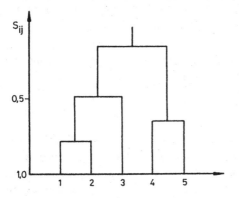

Fig. 2. A dendrogram, constructed of the 5 points of Fig. 1.

Fisher's weights calculation

In case of supervised learning the Fisher's weights calculation can help to select the most significant components (i.e. elements) of the classification, thus elements of low information value can be omitted. The formula for the Fisher's weight calculation, F_i for the i th element between the first (1) and second (2) group:

$$F_i = \frac{[x_{i,1} - x_{i,2}]^2}{S^2_{i,1} - S^2_{i,2}} \quad ....[4]$$

A low F_i value means a lower effectivity of the given element. F_i values < 1.0 are not significant.

Experimental

Materials

Clinker types made in the five Hungarian cement factories, as well as 27 cements made by various Austrian factories and 7 Canadian clinkers and cements have been investigated for this study. The first experiments were done by the 3EA method, using only Sr, Ba and Mn, as dactylogrammatically relevant elements in case of 14 clinker samples (13 Hungarian samples, of the 5 cement factories /BF, B, H, L and D/, and one Austrian sample /A/. Concentrations are shown in Fig. 3.

Characteristic differences can be seen, but a decisive differentiation by this simple method cannot be done; data processing is needed.

Sample preparation

Clinker and cement samples were drawn directly from the kiln at different intervals and from various sites of clinker outlet to ensure a good average sample as far as possible. Clinker samples were crushed to 2 mm, reduced to approximately 20 g and hand-ground in an agate mortar to pass sieve 0.045 mm. In the case of cement samples there was no need of grinding. Small, but hopefully representative samples were used for further tests.

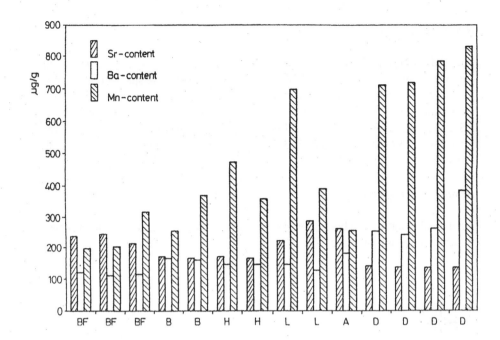

Fig. 3. Sr-, Ba- and Mn content of 14 clinker samples

In the beginning stage approximately 1 g, sample was dissolved in concentrated hydrochloric acid, SiO_2 removed by precipitation (evaporation in the presence of gelatin) and the precipitate washed out several times. The filtrate was analysed by ICP-ES (Inductively Coupled Plasma Emission Spectrography). This method is advantageous when absolute ppm values are needed, as standardization can be done by reagent grade chemicals of known concentration.

Recently the slurry method is being used [3]. For this, the sample, of fineness < 50 μm is suspended in de-ionized water, nebulized in a spray chamber and introduced directly in the plasma unit. This method eliminates tedious sample preparation, but has the drawback that a standardization, with elements in a similar matrix, is needed. (A NIST standard cement sample was used for standardization, although, for our purposes, where only relative values are needed, unstandardized concentration data can be applied too). Comparison between the acid attack vs. slurry methods is in progress.

Analytical method

An ARL/3410 type ICP spectrometer was used, with a mini-plasma torch, using a radiofrequency generator, 27.17 MHz of 650 W power. The spectrometer contains a Czerny-Turner vacuum monochromator. Spectral range 165-800 nm. Computation. EPIC (Evolutionary Program for Instrument Control) software, IBM PS/2 computer, with a PC DOS 3.0 system. The wavelengths used in this study (in nm): Mn = 257.610, Mg = 279.553, Sr = 407.771, Ba = 455.403, Ti = 336.121, Zr = 349.621.

Results and discussion

Analytical results have been published elsewhere [4, 5], and are excluded from this paper, due to lack of space.

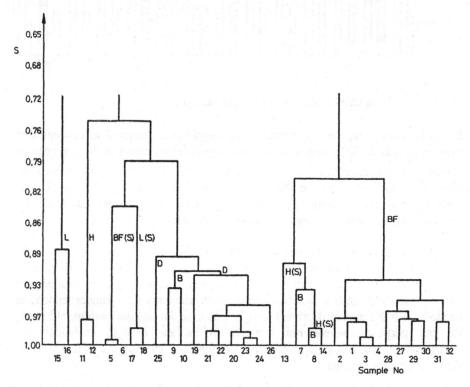

Fig. 4. Dendrogram of 16 Hungarian clinkers (with two parallel analyses) by the 3EA approach

Some results of direct two-dimensional clustering (by the 3EA) were shown abova, in Fig. 3. The dendrogram, constructed of these data is shown in Fig.4. The separation is not satisfactory: although BF samples (with the exception of an SRC clinker) give a good cluster, with S values > 0.9, the others are mixed.

Fisher's weights are shown in Table 1. High differences in the same row, ranging from 0 to > 100 show that a real differentiation between clusters by this method is ineffective.

Table 1.

Significance of elements (Fisher's weights)
of 16 clinkers, manufactured in the 5 Hungarian
cement factories

Factory 1.: **BF** Works (produces OPC and SRC clinkers)
Factory 2.: **B** works (produces OPC clinker)
Factory 3.: **H** works (produces OPC and SRC clinkers)
Factory 4.: **L** works (produces OPC and SRC clinkers)
Factory 5.: **D** works (produces OPC clinker)

	BF/B	BF/H	BF/L	BF/D	B/H	B/L	B/D	H/L	H/D	L/D
Sr	22.7	23.2	0.7	33.1	0.0	93.7	6.2	101.4	6.3	88.7
Ba	8.6	15.2	1.0	27.3	1.0	0.9	2.0	0.2	9.6	4.4
Mn	1.0	0.7	2.8	6.3	0.1	1.5	0.4	0.7	0.0	1.1

Therefore this type of investigations was discontinued, and the 6EA used instead.

As samples analyzed in the beginning stage (Fig. 3. and 4.) were not available, 24 samples coming from two Hungarian factories (two OPC and one SRC clinker) have been investigated.

Results are given in terms of the 6EA approach.

The dendrogram is shown in Fig. 5. An excellent clustering can be seen, as samples from the L plant (OPC clinkers, sample Nos. 11-20, and SRC clinkers, sample Nos. 21-24) as well as from the D plant (OPC clinkers, sample Nos. 1-10) can be easily differentiated.

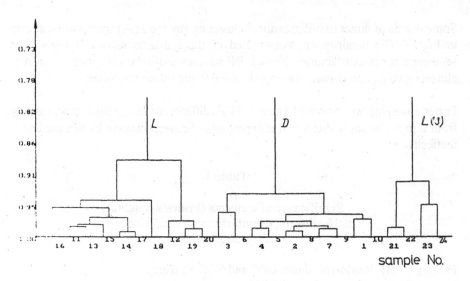

Fig. 5. Dendrogram of 24 Hungarian clinkers by the 6EA approach

Fisher's weights (Table 2.) show, that Mg has the highest significance in all cases, followed by Sr (in two cases) or Ti (in one case)

Table 2.

Significance of elements (Fisher's weights)
of 24 cements, manufactured in 2 Hungarian
cement factories

Grouping: Three groups: **1** Factory D (OPC clinker)   **2** Factory L (OPC clinker)   **3** Factory L (SRC clinker)

	1/2	1/3	2/3
Mg	190.8	46.6	45.8
Zr	1.6	0.6	0.1
Mn	0.2	2.1	3.0
Ba	4.9	6.3	3.0
Ti	4.1	2.5	14.1
Sr	41.4	27.2	4.9

Several Austrian factories have supplied cement samples; 27 different sorts have been received, and, by making parallel tests, 54 samples analyzed. Here the origin of samples was unknown, only codenumbers given. Dendrogram, by the 6EA approach, is shown in Fig. 6. As usual, parallel samples are the closest. Still, five different groups could easily been differentiated, by S values > 0.84. (Groups 1 and 2 merge at $S = 0.82$, groups 3 and 4 at $S = 0.75$, while group 5 is the most different from all others).

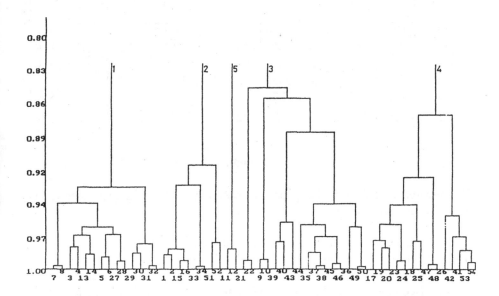

Fig. 6. Dendrogram of 27 Austrian cements (with two parallel analyses) by the 6EA approach

Fisher's weights are shown in Table 3. Again, Mg has the highest significance; Mn is significant in the differentiation of cements of the 1 and 5 as well as in the 4 and 5 groups. Group 5 is differs from all others by its high Zr content. It is quite interesting that Sr, which had a high F value in former samples is absolutely insignificant here.

Table 3.

Significance of elements (Fisher's weights)
of 27 cements, manufactured in Austrian
cement factories

Grouping: Five groups, according to the dendrogram of Fig. 6.

	1/2	1/3	1/4	1/5	2/3	2/4	2/5	3/4	3/5	4/5
Mg	19.4	50.3	17.6	978.4	82.4	40.0	879.0	5.1	16.2	53.3
Zr	1.1	0.0	1.3	8.9	0.6	0.0	52.7	0.7	8.1	292.0
Mn	0.2	0.2	0.1	119.4	0.4	0.0	46.6	0.3	10.3	53.8
Ba	2.3	0.1	0.2	13.4	2.1	1.4	32.8	0.4	5.3	17.2
Ti	0.5	0.2	1.3	4.7	0.0	0.1	12.9	0.3	8.5	38.5
Sr	0.0	0.1	0.0	0.0	0.0	0.1	0.0	0.5	0.1	0.4

Outlook

The research described in this paper is necessarily only a beginning step, to be continued in three directions:

1. Analysis of already investigated and new clinker samples for still more elements, which may be dactylogrammatically relevant.

2. Continuation of research to study cement identication and finally the same in case of concrete (here the separation of the effects of hydraulic additives and aggregates will cause serious problems).

3. The establishment of national, then regional and finally worldwide databases. This sort of research definitely needs international cooperation. RILEM, (Réunion Internationale des Laboratoires d'Essais et de Recherches sur les Matériaux and les Constructions - International Union of Testing and Research Laboratories for Materials and Structures), the prestigious international association has recently established the formation of an international Technical Committee "QIC" (Qualitative Identification of Clinkers and Cements), to harmonize investigationns and create the database, mentioned in # 3. Besides some European countries, North American, South African and some research groups of Asia or the Pacific region are members of this TC, where the first and the second author serve as Chairman and Secretary, respectively.

REFERENCES

[1]   Goguel, R.L. and StJohn, D.A. (1993): Chemical identification of Portland cements in New Zealand concretes, Part I. Characteristic differences among New Zealand cements in minor and trace element chemistry. Cem. & Concr. Res., _23_, 59-68, Part II. The Ca-Sr-Mn plot in cement identification and the effect of aggregates. Cem. & Concr. Res., _23_, 283-293

[2]   Tamas, F.D. (1996): Pattern recognition methods for the qualitative identification of Hungarian clinkers. World Cement/Res. & Development _27_, 75-9

[3]   Goodall, P., Foulkes, M.E. and Ebdon, L (1993): Slurry nebulization inductively coupled plasma spectrometry - The fundamental parameters discussed. Spectrochimica Acta, _48B_ 1563-77

[4]   Tamas, F.D. and Kristof-Mako, E. (1996): Chemical "fingerprints" in portland cement clinker. In.: Advances in Building Materials Science; "Festschrift Wittmann", edited by A. Gerdes. Aedificatio Publishers, Freiburg, Germany, p. 217-28

[5]   Tamas, F.D., Patkai-Horvath, M., Kristof-Mako, E. and Tritthart, J. (1997): Qualitative identification of clinkers and cements - some results and possibilities. Proc. 10th Intl. Congress on the Chemistry of Cements, Göteborg, Sweden. Vol. 3, paper 3v010

ACKNOWLEDGEMENTS

The financial support of OTKA (Hungarian National Scientific Research Fund), Project No. T026307 as well as of theAustro-Hungarian Action (AÖU), Project No. P29Ö11 is gratefully acknowledged.. Further acknowledgement is due to the Vác Factory of the Duna-Dráva Cement Ltd. (Hungary) for sponsoring the senior author's participation at the Diamond Symposium.

FORMATION OF JENNITE AND TOBERMORITE AND THEIR RELATIONSHIP WITH C-S-H GEL IN HYDRATED CEMENT PASTE

Naomichi HARA, Norihiro INOUE and Hiroaki NOMA
Kyushu National Industrial Research Institute
Tosu, Saga-ken, 841-0052, Japan

ABSTRACT

Amorphous siliceous materials, viz. fumed silica, silica glass, volcanic glass and rice husk ash, were hydrothermally reacted with lime at temperatures ranging from 60 to 120°C. Jennite was formed at high Ca/Si mole ratios and 1.4 nm tobermorite at low Ca/Si mole ratios. In the presence of alumina, the formation of 1.4 nm tobermorite was replaced by that of 1.1 nm tobermorite. In addition, an increase of reaction time tended to replace the formation of jennite by that of 1.1 nm tobermorite. The boundary between the formation of jennite and that of 1.4 nm tobermorite was situated at about a Ca/Si mole ratio of 1.0, which coincides with a discontinuous phase change of C-S-H gels from a tobermorite-like structure to a sorosilicate-like structure.

INTRODUCTION

The present authors have been studying hydrothermal syntheses of calcium silicate hydrates using amorphous siliceous materials such as fumed silica, silica glass, volcanic glass, rice husk ash, etc. and have previously reported that jennite and 1.4 nm tobermorite are formed reproducibly in hydrothermal reactions of these amorphous siliceous materials with lime at temperatures between 60 and 120°C (1-7). However, the formations of jennite and 1.4 nm tobermorite were discussed separately in those papers.

Presently, the relationship of jennite and 1.4 nm tobermorite with C-S-H gel in hydrated cement paste are often discussed, but sufficient experimental data are lacking for making a clear image of the structural model of the latter. With the aim of supplementing the present state of knowledge with some new information derived from jennite and 1.4 nm tobermorite, we have reexamined the boundary region between the formation of jennite and that of 1.4 nm tobermorite in relation to C-S-H gels. These results are reported in this paper.

EXPERIMENTAL

The starting materials used for the preparations are listed in Table I. Volcanic glass was obtained by hydraulically separating the glassy part from volcanic ash sampled in Kagoshima Prefecture, South Kyushu, which is locally called 'SHIRASU' (which means 'white sand') . Rice husk ash was prepared by combustion of rice husk at relatively low temperatures to give amorphous silica.

The reaction conditions were selected for the respective siliceous materials according to the aims of each preparation within the following ranges : starting Ca/Si mole ratios = 0.6 - 2.0, reaction temperatures = 60 - 120°C, reaction time = 2 - 300 days, addition of Al_2O_3 = 0.5 - 8.8 wt.% . However, the water/solid weight ratio was mostly set at 20.

The CaO was first mixed with deionized water, then the respective siliceous materials, along with Al_2O_3 when needed, were added and well mixed to give homogeneous suspensions. The suspensions were sealed in stoppered polyethylene bottles and placed in a water bath for reaction temperatures lower than 90°C. For temperatures over 100°C, autoclaves were used. At required intervals, a portion of each preparation was pipetted, filtered and dried in a vacuum desiccator at room temperature. Reaction products were identified by XRD, DTA-TG, SEM, TEM and SED.

Table I. Chemical composition, fineness and preparation of starting materials

Materials	Chemical composition, fineness and preparation
Fumed silica	SiO_2 99.8 % ; 200±25 m^2/g (BET fineness). * Produced by oxidizing $SiCl_4$ (Aerosil 200 : Nippon Aerosil).
Silica glass	SiO_2 99.97 % ; 1550 cm^2/g (Blaine fineness) (Fused quartz).
Volcanic glass	Ig. loss 2.7, SiO_2 74.3, Al_2O_3 12.7, Fe_2O_3 1.3, CaO 1.5, MgO 0.2, Na_2O 4.4, K_2O 3.0 % ; 5950 and 3460 cm^2/g (Blaine fineness).
Rice husk ash	Ig. loss 5.04, SiO_2 91.26, Al_2O_3 0.58, Fe_2O_3 0.60, TiO_2 0.09, CaO 0.54, MgO 0.32, Na_2O tr., K_2O 1.04, P_2O_5 0.45 % *Amorphous; Coarsely ground.
CaO	Prepared by calcining precipitated $CaCO_3$ at 1,000-1,100°C for 4 hours.
Al_2O_3	γ-Al_2O_3 with particle diameter $\leqq 0.05 \mu$m (Micropolish: Buehler).

RESULTS AND DISCUSSION

Reaction products from fumed silica obtained at a reaction temperature of 80°C along with their basal spacings are summarized in Table II.

1.4 nm tobermorite was formed only at a starting Ca/Si mole ratio 0.9. By contrast, jennite was formed over the wide range of starting Ca/Si mole ratios from 1.1 to 1.6. Of these, a Ca/Si mole ratio of 1.15 was optimum, at which well crystalline jennite was formed within a short reaction time. Addition of Al_2O_3 retarded or

Table II. Reaction products from fumed silica and their basal spacings (nm)
(Reaction temperature : 80°C)

Start. Ca/Si.	Al_2O_3 %	Reaction time, days				
		10	20	40	70	100
0.8	0	1.41 C	1.43 C	1.37 C	--	--
0.9	0	1.41 T	1.41 T	1.40 T	1.40 T	1.40 T
1.0	0	1.41-1.25C	1.41-1.25C	1.25-1.18C	1.25-1.16C	1.25-1.13C
1.1	0	1.26 C	1.23 C	1.05 J	1.05 J	1.05 J
	2	1.26 C	1.19 T	1.19 T	1.18 T	1.17T,1.04J
	5	1.17 T	1.19 T	1.18 T	1.17 T	1.18 T
	8.8	1.22 T	1.23 T	1.24 T	1.21 T	1.21 T
1.15	0	1.06 J	1.05 J	1.05 J	1.04 J	1.04 J
	2	1.26 C	1.05,1.34C	1.05 J	1.05J,1.17T	1.04J,1.17T
	5	1.27 T	1.23 T	1.24 T	1.25 T	1.21 T
	8.8	1.28 T	1.30 T	1.26-1.28T	1.27 T	1.25 T
1.2	0	1.41-1.16C	1.05 J	1.04 J	1.04 J	1.04 J
	2	1.32 C	1.05 J	1.19T,1.05J	1.17T,1.05J	1.05J,1.17T
	5	1.28 C	1.28 T	1.25 T	1.26 T	1.26 T
	8.8	1.5-1.7 C	1.5-1.7 C	1.5-1.7 C	0.99 J	0.99 J
1.3	0	C-S-H	C-S-H	1.04 J	1.04 J	1.04 J
	8.8	C-S-H	1.47 C	1.3 C	1.29 T	1.30T,1.01C
1.4	0	C-S-H	C-S-H	1.04 C	1.04 J	1.04 J
	8.8	1.58C, Hy	1.58C, Hy	1.58C, Hy	1.47C, Hy	1.47C, Hy
1.5	0	C-S-H	C-S-H	C-S-H	1.04 C	1.04 J
1.6	0	C-S-H, CH	C-S-H, CH	C-S-H, CH	C-S-H	C-S-H, Af, 1.05 J
1.7	0	C-S-H, CH	C-S-H, CH	C-S-H, CH	C-S-H,CH,Af	Af, CH
2.0	0	C-S-H, CH	C-S-H, CH	C-S-H, CH	C-S-H, CH	C-S-H, CH

T : Tobermorite.   J : Jennite.   C : C-S-H showing weak and diffuse basal reflections.
C-S-H : C-S-H showing no basal reflections.   CH : Ca(OH)$_2$.   Af : Afwillite.
Hy : Hydrogarnet (C$_3$ASH$_4$).   --: Untried.

prevented the formation of jennite, but accelerated the formation of 1.1 nm tobermorite instead. The basal spacing of tobermorite thus formed showed a tendency to increase due to Al-substitution, which was often accompanied by a decrease of the crystallinity of tobermorite. The addition of 8.8 % Al_2O_3 for the preparations with a starting Ca/Si mole ratio of 1.2 gave jennite having the basal spacing 0.99 nm. Hydrogarnet (C_3ASH_4) was formed at a starting Ca/Si mole ratio 1.4 with the addition of 8.8 % Al_2O_3. Starting Ca/Si mole ratios 1.6 and 1.7 led to the formation of afwillite.

In other preparations at starting Ca/Si mole ratios 1.1, 1.15 and 1.2 using

Table III. Reaction products from silica glass, volcanic glass and rice husk ash, and their basal spacings (nm)
(Reaction temperature : 80 and 90°C)

Start. Ca/Si.	Al_2O_3 %	Reaction time, days				
		10	20	40	70	120
Silica glass,	80°C					
0.6	0	1.40T	1.40 T	1.40 T	1.40 T	1.40 T
	2	C-S-H	1.18 T	1.16 T	1.16 T	--
0.7	0	1.40 T	1.40 T	1.40 T	1.40 T	1.40 T
	1	C-S-H	1.21 T	1.21, 1.40T	1.21, 1.40T	--
	2	C-S-H	1.18 T	1.16 T	1.16 T	--
	5	C-S-H, CH	1.00J, CH	1.02 J	1.18 T	--
0.8	0	C-S-H	1.40 T	1.40 T	1.40 T	1.40 T
0.9	0	C-S-H	1.40 T	1.40 T	1.40 T	--
Silica glass,	90°C					
0.6	0	1.40 T	1.40 T	1.40 T	1.40 T	--
0.9	0	1.40 T	1.40 T	1.40, 1.26T	1.40, 1.26T	--
Volcanic glass, 80°C						
0.6	0	C-S-H, CH	0.99 J	1.18T,1.00J	1.18T,1.00J	--
0.7	0	C-S-H, CH	0.99 J	1.18T,1.00J	1.18T,1.00J	--
	#	C-S-H, CH	0.99J, CH	1.00 J	1.00J, 1.23-1.3T	--
Reaction time, days		7	28	91	150	
Rice husk ash, 80°C						
0.8	0	C-S-H	1.05 J	1.40,1.16T	1.40,1.15T	

T : Tobermorite. J : Jennite. CH : Ca(OH)$_2$. -- : Untried.

\# : Volcanic glass with Blaine fineness 3460 cm^2/g was used in this preparation. For the others, that with 5950 cm^2/g was used.

* Reaction time for the preparations from rice husk ash was different from the others.

** Hydrogarnet (C_3ASH_4) was formed in most of the preparations from volcanic glass, but not listed in this table.

fumed silica without the addition of Al_2O_3, we confirmed that jennite was formed in the reaction at 60°C for 70 days and at 100°C for 40 days. However, 120°C was too high to obtain well crystalline jennite (7).

Reaction products from silica glass, volcanic glass and rice husk ash along with their basal spacings are summarized in Table III.

Silica glass was a suitable material to synthesize 1.4 nm tobermorite. At starting Ca/Si mole ratios from 0.6 to 0.9, well crystalline 1.4 nm tobermorite was obtained. Elevation of the reaction temperature to 90°C accelerated its formation, but at longer reaction times 1.4 nm tobermorite gradually degenerated and tobermorite having a shorter basal spacing started to form. This tendency was remarkable at higher Ca/Si mole ratios. Reaction temperatures of 70 and 100°C were also tried using silica glass (4). At 100°C, 1.4 nm tobermorite was formed in 2 days, but 1.1 nm tobermorite emerged as a dominant phase in 7 days at a starting Ca/Si mole ratio 0.9. At 70°C, 1.4 nm tobermorite was formed but its formation was delayed considerably.

From volcanic glass, jennite having the basal spacing 0.99-1.00 nm was first formed and then replaced by the formation of Al-substituted 1.1 nm tobermorite. This jennite was a little different from ordinary jennite prepared from fumed silica without the addition of Al_2O_3, but similar to that prepared with the addition of 8.8 % Al_2O_3, which had the basal spacing 0.99 nm as shown in Table II. In the latter preparation, the Al_2O_3 content of 8.8 % nearly corresponds to that of volcanic glass. Jennite of this kind was obtained from silica glass with the addition of 5 % Al_2O_3, too, as shown in Table III. We tentatively designate this jennite as 'Al-bearing jennite' and characteristics of both ordinary and Al-bearing jennite are compared elsewhere (6). The rate of reaction was affected by the fineness of volcanic glass, viz. the finer glass reacted faster.

From rice husk ash, ordinary jennite was first formed and then replaced by the formation of 1.4 nm tobermorite along with less 1.1 nm tobermorite.

SEM images of jennite and 1.4 nm tobermorite are given in Fig. 1 and 2. The jennite was blade-shaped or lath-like crystals well elongated along *b* with (001) cleavage. The 1.4 nm tobermorite was platy with (001) cleavage and had a tendency to be elongated along *b*. The precursor of jennite was fibrous or fibre bundle-like, which looked similar to C-S-H(II) (8) and 'Type I' C-S-H gel (9). The precursor of 1.4 nm tobermorite was rather foil-like, which looked similar to C-S-H(I) (8). There was no distinct difference in morphology between 1.4 nm tobermorite and 1.1 nm tobermorite, the latter of which was prepared with the addition of Al_2O_3 or from volcanic glass.

Fig. 3 shows a TEM image of the preparation from rice husk ash where jennite and 1.4 nm tobermorite along with their precursors coexisted. A crystal designated by 'T' in Fig. 3 was 1.4 nm tobermorite lying on (100), which was a very rare case, and its SED pattern is given in Fig. 4. Reflections with odd values of k gave spots but did not show continuous streaks along c^* such as observed in 1.1 nm tobermorite from Loch Eynort (8). This seems to suggest the absence of stacking

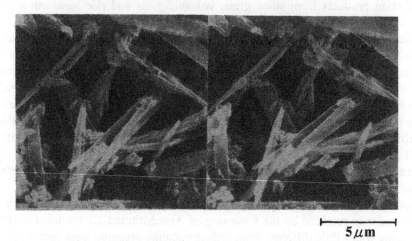

Fig. 1. SEM stereoscopic image of jennite prepared from fumed silica.
(Starting Ca/Si = 1.4,   80°C,   100 days)

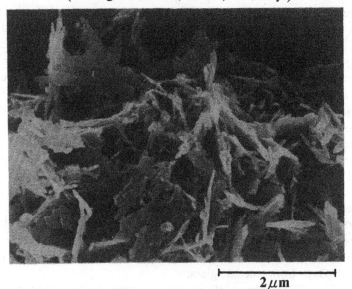

Fig. 2. SEM image of 1.4 nm tobermorite prepared from silica glass.
(Starting Ca/Si = 0.7,   100°C,   2 days)

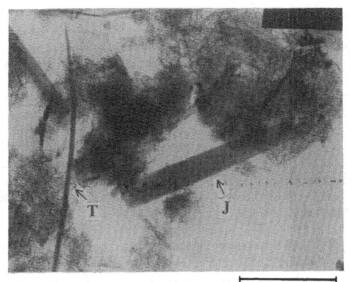

Fig. 3. TEM image of reaction product from rice husk ash.
(J : Jennite,   T : 1.4 nm Tobermorite)

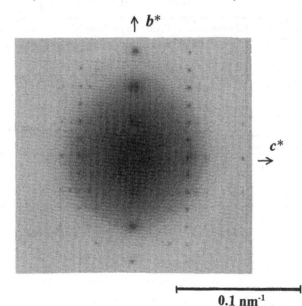

Fig. 4. SED pattern of 1.4 nm tobermorite lying on (100).

disorder along *c*. Many preparations were examined by TEM and SED, but coexistence of jennite and tobermorite within one crystal could not be observed. Even in the preparations where XRD showed coexistence of both phases, jennite and tobermorite were always observed as separate crystals showing different morphologies and SED patterns.

Ca/Si mole ratios determined by energy dispersive X-ray microanalysis of tobermorite and jennite crystals including their precursors in the preparations from silica glass and volcanic glass are summarized in Tables IV and V.

The 1.4 nm tobermorite as well as its precursor showed low Ca/Si mole ratios compared with 1.1 nm tobermorite. Jennite had a higher Ca/Si mole ratio than tobermorite. In terms of Ca/(Si+Al) mole ratio, its ratio was still high. The Al/(Si+Al) mole ratio of 1.6-1.8 of reaction products from volcanic glass was similar to that of volcanic glass, i.e. 1.7.

Table IV. Ca/Si mole ratios of tobermorites prepared from silica glass
(Starting Ca/Si = 0.7,   Reaction temperature : 80°C)

Reaction product	1.4 nm Tobermorite				1.1 nm Tobermorite*
Reaction time, days	5	10	20	40	20
Foil	1.00	0.82	0.70	0.65	1.38
Plate	#	0.81	0.78	0.78	0.89

Foil : Precursor of tobermorite.   Plate : Crystalline tobermorite.   # : Unobserved.
* 2 wt. % of γ-Al_2O_3 (vs. solids) was added to this preparation.

Table V. Ca/Si mole ratios of jennite and tobermorite prepared from
volcanic glass   (5950 cm^2/g)
(Starting Ca/Si = 0.8,   Reaction temperature : 80°C)

Reaction product	Jennite		1.1 nm Tobermorite		
Reaction time, days	20	30	20	30	40
Ca/Si	1.38#	1.15	1.75*	0.94**	0.93
Ca/(Si+Al)	1.13	0.95	1.46	0.78	0.79
Al/(Si+Al)	0.18	0.17	0.17	0.17	0.16

Fibre bundle : Precursor of jennite.
* Crumpled foil and ** Thin plate : Precursor of tobermorite.

CONCLUSIONS

1. Both jennite and 1.4 nm tobermorite are formed at similar temperature ranges below 100°C. In order to obtain well crystalline phases, the reaction temperature 80°C is recommended but it should be noted that both phases are semi-stable at these temperatures.

In terms of stability, jennite is more stable at high Ca/Si mole ratios. On the contrary, 1.4 nm tobermorite is more stable at low Ca/Si mole ratios. The stability of 1.1 nm tobermorite is situated between them.

2. The presence of Al_2O_3 accelerates the formation of 1.1 nm tobermorite at low starting Ca/Si mole ratios and that of Al-bearing jennite at high Ca/Si mole ratios. Similar relations between ordinary jennite and 1.4 nm tobermorite are observed between Al-bearing jennite and Al-substituted 1.1 nm tobermorite.

3. Both phases are formed by a dissolution-precipitation process. Therefore, the particle size of the siliceous material markedly affects the total progress of the reaction. In the preparations using fumed silica (average particle diameter = about 12 nm), the dissolution of silica and the formation of C-S-H, where the Ca/Si mole ratio of C-S-H seems to be near to the initial starting Ca/Si mole ratio, proceeds within a short period of time and then crystallization follows. In the case of silica glass, volcanic glass and rice husk ash, the dissolution of siliceous material continues for a longer period of time. At an early age, C-S-H with a Ca/Si mole ratio higher than that of the starting Ca/Si mole ratio is formed and then the supply of silicate anions due to the dissolution of the un-reacted part of the siliceous material continues for a longer period of time to decrease the Ca/Si mole ratio of C-S-H along with the progress of crystalliza-tion. It seems that the latter condition is favourable for the formation of to-bermorite.

4. Morphologically, the precursor of jennite looks very similar to C-S-H(II) and 'Type I' C-S-H gel, and that of tobermorite looks similar to C-S-H(I). In SED pattern, 1.4 nm tobermorite lying on (100) shows reflections with odd values of k as spots. Continuous streaks along c^* such as observed in 1.1 nm tobermorite from Loch Eynort are not observed. This seems to suggest the absence of stacking disorder along c.

5. It is noticeable that the boundary between the formation of jennite and that of 1.4 nm tobermorite is situated at about a Ca/Si mole ratio of 1.0, which coin cides with a discontinuous phase change of C-S-H gels from a tobermorite-like structure to a sorosilicate-like structure reported by Grutzeck et al (10). Further investigation on the crystallization process from precursors to both jennite and tobermorite especially at lower temperatures is needed to clarify these relations.

REFERENCES

[1]N. Hara and N. Inoue, "Formation of 14 Å Tobermorite from Silica Glass", *Yogyo-Kyokai-Shi (Journal of the Ceramic Society of Japan)*, **84** [4] 181-85 (1976).

[2]N. Hara, N. Inoue and O. Matsuda, "Hydrothermal Reaction of Volcanic Ash

(Shirasu) with Lime at 80° and 100°C", *Yogyo-Kyokai-Shi (Journal of the Ceramic Society of Japan)*, **87** [2] 86-94 (1979).

[3]N. Hara and N. Inoue, "Hydrothermal Reaction of Silica Glass with Lime at 70°-100°C", *Yogyo-Kyokai-Shi (Journal of the Ceramic Society of Japan)*, **87** [3] 134-41 (1976).

[4]N. Hara and N. Inoue, "Formation of 10Å and 14Å Tobermorite from Pozzolanic Glassy Silica", *Proceedings of the 7th International Congress on the Chemistry of Cement*, Paris 1980, Vol. **3**, IV-13-18 (1980).

[5]N. Hara and N. Inoue, "Formation of Jennite from Fumed Silica", *Cement and Concrete Research*, **10**, 677-82 (1980).

[6]N. Hara and N. Inoue, "Formation of Jennite and its Thermal Behaviour", *Proceedings of the 1st International Symposium on Hydrothermal Reactions*, Yokohama 1982, 849-58 (1982).

[7]N. Hara, N. Inoue, H. Noma and T. Hasegawa, "Formation and Characterization of Jennite", *Proceedings of the 8th International Congress on the Chemistry of Cement*, Rio de Janeiro 1986, Vol. **3**, 160-66 (1986).

[8]J. A. Gard, "Electron Microscopy and Diffraction"; pp. 243-270 in *The Chemistry of Cements, Vol. 2*. Edited by H. F. W. Taylor. Academic Press, London, 1964.

[9]S. Diamond, "Cement Paste Structure–An Overview at Several Levels"; pp. 2-30 in *Hydraulic Cement Pastes: Their Structure and Properties*. Cement and Concrete Association, Slough, U.K., 1976.

[10]M. Grutzeck, J. LaRosa-Thompson and S. Kwan, "Characteristics of C-S-H Gels", *Proceedings of the 10th International Congress on the Chemistry of Cement*, Gothenburg, Sweden 1997, Vol. **2**, 2ii067 (1997).

PORTLAND CEMENT WITHOUT GYPSUM

Wiesław Kurdowski,
Sywester Duszak,
Barbara Trybalska
University of Mining and Metallurgy,
Al. Mickiewicza 30, 30-059 Kraków, Poland

ABSTRACT

Among many other papers Prof. Diamond published the results of research on gypsum free cement containing sodium lignosulfonate and sodium bicarbonate in 1980 in World Cement. We undertook similar studies using portland cement rich in silicates. The results showed that the content of sodium bicarbonate added together with sodium lignosulfonate had a very significant effect on the setting time. In the absence of gypsum the cement paste presented the behaviour of a classic Newtonian liquid. The content of water addition can be easily diminished to 0,37. Consequently, the porosity of the paste was much lower, for example for 90 days of hydration it reached about 18 mm^3/g and 58 mm^3/g for cement without and with gypsum, respectively. However, after 2 days of hydration, the porosity of the paste from cement without gypsum was a little higher. It is caused by the diminished velocity of hydration by lignosulfonate, which is confirmed also by the X-ray analysis. After 2 days of hydration no lines of $Ca(OH)_2$ can be found and after 90 days relatively high quantities of unhydrated calcium silicates were present in the paste.

In spite of the low velocity of hydration the strengths after 28 days and 90 days were higher in the presence of lignosulfate.

INTRODUCTION

It is well known that calcium sulphate, normally used as a set regulator, causes the flocculation of the cement paste suspension [1]. These flocs are composed of many cement grains and water is trapped between them. As a result the cement water demand is high and has the negative consequences on the behaviour of concrete, namely lowering strength and durability of this material.

It was the main reason of many works done principally in the US, which developed another alternative of the cement set regulator, eliminating gypsum [2-6].

Basing on Rebinder's paper [7] Brunauer and associates [2-6] elaborated the combined regulator composed of calcium lignosulfonate and potassium carbonate. To obtain the fine ground clinker the grinding aids were also used [2-6]. This combined regulator acted simultaneously as a set-regulating and cement paste dispersing agent. Potassium carbonate counterbalanced the retarding action of lignosulfonate and accelerated the cement hydration. In these conditions the cement paste can be prepared with low w/c ratio and, consequently, the porosity of the hardened paste will also be low. Brunauer [2] called this gypsum-free cement „low porosity cement".

Odler [8] summarized the results of Brunauer school at the 6th International Congress on Cement Chemistry, namely the addition of calcium lignosulfonate must lie between 0,25-1,5%, alkali carbonate between 0,25-1,25% and grinding aid between 0,1-0,5%.

In his later work Odler [9] used sodium lignosulfonate and sodium carbonate to control the setting process of the portland cement.

Diamond [10, 11] presented very interesting papers on this subject in 1978 explaining clearly the basic properties of „low porosity" portland cement pastes.

In the light of a quick development of High Performance Concrete in the last decade the problem of water demand of the cement paste became even more up-to-date and it can be foreseen that the importance of „low porosity cement" will increase considerably.

A program related to the Technology of High Performance Concrete with the use of „low porosity cement" is being carried out in our laboratory.

The technology of „low porosity cement" is the first topic of this program.

This paper presents the preliminary results on this gypsum free cement.

MATERIALS AND METHODS

The industrial clinker, rich in silicates, was used. Its analysis and mineralogical composition, calculated according to Bogue, are presented in Table I

The following admixtures were used as a set-regulator: pure sodium lignosulfonate, produced by Grace, and as a source of carbonate ions alkali bicarbonates and carbonates, all chemically pure reagents.

Pure gypsum was added for comparison and chemically pure sodium and potassium chlorides were also tested.

Clinker was ground in laboratory mill to the specific area of about 3000 cm^2/g (Blaine) and was mixed with additions for 10 minutes in a small mill. Vicat needle was used for setting time determinations and a Vicat rod for water demand. ISO method was adopted for strength measurements, although the water - cement ratio was not kept constant and is given for each sample.

TABLE I
Chemical and mineralogical composition of clinker

Chemical composition		Mineralogical composition	
loss of ign.	0,28%	C_3S	66%
SiO_2	22,62%	C_2S	15%
Al_2O_3	4,76%	C_3A	9%
Fe_2O_3	2,09%	C_4AF	6%
CaO	68,22%		
MgO	0,82%		
SO_3	0,38%		
Na_2O	0,10%		
K_2O	0,42%		
Free CaO	0,80%		

The rheological properties of all pastes were studied by a coaxial cylinder viscosimeter „Rheotest". The porosity of selected pastes was measured by a mercury intrusion porosimeter.

The kinetic of cement hydration was followed by microcalorimetry and cement paste composition and microstructure were studied by X-ray and SEM.

RESULTS AND DISCUSSION

The pastes with addition of 0,8 lignosulfonate presented a very long setting time (Table II). The decrease of this admixture to 0,5% with a simultaneous addition of 1% or 1,5% of alkali bicarbonate or carbonate gave pastes with a good setting time. However, the difference between the initial and final set is relatively short and lasts 16-20 minutes. The pastes with sodium bicarbonate (samples 3 and 4) are exceptions for which this difference lasts from one to two hours.

The rheological tests showed that all the pastes with carbonate ions behaved almost as Newtonian fluids with a very small yield stress in the range of 2,1 to 12,8 Pa. The higher values are typical for bicarbonates, the lower for carbonates. The flow curves are reversible in case of carbonates and present a slightly thixotropic behaviour in case of bicarbonates (Fig 1).

TABLE II

TABLE II
Setting time

Sample	Admixtures	Initial, h, min	Final, h, min	w/c ratio
1.	5% gypsum,	2^{24}	4	0,31
2.	0,8% lgns 1,4% NaHCO$_2$,	8	-	0,28
3.	0,5% lgns 1,0% NaHCO$_3$	3^{15}	5^{03}	0,28
4.	0,5% lgns 1,5% NaHCO$_3$	6^{05}	7^{10}	0,25
5.	0,5% lgns 1,5% Na$_2$CO$_3$	2^{40}	2^{58}	0,22
6.	0,5% lgns 1,5% KHCO$_3$	2^{42}	2^{58}	0,25
7.	0,5% lgns 1,5% K$_2$CO$_3$	2^{25}	2^{45}	0,22
8.	0,8% lgns 1% NaCl	0^{03}	0^{05}	0,25
9.	0,5% lgns 1% NaCl	0^{03}	0^{05}	0,25
10.	0,5% lgns 1,5% KCl	0^{03}	0^{05}	0,23
11.	0,5% lgns 0,5% KCl	0^{03}	0^{05}	0,22

The samples with chlorides presented a flash set .

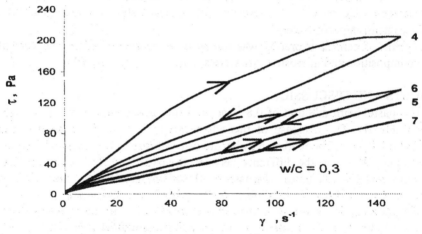

Fig 1. Flow curves for selected pastes

The yield stress increases with time but the change is extremely small in the first hour for sodium compounds and K$_2$CO$_3$ but this increase is much higher for potassium bicarbonate (Fig 2).

As it could be expected, the rheological properties of the sample with NaCl can be tested only with a higher w/c ratio, equal to 0,5. The yield value is about 52 Pa. This paste has a antitixotropic character which can be explained by a quick set.

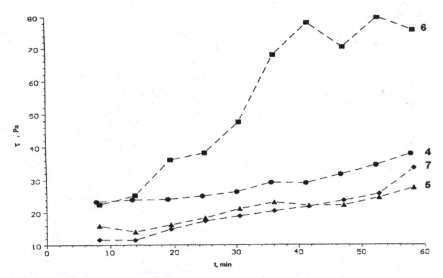

Fig 2. Stress variation vs time for selected pastes

According to microcalorimetric curves (Fig 3) the pastes can be divided into three groups: chlorides, bicarbonates and carbonates.

With the addition of NaCl, immediately after mixing of cement with water, a great peak appears after which a relatively long induction period is marked, followed by a lengthened hydration process of silicates.

In the presence of bicarbonates three peaks appear on the microcalorimetric curve. The hydration of silicates starts after a second, long dormant period. It is a result of a strong retardation of cement hydration by Na-lignosulfonate.

The carbonates belong to the third group. There is a short hydration, practically without the induction period, but the heat of hydration is small (Table III).

Assuming that the degree of hydration is proportional to the heat of hydration we can estimate that in case of K_2CO_3 the hydration after 70 hours presents only 30% in comparison to the standard (in case of Na_2CO_3 even lower). $KHCO_3$ counterbalanced in the best manner the retardation effect of Na-lignosulfonate and the heat of hydration after 72 hours is on the level of 63% of the standard. However, it is not expressed in the strength which was higher after 3 days for sample 7, containing K_2CO_3 (Table IV).

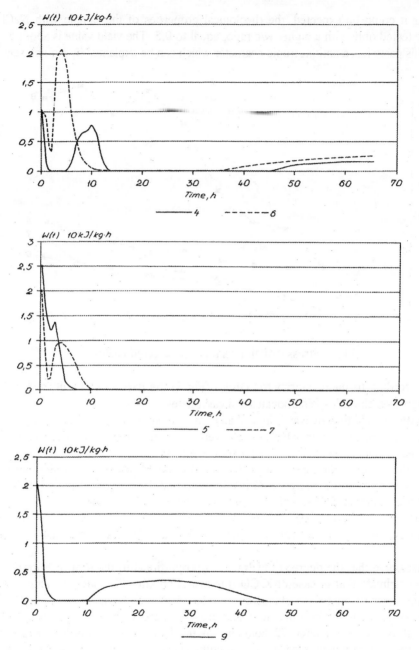

Fig 3. Microcalorimetric curves for selected pastes

TABLE III
Heat of hydration of pastes

Sample	Admixture[1]	Heat, kJ/kg after hours	
		24	72
9	1% NaCl	92	154[2]
4	1,5% NaHCO$_3$	58	95
6	1,5% KHCO$_3$	103	198
5	1,5% Na$_2$CO$_3$	68[3]	
7	1,5% K$_2$CO$_3$	106[4]	
1	5% gypsum	245	315

[1] all samples with 0,5% Na-lignosulfonate
[2] after 45 h
[3] after 7 h
[4] after 11 h

TABLE IV
Strength of the samples

Sample	Admixt.	w/c ratio	Strength in MPa							
			Bending after days				Compressive after days			
			3	7	28	90	3	7	28	90
1	gypsum	0,5	7,7	9,3	7,9	10	25,4	36,9	41,0	44,3
3	NaHCO$_3$	0,37	2,1	5,9	8,9	10	8,6	29,5	45,4	55,5
6	KHCO$_3$	0,37	2,5	6,4	9,6		10,9	28,8	51,0	
7	K$_2$CO$_3$	0,37	3,2	5,9	10,5		17,5	28,3	51,6	

The best strength after 3 days was presented by the sample with addition of K$_2$CO$_3$ but much lower than the standard. The strength higher than the standard is achieved by the pastes with Na-lignosulfonate after 28 days and significantly higher after 90 days. It is evident that the retardation effect of Na-lignosulfonate gives a relatively slow increase of strength in the first seven days.

The porosity measurements showed that the pastes with Na-lignosulfonate had a much lower porosity than the standard sample with gypsum. The porosities of sample 3 and the standard sample are presented in figure 4 as an example.

After 90 days the total porosity of sample 3 (with 1% of NaHCO$_3$) is about 18 mm^3/g and for the standard 58 mm^3/g. The maximum is at 38 nm for the standard paste and at 9 nm for sample 3. However, the capillary porosity in the range of 100 to 1000 nm is very slightly higher for sample 3.

The X-ray examination of the pastes showed that the addition of Na-lignosulfonate retarded the hydration of silicates very significantly. Portlandite is not found after 2 days and its lines appear on the X-ray pattern only after 7 days (Fig 5).

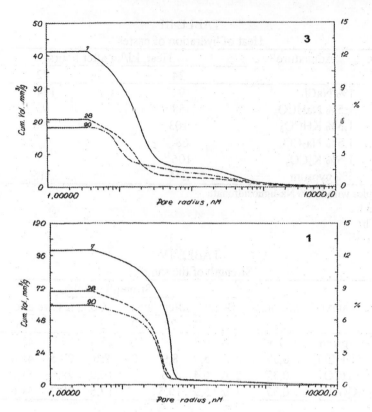

Fig 4. Porosities of cement pastes 3 and standard for a different time of hydration

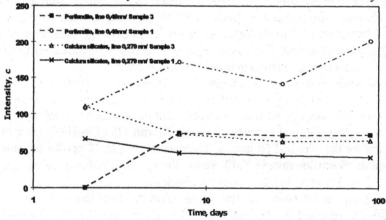

Fig 5. Intensity of portlandite and silicates lines for sample 3 and the standard with the time of hydration

Fig 6. SEM image of paste 3 after 7
days of hydration. Region rich in
carboaluminate crystals

Fig 7. SEM image of paste 3 after 2
days of hydration; weakly bound grains

After 28 days of hydration the microstructure is very compact with type III C-S-H phase [12] (Fig 8).

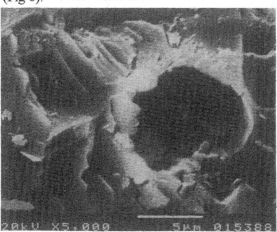

Fig 8. SEM image of paste 3 after 28 days of hydration; very compact
microstructure

The same very compact microstructure of the paste is also typical after 90 days of hydration.

The examination of the paste with the NaCl addition confirmed that the cause of the flash set is the formation of Friedel salt (Fig 9).

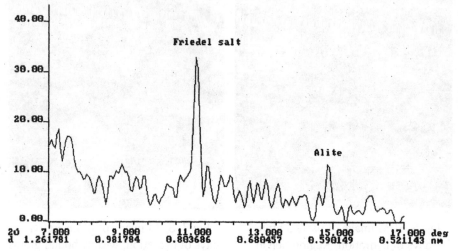

Fig 9. X-ray pattern of sample 9 after 3 hours of hydration; strong Friedel salt line

In the presence of lignosulfonate the crystals of this salt form needles (Fig 10). It is evident from the SEM examination that these crystals link together the cement grains, causing flash set (Fig 11).

Summing up it can be concluded that the combined addition of Na-lignosulfonate and alkali bicarbonate or carbonate gives the paste a very low viscosity and a low water demand. The decrease of water is of the order of 30%. This decrease of water enables the mortar of very low porosity and high strength to be obtained. The microstructure of the paste is very compact. However, the hydration of silicates is very slow which has the consequences in a relatively slow strength development. The strengths after 3 and 7 days are lower than for the standard. In this respect the utilisation of potassium bicarbonate seems to be more promising.

One of the possible solutions of this unfavourable feature of lignosulfonate addition is the increase of cement fineness and the application of grinding aids [1, 2, 8]. Also the optimum quantity of lignosulfonate must be found for each clinker.

Fig 10. Paste 9 after 3 hours of
hydration, needles of Friedel salt

Fig 11. Paste as on fig 10; needles of
Friedel salt linking cement grains

REFERENCES
[1]. S. Diamond, World Cement Technology, April 1980, p. 116.
[2]. S. Brunauer et al., Cem.Concr.Res. vol 2, 1972, p. 313, 331, 463, 577, 731, vol. 3, 1973, p. 129, 279.
[3]. I. Odler, Cem and Concr. Res. vol 1, 1971, p. 225.
[4]. J. Skalny, J.C. Philips, D.S. Cahn, Cem. and Concr.Res. vol 3, 1973, p. 29.
[5]. E.E. Bodor J. Skalny, S. Brunauer, J. Hagymassy, M. Yudenfreund, J. Colloid Interface Sci., vol 38, 1972, p. 560.
[6] J. Odler, J. Hagymassy, M. Yudenfreund, K.M. Hanna, S. Brunauer, J. Colloid Interface Sci., vol 38, 1972, p. 256.
[7] O.I. Luk'yanova, E.E. Segalova, P.A. Rehbinder, Kolloid Zhurnal, 19, 1957, p. 82.
[8] I. Odler, J. Skalny, S. Brunauer, 6th Inter. Cong. on the Chemistry of Cement, vol II, book 2, p. 30, Moscow 1976.
[9] I. Odler, U. Duckstein, Th. Becker, Cem.Concr. Res. vol 8, 1978, p. 469.
[10] S. Diamond, C. Gomez-Toledo, Consistency, Setting, and Strength Gain Characteristic of a „Low Porosity Portland Cement Paste, Cem.Concr. Res. vol 8, 1978, p. 613.

[11] S. Diamond, C. Gomez-Toledo, The Microstructure of Low Porosity Portland Cement Paste, Il Cemento, vol. 75, 1978, p. 189.

[12] S. Diamond, In: Hydraulic Cement Paste: Their Structure and Properties, Proceedings of a Conference at University of Sheffield, April 1976, Cement and Concrete Association, Wexham Springs, 1976, p. 2.

EFFECT OF THE SYNTHESIS TEMPERATURE ON THE HYDRATION REACTION OF FLY ASH-BELITE CEMENT

S. Goñi, A. Guerrero. A. Macías and M.P. Luxán
Institute of Construction Science "Eduardo Torroja" CSIC
Serrano Galvache s/n 28033 Madrid. Spain

ABSTRACT
    Fly Ash-Belite Cement (FABC) is being synthesised in our laboratory by means of a low-energy method which is based on a previous hydrothermal activation of the fly ash pozzolanic reactivity in the presence of lime and water. In the present paper the effect of the synthesis temperature on the hydration reaction of the FABC is presented. The temperatures studied after hydrothermal treatment were 700°C, 800°C and 900°C. Pastes were prepared with FABC (fabricated with two kind of Spanish fly ashes (ASTM Class F) of different granulometry), at a demineralized water to cement ratio of 0.75. The pastes were kept at >90% rh and 21°C±2°C during a period of 200 days from mixing. The evolution of the hydration reaction was followed by the combined water content which was evaluated from thermogravimetry (TG) analysis, X-ray diffraction (XRD) and infrared spectroscopy were used to the characterisation of the solid phase.

INTRODUCTION
    The hydrothermal process is a very useful way to activate the reactions whose mechanisms are via dissolution, as is the case of the fly ash pozzolanic reaction. During this process the acid components of the fly ash are firstly dissolved, reacting with the basic Ca^{2+} and OH^- ions from the CaO. The hydrated calcium silico-aluminate gels formed are the precursors of a new kind of low-energy cements called fly ash belite cements (FABC).
    Jiang and Roy (1) were the first investigators who synthesised such FABC from a mixture of fly ash (ASTM class F), lime and water, by applying the hydrothermal process and posterior heating at 900°C during 4 hours of the mixture. On the basis of that work, is being developed in our laboratory an

extensive investigation project to optimise the parameters of the FABC synthesis, and to establish the lower heating temperature and the best cementitious product.

Mixtures of two kind of Spanish fly ashes (called A and B), of different granulometry and alkali contents, CaO and water, are being firstly hydrothermally treated at 200°C and 1.24 MPa pressure during 4 hours. Subsequently, the dried mixtures are heated at 700°C, 800°C, and 900°C.

In a previous work (2) the microstructural study of hydration reactions of the FABC obtained with the fly ash-A at the heating temperature of 900°C was presented. The main crystalline compounds of that FABC were: C_2AS (gehlenite), α'-C_2S, β-C_2S and C_3A.

From those results it was considered to reduce the heating temperature of the FABC synthesis to avoid the formation of gehlenite (of slow hydraulic activity).

In the present work, the effect of the FABC synthesis temperature on the hydration reactions is studied during a period of 200 days from mixing. The evolution of the reaction was followed by the combined water content which was evaluated from thermogravimetric (TG) analysis, X-ray diffraction (XRD) and infrared (IR) spectroscopy analyses were carried out to characterise the new hydrated solid phases.

EXPERIMENTAL PROCEDURE

Two kind of Spanish fly ashes (ASTM class F) called A and B, of different granulometry and alkali content were used. Their chemical compositions and particle size distribution are given in Table I and Fig. 1 respectively.

As can be seen in Fig. 1, the particle sizes of the Fly ash B (FAB) are more homogeneous being the size of the majority of particles ≈ 40 μm. Nevertheless, the fly ash A (FAA) has two maxima at ≈ 100 μm and ≈ 10 μm.

Both fly ashes were mixtured with CaO (analytical reagent grade) at a Ca/Si molar ratio of 2 and demineralized water-to-solids ratio of 5. After 4 hours of hydrothermal treatment at 200°C and 1.24 MPa pressure, the solids were filtered and dried at 80°C. The dried products were then heated at a rate of 10°C/min. (up to 600°C) and 5°C/min. from 600°C to 700°C, 800°C and 900°C.

Six series of pastes were prepared with the two kind of FABC called FABC-A and FABC-B at a deminealized water to FABC ratio of 0,75. The pastes were kept at >90% rh and 21°C±2°C during a period of 200 days from mixing.

Table I. Chemical Composition of the two fly ashes (% by weight)

(%)	Fly Ash-A	Fly Ash-B
IL	5.6	8
IR	0.3	0.2
CaO (total)	4.65	6.75
SiO_2	48.8	45.2
CaO (free)	0.25	0.14
Fe_2O_3	7.45	9.1
Al_2O_3	26.8	25.3
MgO	1.9	1.7
SO_3	0	0
Na_2O	0.67	0.48
K_2O	3.65	3.25
SiO_2 (reactive)	35.8	35.3

IL = ignition loss
IR = insoluble residue

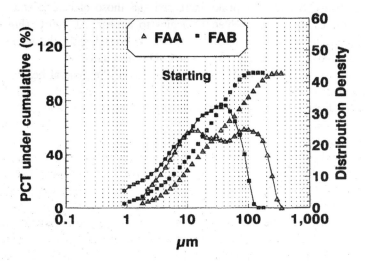

Figure 1. Granulometric study of fly ash A (FAA) and fly ash B (FAB).

The hydrothermal treatment was carried out with a Parr pressure reactor equipment Model 4522 (100 mL bomb with continuos stirring, split-ring closure and a 4842 temperature controller). XRD patterns were recorded on a Philips PW

1730 diffractometer with a graphite monochromator and Cu $K\alpha_1$ radiation. The thermal analyses were recorded using a Netzsch equipment with STA 409 simultaneous analyses system using 50 mg samples and a dynamic nitrogen stream (flow rate = 100 cm^3/min) at a heating rate of 10°C/min. IR study was carried out on a Perkin Elmer 783 instrument and KBr pellets contained 0.3% of sample. The particle size analyses were made with a SYMPATEC ELOS equipment.

RESULTS AND DISCUSSION

X-Ray Diffraction (XRD)

As shown in Fig. 2 and 3, the XRD patterns of the samples, after the hydrothermal treatment, are similar (compare Fig. 2(a) and Fig. 3(a)). The main crystalline compounds detected are the following: C_3ASH_4 (katoite silicaton), C_2SH, $CaCO_3$ (calcite) and $Ca(OH)_2$ (portlandite). It seems that, in the case of the FA-B the katoite crystallinity is higher in comparison with that of FA-A.

The hydrated precursors disappeared at 700°C (Fig. 2(b)). At this temperature some broad new peaks are detected which could be attributed to α'-C_2S, ß-C_2S, $C_{12}A_7$ (mayenite), C_3A and C_3AS_3. At 800°C (Fig. 2(d)), the intensity of the α'-C_2S, ß-C_2S and C_3A peaks increased and those of calcite and C_3AS_3 decreased. Besides the C_2AS (gehlenite) begins to appear. At 900°C (Fig. 2(f)) mayenite and calcite almost disappeared, increasing considerably the intensity of the gehlenite, α'-C_2S and ß-C_2S peaks; some CaO is also detected.

In the case of the FABC-B (Fig. 3) eventhough the general behaviour on heating is similar, to that of FABC-A, nevertheless some differences can be shown: (i) mayenite is almost not formed, (ii) the XRD patterns at 700°C and 800°C show the broad α'-C_2S and ß-C_2S peaks much less intense and (iii) the gehlenite formed at 900°C has much higher intensity.

The new hydrated compounds formed during the hydration depended on the FABC heating temperature and the kind of the fly ash used. As an example in Fig. 2 and Fig. 3 the XRD patterns of the FABC-A and B hydrated during 200 days are presented.

In the case of the FABC-A, the hydrated calcium carboaluminates: $C_4Ac_{0.5}H_{12}$ and C_4AcH_{11} are well detected irrespective of the heating temperature. The katoite (C_3ASH_4) is only present in the FABC-A heated at 900°C (see Fig. 2(g)), at the others temperatures (Fig. 2 (c) and (e)) the $C_3AS_3c_3H_{1.5}$ could be present. Vertumnite ($C_4A_2S_4H_{15}$) seems to insinuate in the samples heated at 800°C and 900°C (Fig. 2 (e) and (g)). The tobermorite ($C_5S_6H_5$), CSH gel ($C_{1.5}SH_x$) and $C_2SH_{0.35}$ could be also present in all the samples.

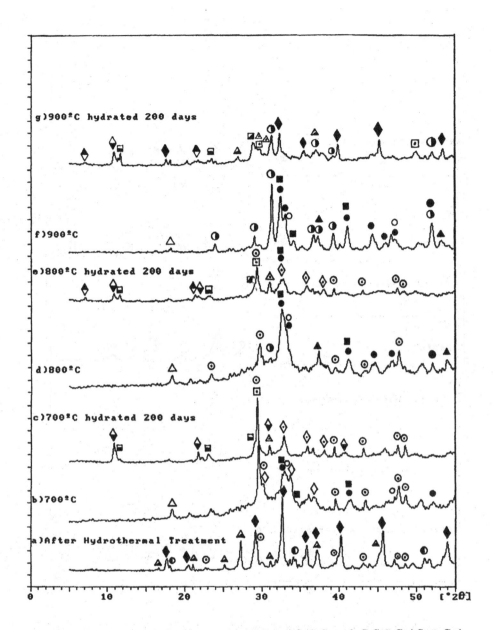

Figure 2. X-ray Diffraction patterns of FABC-A: ■ β-C_2S; ● α`-C_2S; ◑ C_2AS; ○ C_3A; ◇ C_3AS_3; ⊙ Cc; ▲ CaO; △ C_2SH; ◒ CH; ◆ C_3ASH_4; ▽ $C_4A_2S_4H_{15}$; ◆ $C_4Ac_{0.5}H_{12}$; ◈ $C_3AS_3c_3H_{1.5}$; ⊡ C_4AcH_{11}; ▣ $C_5S_6H_5$; ⊡ CSH; △ $C_2SH_{0.35}$; △ $C_{12}A_7$

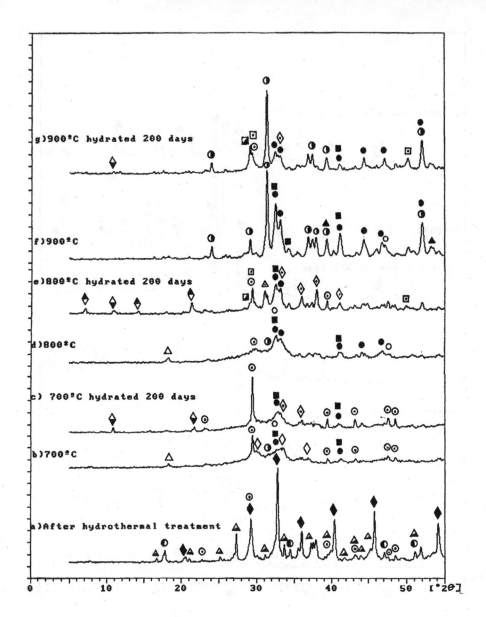

Figure 3. X-ray Diffraction patterns of FABC-B:■ β-C₂S;● α`-C₂S;◐ C₂AS;○ C₃A; ◇ C₃AS₃; ⊙ Cc; ▲ CaO; ▵ C₂SH;◑ CH; ◆ C₃ASH₄; ⬧ C₄A₂S₄H₁₅; ⬦ C₄Ac₀.₅H₁₂; ◈ C₃AS₃c₃H₁.₅;⊟ C₄AcH₁₁;◪ C₅S₆H₅; ⊡ CSH;◮ C₂SH₀.₃₅; △ C₁₂A₇

Nevertheless, for the FABC-B, in general, the hydrated calcium-carboaluminates peaks appeared much less intense in comparison to those of FABC-A. The katoite phase does not appear, but the $C_3AS_3c_3H_{1.5}$ seems to insinuate in all the cases. The vertumnite is clearly detected at 800°C (Fig. 3(e)), together with the tobermorite the CSH gel, and $C_2SH_{0.35}$ phases.

As a result of the hydration, the intensity of the $\alpha`$-$C_2S$, $\beta$-$C_2S$ and gehlenite peaks decreased, so much for the FABC-A heated at 900°C. In the case of the FABC-B, the gehlenite does not change practically. The semiquantitative hydration evolution of the $\alpha`$-C_2S and β-C_2S with time appeared in Fig. 4 for the FABC-A and FABC-B.

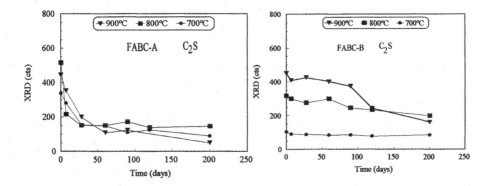

Figure 4. Semiquantitative hydration evolution of C_2S with time

In this Figure, the intensity of the maximum peak at 32.6 2θ angular zone of the both C_2S is represented versus hydration time. As shown, in the case of FABC-A the C_2S decreased up to 28 days of hydration developing a plateau thereafter. The hydration is the fastest during the first 7 days for the FABC-A heated at 800°C. After 200 days, the C_2S hydration degrees, calculated from the values of Fig. 4 were: 0.89, 0.72 and 0.73 for the FABC-A heated at 900°C, 800°C and 700°C respectively. These values were corroborated from the combined water contents, which will be shown later.

In the case of the FABC-B, the hydration of the C_2S is slower in comparison with that of the FABC-A. An induction period of about 90 days is shown for the FABC-B heated at 900°C. The hydration is more or less continuous for the sample heated at 800°C, and is not produce for that heated at 700°C. The corresponding C_2S hydration degree values after 200 days were: 0.64, 0.37 and 0.17 for the FABC-B heated at 900°C, 800°C and 700°C, respectively. The hydration of gehlenite for the FABC-A (Fig. 5) heated at 900°C shows an induction period of 28 days, and is practically not produced for the FABC-B.

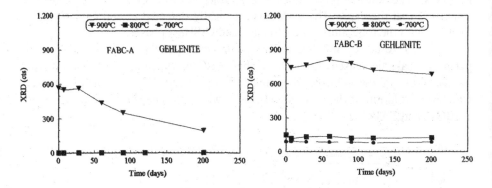

Figure 5. Semiquantitative hydration evolution of Gehlenite with time

One of the reasons to explain the higher and faster FABC-A hydration, in spite of the apparently higher particle size of the starting fly ash, could be the alkali content which is higher for the fly ash-A (Na$_2$O equivalent 3.1% and 2.6% for fly ash-A and B respectively). As is known, in absence of SO$_3$, the major portion of alkalis is incorporated in the belite, absorbing higher concentrations of K$_2$O than Na$_2$O. Increasing amounts of alkalis incorporated in the belite lattice result in a more extensive stabilisation of the hydraulic activity of the α'-C$_2$S and ß-C$_2$S (3-6). In fact the alkalinity and specially the K$^+$ concentrations of the pore solutions extracted from mortar samples fabricated with FABC-A were higher than those fabricated with FABC-B.

Very interesting from an economical point of view are the results obtained with the FABC-A heated at 700°C. This sample seems to react with water reaching the same hydration degree that the FABC-A heated at 900°C (the hydraulic activity was confirmed from the combined water content and mechanical strength development). The energy reduction in the fabrication process of this kind of cements is dramatic in comparison with that necessary in a conventional process (1350°C).

Thermal Analysis

Before to make the thermal analysis, the samples were dried at room temperature in desiccators with silica gel to eliminate the evaporable water. The profile of TG and DTG curves for the FABC-A and FABC-B hydrated during 200 days are given in Figs. 6 and 7, respectively.

A weight loss between 25°C and 1000°C is produced for the FABC-A, which is due to the release of the water molecules from the hydrated products aforementioned. In the corresponding DTG two maxima centred at 130°C and 170°C can be shown. In the case of the FABC-A heated at 800°C and 700°C, the TG curves show a second weight loss between 660°C and 1000°C, due to the release of CO_2 molecules from calcite, and carboaluminates decomposition which produce a DTG peak centred at 740-750°C. A part of that $CaCO_3$ content was already present in the anhydrous FABC with the exception of those heated at 900°C.

The hydrated FABC-B samples (Fig. 7) have similar TG and DTG profiles to those of FABC-A but the maximum centred at 100°C is much less marked.

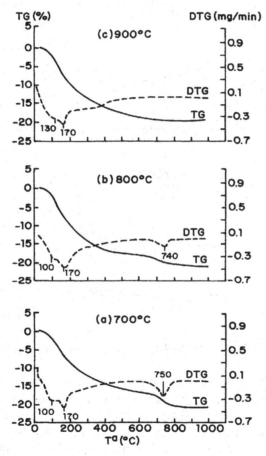

Figure 6. TG and DTG curves of FABC-A after 200 days of hydration

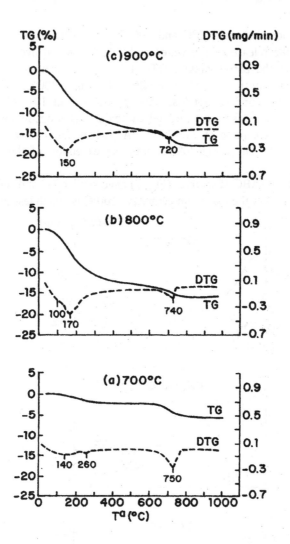

Figure 7. TG and DTG curves of FABC-B after 200 days of hydration

The evolution of the combined water with hydration time for both FABC-A and FABC-B is given in Fig. 8. The curve profiles are opposite to those of the C_2S evolution (Fig. 4). The FABC-A heated at 800°C shows the fastest hydration rate during the first 7 days, being the combined water content the highest up to 60 days. The combined water contents for FABC-A heated at 700°C and 900°C are similar up to 60 days of hydration. At 200 days the values ranged from 23% to 26% (on residue). For the FABC-B an induction period of 90 days similar to that of the C_2S (Fig. 4) can be seen for the sample heated at 900°C. The sample heated at 800°C has the fastest hydration during the first 7 days similarly to what happened in the FABC-A. At 200 days the combined water content of FABC-B-900°C and 800°C were similar (18%) whereas that of the sample heated at 700°C was 3.6%.

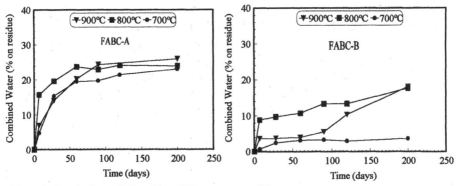

Figure 8. Evolution of Combined Water versus Time

Infrared Spectroscopy (IR)

When the FABC samples were heated at 900°C (Figs. 9 and 10 (f)), the following absorption bands appeared below 1200 cm^{-1}: a strong band with 4 maxima at 1000, 910, 880 and 820 cm^{-1} two medium bands centred at 720 and 430 cm^{-1}, a strong band centred at 530 cm^{-1}. These bands could correspond to the β-C_2S variety according to the previously IR study reported by Vázquez (7). The aluminic component of those samples (gehlenite) can not be distinguished because of its absorption bands are overlapped with those of SiO_4^{4-} groups (8, 9): 819-800 cm^{-1} (υ_1); 340 cm^{-1} (υ_2); 1050-956 cm^{-1} (υ_3); 527 cm^{-1} (υ_4). As the heating temperature decreased (Figs. 9 and 10 (b) and (d)) and amorphization of the β-C_2S crystallinity seems to be produced (the broadness of the aforementioned bands increased disappearing the 4 maxima of the principal band located between 1200 and 750 cm^{-1}), in a good agreement with the XRD data of Figs. 2 and 3.

The presence of calcite produces a wide band (υ_3) with two maxima at 1500 and 1430 cm^{-1}; at 890-800 cm^{-1} (υ_2) and 745-670 cm^{-1} (υ_4), the intensity of which decreased with the heating temperature. The lattice water molecules are evidenced by the wide band appeared in the stretching 3800-3200 cm^{-1} region and the HOH bending vibration at 1660-1640 cm^{-1}, which also decreased with the heating temperature. In the case of the samples after the hydrothermal treatment

(Figs. 9 and 10 (a)) the two maxima located at 3700 cm^{-1} and 3560 cm^{-1} are due to the OH$^-$ groups of the hydrated compounds.

After 200 days of hydration the intensity of the bands appeared below 1200 cm^{-1} decreased strongly, with the exception of the FABC-B heated at 700°C (see Fig. 10 (b) and (c)). In this sample the combined water content was 3.6%. In the principal band located between 1200 and 750 cm^{-1} a maximum at 980 cm^{-1} with a shoulder at 880 cm^{-1} is produced for the FABC-A (Fig. 9 (c), (e) and (f)). This band could be due to the CSH and tobermorite gel (10-12) formed during the hydration, which were insinuated from XRD results.

The shoulder at 880 cm^{-1} could be due to both: the υ_2 vibration of calcite and the hydrated calcium carboaluminates. In the case of FABC-B (Fig. 10) the different maxima aforementioned do not disappeared totally. The intensity of the band in the stretching 3800-3200 cm^{-1} region increased strongly as a result of the hydration. The inflections at 3720 and 3600 cm^{-1} could be attributed to the OH$^-$ groups.

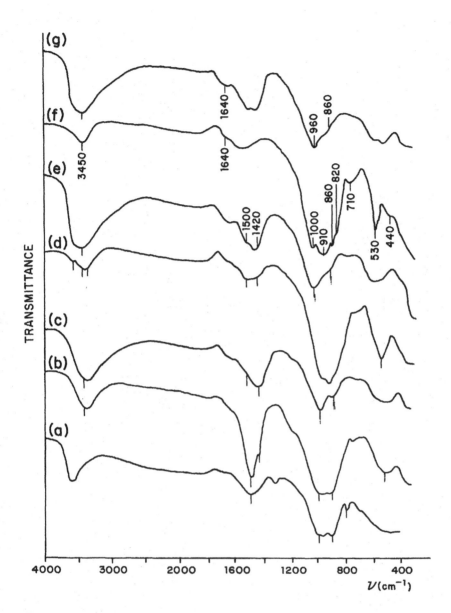

Figure 9. IR Analysis of the Hydrated Fly Ash-Belite-Cement-A (FABC-A): (a) after hydrothermal treatment; (b) 700°C; (c) 700°C hydrated 200 days; (d) 800°C; (e) 800°C hydrated 200 days; (f) 900°C; (g) 900°C hydrated 200 days

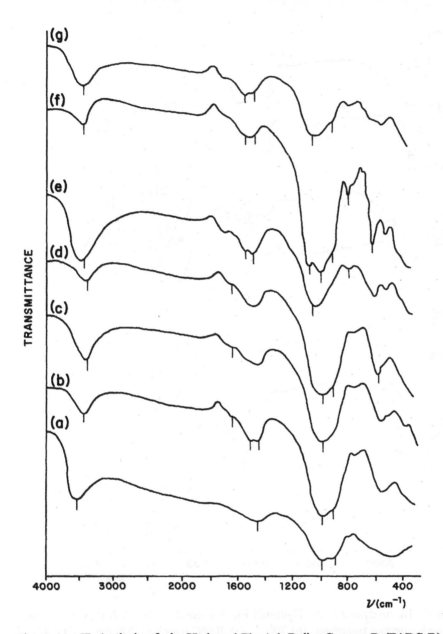

Figure 10. IR Analysis of the Hydrated Fly Ash-Belite-Cement-B (FABC-B): (a) after hydrothermal treatment; (b) 700°C; (c) 700°C hydrated 200 days; (d) 800°C; (e) 800°C hydrated 200 days; (f) 900°C; (g) 900°C hydrated 200 days

CONCLUSIONS

The main crystalline phases detected in both FABC samples were α'-C_2S and β-C_2S varieties, whose crystallinities increased with the heating temperature. Gehlenite (C_2AS) is formed at 900°C so much for the FABC-B.

The semiquantitative evolution of the C_2S hydration, calculated from the XRD results, showed important differences for the two FABC-A and FABC-B. In the case of the FABC-A, the C_2S hydration is produced during the first 28 days, whereas for the FABC-B heated at 900°C the C_2S hydration is produced after an induction period of about 90 days. The gehlenite hydration is not produced for the FABC-B and after an induction period of 28 days for the FABC-A.

These XRD results were confirmed from the combined water content.

The higher hydraulic activity of the FABC-A, even for the sample heated at 700°C, could be due to the higher alkali content of the starting FAA. The inclusion of K^+ in the C_2S crystallographic lattice is one of the reason of its higher solubility in water.

ACKNOWLEDGMENT

The authors gratefully acknowledge the financial support of the CICYT (Project n° MAT 95-0054) and the Powder Stations of Soto de Ribera and Velilla de Río Carrion for the fly ash supplied.

REFERENCES

[1] W. Jiang and D.M. Roy, "Hydrothermal Processing of New Fly Ash Cement". Ceramic Bulletin, **71** (4), 642-647 (1992).

[2] S. Goñi; M.P, Luxán; A. Guerrero; M.S. Hernández and A. Macías, "Microstructural Study of Hydration Reactions of a Fly Ash-Belite Cement". Sixth CANMET/ACI International Conference on Fly Ash, Silica Fume, Slag and Natural Pozzolans in Concrete. Bangkok, Thailand (1998) (to be present).

[3] A. Gies and D. Knöfel, "Influence of Alkalies on the Composition of Belite-Rich Cement and the Technological Properties of the Resulting Cements". Cem. and Concr. Res., **16**, 411-422 (1986).

[4] F. Lampe and R. Seydel, "On a new form of β-Belite", Cem. and Concr. Res., **19**, 509-518, (1989).

[5] W. Nocun-Wczelik, "Effect of Some Inorganic Admixtures on the Formation and Properties of Calcium Silicate Hydrates produced in Hydrothermal Conditions". Cem. and Concr. Res., **27** (1), 83-92 (1997).

[6] W. Kurdowski; S. Duszak and B. Trybalska, "Belite Produced by Means of Low-Temperature Synthesis". Cem. and Concr. Res., **27** (1), 51-62 (1997).

[7]T. Vázquez, "Spectrocopie Infrarouge de quelques composés qui out trait à la Chimie du ciment". Cahiers de Recherche IETcc, n° 22, (1969).

[8]K. Nakamoto, "Infrared and Raman Spectra of Inorganic and Coordination Compounds". Fourth Edition, Ed.: John Wiley & Sons., 1986.

[9]A. Nyquist, and R. Kagel, "Infrared Spectra of Inorganic Compounds", Ed.: Academic press, New York and London, 1971.

[10] R. Gabrovsek, B. Kurbus, D. Mueller and W. Wieker, "Tobermorite Formation in the System CaO, C_3S-SiO_2-Al_2O_3-NaOH-H_2O under Hydrothermal Conditions", Cem. and Concr. Res., 23, 321-328 (1993).

[11]N.S. Bell, S. Venigalla, P.M. Gill and J.H. Adair, "Morphological Forms of Tobermorite in Hydrothermally Treated Calcium Silicate Hydrate Gels", J. Am. Ceram. Soc., 79 (8), 2175-2178 (1996).

[12]W. Ma and P.W. Brown, "Hydrothermal Synthesis of Tobermorite from Fly Ashes", Adv. Cem. Res., 9 (33), 9-16 (1997).

THE DEGREE OF CEMENT HYDRATION DETERMINED BY BACKSCATTERED ELECTRON IMAGING

João Castro Gomes
Civil Engineering Department
University of Beira Interior
6200 Covilhã
Portugal

Joseph Cabrera
Civil Engineering Materials Unit
University of Leeds
LS2 9JT Leeds
UK

Said Jalali
Civil Engineering Department
University of Minho
4800 Guimarães
Portugal

ABSTRACT

The degree of hydration of Portuguese composite cement pastes mixed with 0.25, 0.35, 0.45, 0.55 and 0.65 w/c ratios and cured at 5°C, 20°C and 38°C was determined by backscattered electron imaging. Specimens for microscopic study were prepared at the ages of 4, 12 and 24 hours, 2, 3, 7 and 28 days and 11 months hydration time. From each specimen, two random image files of two random 'fields' were obtained at 120 times magnification. Image files were enhanced and the percentage of unhydrated cement, in relation to the total amount of pixels of the image file, was clearly distinguished and estimated by histogram thresholding. The degree of hydration was calculated for each different age from the percentage of unhydrated cement measured by a proposed method that does not require measurements from fully hydrated samples; i.e. it only uses the percentage of unhydrated cement at each age and the w/c ratio of the mix. The degree of hydration obtained by backscattered electron imaging was compared with the degree of hydration obtained from thermogravimetric analysis and from specific gravity measurements. The comparison indicate that a good relation exist.

INTRODUCTION

The process of hydration of cement is of fundamental interest to the durability of concrete. Engineering properties such as permeability, absorption and diffusion in concrete depend on the microstructural parameters, like porosity, pore size distribution and nature and amount of hydrated products of the cement paste present.(1) The amount of hydrated products depend, in turn, on the mixing and curing conditions, specially the w/c ratio, the curing temperature and the degree of hydration.

Models based on microstructural parameters, and on particularly on the degree of cement hydration, have been proposed, to form the basis for predicting properties related to the performance and durability of concrete.(2,3,4) However, the degree of cement hydration has been estimated by measuring related parameters which are believed to give only indications or approximations of the actual degree of hydration. Most experimental techniques, for example, measuring the liberated heat of hydration, the amount of $Ca(OH)_2$ or chemically combined water, measuring the specific gravity of the paste or the strength development, have limitations and the results are only of relative value. Furthermore, most techniques require comparing each measured parameter with the measured parameter for fully hydrated paste, normally obtained only after two years time.

In this article, a new method to calculate the degree of hydration is proposed. This method allows measuring directly the amount of unhydrated cement and does not require measurements from fully hydrated samples to estimate the degree of hydration at any given age. With this method the degree of hydration can be estimated by measuring only the percentage of unhydrated cement determined by backscattered electron imaging for a given w/c ratio. Furthermore, this method may give a very close approximation of the actual degree of hydration, if not, the actual degree of hydration.

CEMENT CHARACTERISTICS

The cement used in this research is produced in Portugal by CIMPOR. A composite cement type II 32.5 was used. It consists of Portland cement blended with 10 percent of limestone and a small percentage, about 1 to 2%, of Portuguese PFA. Based on the chemical analysis of this type of cement, as supplied by the manufacturer, the percentage of each cement compound was estimated using Bogue equations, as follows: $C_3S=42.6\%$, $C_2S=22\%$, $C_3A=9.6\%$, $C_4AF=9.6\%$.

The Portuguese PFA contains 40% to 60% SiO_2, 20% to 35% Al_2O_3, 5% to 10% Fe_2O_3 and small quantities of CaO, MgO, Na_2O, P_2O_5 and SO_3. The limestone filler contains mainly 85% to 95% of $CaCO_3$ and 5% to 15% SiO_2.

EXPERIMENTAL PROGRAM OF RESEARCH

The scope of the experimental program of research work included obtaining backscattered electron image files of cement microstructure, for quantification of unhydrated cement. In this study(3) five different w/c ratios were used: 0.25, 0.35, 0.45, 0.55 and 0.65 and three curing temperatures were adopted, i.e. 5°C, 20°C and 38°C. Considering that the microstructural development due to hydration, changes rapidly in the first few days, relatively slowly after the first week, and very slowly afterwards, it was decided to observe specimens at ages of 4, 12 and 24 hours, 2, 3, 7 and 28 days, and 11 months. For each condition, i.e. w/c ratio, curing temperature and age, one specimen was produced for microscopic observation.

The specimen for microscopic observation consisted of a hardened block of small particles of crushed cement paste embedded in LR white resin (Hard Grade) obtained from Agar Scientific. This resin has very low viscosity and allows good penetration. Figure 1 shows specimens for the microscopic study after surface polished. The specimens were coated with carbon which under BSE detectors is known to give better results when compared with gold coating. To increase conductivity and to reduce 'charging' the specimens were wrapped in aluminium foil, as shown in Figure 2.

Figure 1. Specimens for the microscopic study after surface polished

Figure 2. Specimens for the microscopic study after carbon coating and wrapped in aluminium foil

Experimental Procedure for Thermogravimetric Analysis

For each specimen made for the microscopic study a certain amount of cement paste was submitted to thermogravimetric analysis in a Stanton Redcroft model TG-760 apparatus. Cement particles were grounded to powder and passed through a 75 μm sieve and during TG analysis the heat was raised from 20°C to 1000°C at a rate of 20°C per minute. A weight loss curve was obtained by plotting the cumulative weight loss against temperature. Figure 3 is an example of the type of output produced, in which, the first derivative curve is obtained from the weight losses curve by plotting the percentage of weight loss per degree of temperature, i.e. dw/dT.

The percentage of total amount of $Ca(OH)_2$ was determined from the thermogravimetric curves using the following equation:

$$\%Ca(OH)_2 = A\left(\frac{74}{18}\right) + C\left(\frac{74}{44}\right) \qquad (1)$$

where A is the percentage of weight loss between point 1 and point 2, as shown in Figure 3, which corresponds to the "beginning" and "ending" points of the middle peak on the first derivative curve. The first part of equation (1) gives the total weight loss corresponding to the dehydroxylation of $Ca(OH)_2$.

The second part of equation (1) corresponds to the decarbonation of calcium carbonate (CaCO₃) but also includes 10% of limestone (95% of CaCO₃). Therefore, the amount of carbonation obtained from the right peak on the first derivative curve also includes 10% of limestone. Thus,

$$C = \left(B \, \frac{100}{44} \right) - 10 \qquad (2)$$

where B is the percentage of weight loss between point 3 and point 4, as shown in Figure 3, which corresponds to the limiting points of the right peak on the first derivative curve.

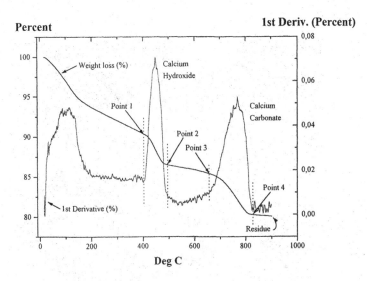

Figure 3. Typical thermogravimetric analysis plot

Experimental Procedure for the Specific Gravity Measurements

To measure the specific gravity of the cement specimens a micromeritics AccuPyc 1330 pycnometer was used with helium as a displacement medium. The instrument is fully automated and can measure the volume of solid objects of irregular shape whether in small pieces or in one piece. To run the specific gravity test, a certain amount of cement particles (about 10 g) from each specimen was weighed and placed into the cell chamber. The air and moisture are then removed from the cell chamber, containing the specimen, automatically by the pycnometer which performs twenty purge cycles of charging and discharging helium into the chamber. Five runs were carried out after that, at each run the pressure change of the helium in the calibrated cell and expansion volumes were used for measuring the sample volume and its density was estimated. The average of these five runs was accepted as the volume of the sample, the pycnometer then produced the relative density (specific gravity) of the sample.

QUANTIFICATION OF UNHYDRATED CEMENT BY BACKSCATTERED ELECTRON IMAGING

The appearance of the hydrated cement particles embedded in resin is shown in the backscattered electron micrograph of Figure 4, where the resin appears completely black. This type of resin has very low mean atomic number thus, absorbing and scattering very few electrons, reason why it appears black in backscattered electron mode.

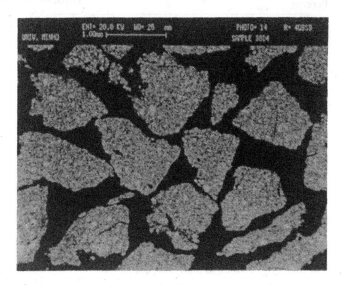

Figure 4. Backscattered electron micrograph showing cement particles embedded in resin

For each specimen two particles were randomly chosen and from each particle two image files were obtained at 120 times magnification. The decision as to which magnification to use involve a trade-off between good resolution and the number of fields required to provide a good representative sampling. This criterion has been used by other researchers.(5) For a given magnification, the number of frames required for a 95% degree of confidence that the area percentage of a determinate phase is within δ of the true average is determinate by the percentage of that area. Thus, phases with small area percentage require large number of frames for a given magnification, as is the case of unhydrated cement phase. However, in this research work it was not possible to produce a large number of image frames and at the same time to have several variables of study. Thus, the 'best' solution was to choose the lowest possible magnification for a reasonable resolution, i.e. around 1μm per image pixel. This was possible with 120 times magnification. With this magnification two frames per specimen give a 95% confidence level that the area percentage of unhydrated cement phase is within 15% of the true average. Figure 5 shows a typical backscattered electron micrograph obtained at 120 times magnification.

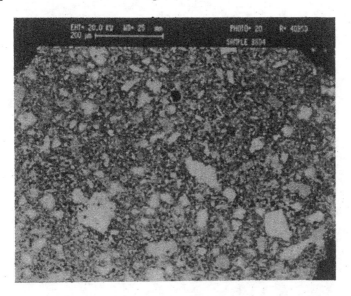

Figure 5. Typical backscattered electron micrograph obtained at 120 times magnification

On the other hand, such low magnification has the advantage of covering a larger area of microstructure, thus being more representative of the features. An obvious example of this advantage is presented in Figure 6 where a large agglomeration of calcium hydroxide-CH is observed. It is possible that with higher magnifications such agglomeration would not be detected. It was reported that in limestone filled pastes larger regions of CH are more unevenly distributed throughout the paste while in OPC pastes small regions of CH are evenly distributed.(6) This was also found in this research work and will be published with more detail elsewhere.

After collecting image files some image processing is required to enhance the image due to electronic noise that occurs during image acquisition.(7) Multiple thresholds were used to divide the image according to bands of grey level corresponding to each hydration phase, unhydrated cement and porosity. Each threshold grey level was found to be in "valleys" between histogram peaks. The image processing steps and the multiple threshold method used are presented elsewhere.(3) Figure 7 shows the corresponding backscattered image file of Figure 6 after image processing and multiple thresholding.

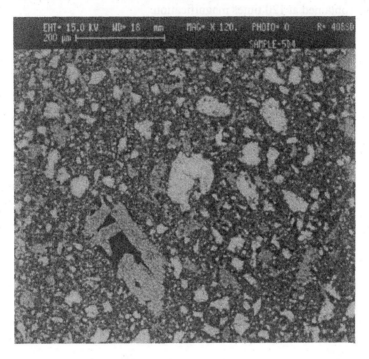

Figure 6. Typical backscattered electron image file as obtained from electron microscope at 120 times magnification

Essentially, three major phases of hydrated cement were clearly identified in this study, namely: calcium hydroxide-CH, inner product calcium silicate hydrate -IP and other hydration products-OHP (also including the cement fillers). Unhydrated cement-UH was also clearly distinguished. In Figure 7, the white areas correspond to UH, the light grey areas correspond to CH, porosity appears black and OHP dark grey.

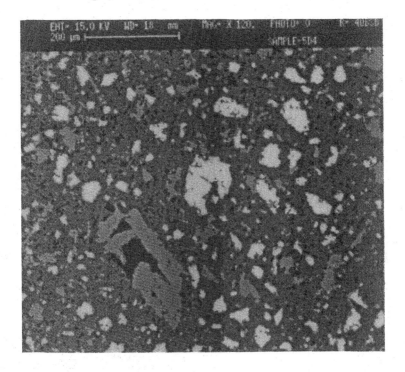

Figure 7. Backscattered electron image file after image processing and multiple thresholding

After positioning each threshold level the percentage of UH was estimated by counting the number of pixels between thresholds and dividing this value by the total number of pixels in the image file. Figure 8 presents a bar chart representation of the percentage of unhydrated cement obtained at 20°C curing temperature for the w/c ratio variables of this study. As expected, the percentage of UH is smaller for higher w/c ratios and, in the case of w/c=0.25 the percentage of UH at later ages is 2 to 3 times higher than in the other w/c ratio mixes. Figure 9 presents a bar chart representation of the percentage of unhydrated cement obtained for a w/c=0.45 and for the curing temperatures of this study. It is shown that as the time of hydration increases the percentage of UH at lower curing temperatures reduces faster and tends to approximate the same values obtained for higher curing temperatures or even to the opposite trend. The results obtained for 5°C and 38°C are presented elsewhere.(3)

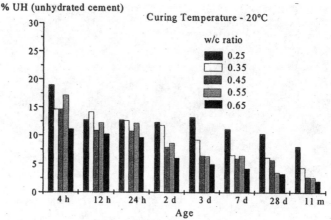

Figure 8. Bar chart representation of the percentage of unhydrated cement obtained at 20°C curing temperature for different w/c ratios

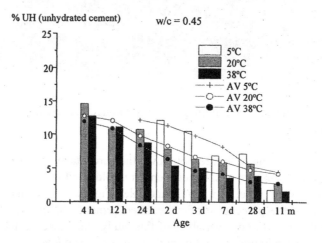

Figure 9. Bar chart representation of the percentage of unhydrated cement obtained for a w/c=0.45 and for the curing temperatures of this study

THE DEGREE OF CEMENT HYDRATION DETERMINED BY BACKSCATTERED ELECTRON IMAGING

By definition, the degree of cement hydration is the ratio between the amount of cement that has hydrated at a given time, and the original amount of cement. Using backscattered electron imaging quantification the percentage of unhydrated cement can be obtained directly at a given time. The percentage of unhydrated cement obtained refers to the total amount of pixels contained in the image file. This percentage is also referred to the total area defined by the image file, or to the total volume considering that the height is one unit. Furthermore, the percentage of volume of unhydrated cement can be estimated from the total volume occupied by the solid phases.

Having an estimation of the percentage of unhydrated cement used in a mix, $\%UH_{(mix)}$, (immediately after mixing and as percentage of volume of the paste) and the percentage of unhydrated cement obtained by backscattered electron imaging, $\%UH$, (at a certain time and as percentage of the total volume of solid phases), then the degree of hydration may be obtained by;

$$DH(\%) = 100 - \frac{\%UH}{\%UH_{(mix)}} x100 \tag{3}$$

The percentage of volume of the paste of unhydrated cement used in a mix can be obtained as follows: A unit volume of a paste is the sum of the unit volumes of water, cement, and air;

$$w + c + a = 1 \tag{4}$$

or,

$$\frac{W}{\rho_w} + \frac{C}{\rho_c} + a = 1 \tag{5}$$

where; W, is the weight of water for a unit volume of the paste; C, is the weight of cement for a unit volume of the paste; and ρ_w and ρ_c are the relative density of water and cement, respectively.

The percentage of air present in a paste is usually quite small (2 to 3%) and may be ignored in calculations. Finally, knowing the weight ratio between the water and the cement (usually referred as w/c), the relative density of water (usually taken as 1) then, the weight of water, the weight of cement, the volume of water, the volume of cement and thus the volume percentage of cement can be estimated solving the following system of equations;

$$\begin{cases} \dfrac{W}{\rho_w} + \dfrac{C}{\rho_c} = 1 \\ \dfrac{W}{C} = (w/c) \end{cases} \qquad (6)$$

the volume percentage of unhydrated cement used in a mix, $\%UH_{(mix)}$, can be obtained for different w/c ratios, as shown in Figure 10. (the relative density of the cement used in this research work is 3.12)

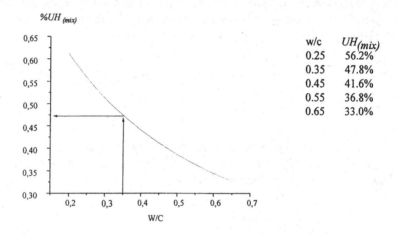

w/c	$UH_{(mix)}$
0.25	56.2%
0.35	47.8%
0.45	41.6%
0.55	36.8%
0.65	33.0%

Figure 10. Volume percentage of unhydrated cement in a mix for different w/c ratios

Kjellsen et al.(8) used backscattered electron imaging to estimate the percentage of unhydrated cement, in a cement paste with 0.50 w/c ratio, at different curing temperatures of 5°C, 20°C and 50°C at approximately 70% hydration. The degree of hydration was determined from the amount of non-evaporable water.

Table I shows good agreement between the degree of hydration values measured by Kjellsen et al. from non-evaporable water and calculated degrees of hydration using the percentage of UH obtained by Kjellsen et al., referred to as the solid phases and using equation (3) proposed in this study.

Table I. Degrees of Hydration measured by Kjellsen et al. from Non-evaporable water, %DHnew, and Authors calculations using Equation (3), %DHuh, from Kjellsen et al. %UH results

	Kjellsen et al. %UH Results and DHnew			Authors calculations	
Temp. (° C)	%UH	%Porosity	%DHnew	%UHsolid	%DHuh
5	11.25	4.27	73	11.75	70
20	10.65	10.93	72	11.96	70
50	10.23	15.11	71	12.05	69

TABLE II. Percentage of Unhydrated Cement, %UH, for 20°C Curing Temperature

W/c → Time	0.25	0.35	0.45	0.55	0.65
4 h	21.8	18.5	17.8	20.8	13.7
12 h	15.1	17.7	13.0	14.4	11.7
24 h	14.8	15.2	12.3	13.2	11.0
2 d	13.9	13.3	9.1	9.5	6.8
3 d	14.4	10.1	7.5	6.8	5.4
7 d	12.3	7.2	6.3	6.9	5.0
28 d	11.5	6.4	6.1	3.8	3.5
11 m	9.0	4.5	3.0	2.8	2.4

Table II gives the percentage of unhydrated cement referred to the total volume of solid phases for 20°C curing temperature and Table III gives the corresponding degrees of hydration obtained using this method. The results obtained for 5°C and 38°C are presented elsewhere.(3)

TABLE III. Degrees of Hydration using Backscattered electron Imaging Results and the Proposed Method

w/c → Time	0.25	0.35	0.45	0.55	0.65
4 h	60.9	61.0	56.8	43.0	58.1
12 h	72.9	62.7	68.4	60.5	64.2
24 h	73.5	67.9	70.1	63.8	66.7
2 d	75.1	71.9	77.9	74.0	79.2
3 d	74.2	78.7	81.8	81.4	83.5
7 d	77.9	84.8	84.7	81.1	85.6
28 d	79.4	86.5	85.2	89.6	89.3
11 m	83.9	90.5	92.7	92.3	92.7

COMPARISON BETWEEN THE DH USING BSE IMAGING RESULTS AND THE DH OBTAINED BY TG ANALYSIS AND SG MEASUREMENTS

The degree of hydration was calculated from calcium hydroxide content using the following equation:

$$DH_{CH(i)} = \frac{Ca(OH)_{2\ (i)}}{Ca(OH)_{2\ (fh)}} x100 \tag{7}$$

where: *(i)* age of hydration; *(fh)* full hydration; *CH* calcium hydroxide.

The degree of hydration from the specific gravity was calculated using equation 8. A sample of cement paste with 0.45 w/c ratio, kept at 20°C and 99%RH for two years was taken as fully hydrated. The CH content measured in this sample was, $Ca(OH)_2=25\%$. The specific gravity measured corresponding to full hydration was 2.14.

$$SG_{(i)} = \frac{100}{\dfrac{DH_{(i)}}{SG_{(fh)}} + \dfrac{100 - DH_{(i)}}{SG_{(uh)}}} \tag{8}$$

where: *(i)* age of hydration; *(fh)* full hydration; *(uh)* unhydrated; *SG* specific gravity. Figures 11, 12 and 13 present bar chart histograms of the degree of hydration calculated from TG analysis and SG measurements, and the degrees of hydration obtained using BSE imaging results. At 5°C curing temperature and at early ages i.e. under 2 days curing, it was not possible to obtained good quality files, as in these cases the paste was not hard enough to provide a good polished surface.

The degrees of hydration obtained with the proposed method at earlier ages are higher than those obtained with the other methods. This may be due to the fact that the proposed method measures the actual degree of reaction of the cement particles with water while the other methods are only indications of the growth of a certain hydration phase. For example in the case of calcium hydroxide, content increases more significantly after 12 to 24 hours of hydration. The specific gravity measurements are also influenced by calcium hydroxide content. On the other hand, the degrees of hydration measured by other methods depend on parameters for fully hydrated cement, normally considered after two years. However it is not clear yet that the cement does not continue to hydrate after two years. By considering fully hydrated paste at that period of time one can underestimate the degrees of hydration obtained. This is probably the main reason why the degrees of hydration obtained by the proposed method are higher than by the other two methods. Finally, it should be noted too, that the estimation of the degree of hydration using calcium hydroxide content is influenced by the presence of PFA, reason why the degree of hydration reduces at later ages. Although the percentage of PFA is low the consumption of calcium hydroxide by the pozolanic reaction is notorious at later ages of hydration and higher for higher curing temperatures.

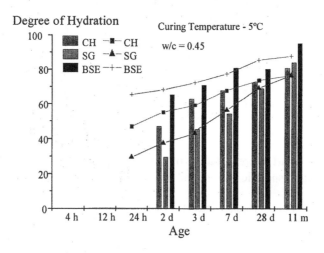

Figure 11. Bar chart representation of the degree of hydration obtained from calcium hydroxide, specific gravity and by backscattered electron imaging; 5°C

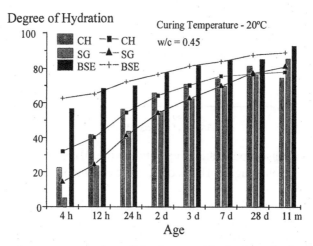

Figure 12. Bar chart representation of the degree of hydration obtained from calcium hydroxide, specific gravity and by backscattered electron imaging;20°C

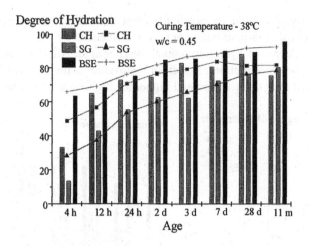

Figure 13. Bar chart representation of the degree of hydration obtained from calcium hydroxide, specific gravity and by backscattered electron imaging;38°C

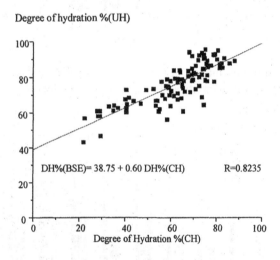

Figure 14. Linear regression relationship between the DH obtained by backscattered electron imaging, DH%(UH), and CH content from TG, DH%(CH)

Nevertheless, significant linear regression correlation's are found between the degrees of hydration obtained with the proposed method and the other methods as show in Figures 14, 15 and 16. Naturally, the results obtained with the proposed method are of relative statistical value and comparisons might not be sufficiently accurate.

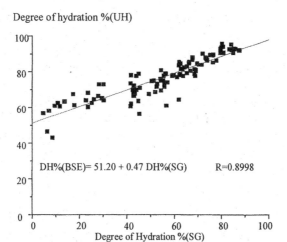

Degree of hydration %(UH)

DH%(BSE)= 51.20 + 0.47 DH%(SG)   R=0.8998

Degree of Hydration %(SG)

Figure 15. Linear regression relationship between the DH obtained by backscattered electron imaging, DH%(UH), and SG measurements, DH%(SG)

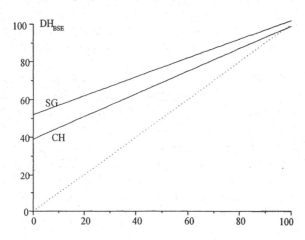

Figure 16. Comparison between DH_{BSE} with DH_{CH} and DH_{SG}.

CONCLUSIONS

The backscattered electron imaging quantification gives directly the degree of hydration since this technique measures the percentage of unhydrated cement. In 'traditional' techniques, such as thermogravimetric analysis and specific gravity measurements the degree of hydration is estimated from indirect parameters. Furthermore, the percentage of unhydrated cement is estimated from the total volume occupied by the solid phases while other parameters are based on weight measurement.

The proposed method for estimating the degree of hydration has the advantage of not needing the parameters corresponding to full hydration. It only requires the w/c ratio of the mix.

When comparing the proposed method with 'traditional' methods, good statistical linear relationships were found between the degrees of hydration. The best relationship was found between the degrees of hydration obtained by specific gravity measurements and backscattered imaging giving a determination coefficient of 0.89.

REFERENCES

[1] N. Gowripalan, J.G. Cabrera, A.R. Cusens, and P.J. Wainwright, "Effect of Curing on Durability", *Concrete International*, February 47-54 (1990).

[2] K. van Breugel, "Numerical Simulation of Hydration and Microstructural Development in Hardening Cement Based Materials – (I) Theory", *Cement and Concrete Research*, **25** [2] 319-331 (1995).

[3] J.P. Castro Gomes, "Modelling the Degree of Hydration"; pp.277-320 in *Mathematical Models for Assessing Hydration and Microstructure of Cement Pastes*, unpublished PhD Thesis, Civil Engineering Department, The University of Leeds, UK, 1997.

[4] J.G. Cabrera, S. Jalali and J.P. Castro Gomes, "Characterisation of Cement Microstructure and Concrete Durability using Fractal Geometry", to be published in the *12th International Congress of the European Ready Mixed Concrete Organization*, 23-26 June, Lisbon, (1998).

[5] H. Zhao, and D. Darwin, "Quantitative Backscattered Electron Analysis of Cement Paste", *Cement and Concrete Research*, **22** 695-706 (1992).

[6] A.P. Barker and H.P. Cory, " The early Hydration of Limestone-filled Cements", *Blended Cements in Construction*, Edited by R.N. Swany, Elsevier Applied Science, 107-116 (1992).

[7] K. Scrivener, "The Use of Backscattered Electron Microscopy and Image analysis to Study the Porosity of Cement Paste", *Materials Research Society Symposium Proceedings*, Materials Research Society, **137** 129-140 (1989).

[8] K.O. Kjellsen, R.J. Detwiler, and O.E. Gjørv, "Backscattered Electron Imaging of Cement Pastes hydrated at Different Temperatures", *Cement and Concrete Research*, **20** 308-311 (1990).

THE STRUCTURE OF ETTRINGITE

R. Berliner
Research Reactor Center
University of Missouri
Columbia, MO 65211

ABSTRACT

Neutron powder diffraction measurements at temperatures between 20 K and 300 K have been performed on specimens of ettringite grown by precipitation from solutions of $Ca(OH)_2$ and Al_2SO_4. H-D substitution was performed by soaking the specimens in D_2O. The starting point for the structural analysis by profile refinement was the X-ray description of the positions of the heavy ions which are known to form a crystal of symmetry P31c with triads of Ca-ions separated by $Al(OH)_6$ octahedra along the c-axis. Adjacent to the c-axis columns are channels containing triplets of SO_4-tetrahedra alternating with one site containing three waters which is 2/3 occupied. The neutron measurements show that the four waters coordinating each Ca-atom are on the outside of the $Ca-Al(OD)_6$ column with those above and below the Ca-atom plane arranged with their molecular plane approximately horizontal. The water molecules in the plane of the Ca-atom triad alternate between a vertical and horizontal orientation. The deutero-oxide pairs of the $Al(OD)_6$ octahedra are tilted toward the plane of the Al-ion with their orientation otherwise symmetrically radial with the O-atom toward the inside of the column.

INTRODUCTION

Ettringite is a hydrated calcium aluminate sulfate, $Ca_6[Al(OH)_6]_2(SO_4)_3 \cdot xH_2O$ with x ~ 26, that appears as one of the important hydration products of tricalcium aluminate $((CaO)_3Al_2O_3)$ and gypsum $(CaSO_4 \cdot nH_2O)$. This compound is found rarely in nature but is implicated in a number of important processes relevant to the setting of practical portland cements and to different avenues of cement deterioration. (Needle like crystals of ettringite were reported in the electron microscopy studies of portland cement hydration by Scrivener.[1]) The voluminous literature on the hydration of the aluminates and the significance of ettringite for cement and concrete durability is reviewed in the recent comprehensive book by Taylor.[2]

The first determinations of the structure of ettringite were made by Bannister, Hey and Bernal in 1936.[3] Several X-ray determinations were made in subsequent years concluding with the definitive result of Moore and Taylor[4] who reported that the true space group symmetry for ettringite was trigonal (P31c) with a = 11.26 Å and c = 21.48 Å. In their structural model, a column of Ca-ions and $Al(OH)_6$ groups alternate along the c-axis. Each of the Ca-ions is coordinated by four waters. They were able to show that the channels on each side of the Ca and $Al(OH)_6$ core were occupied by sulfate tetrahedra alternating with water groups. There are three SO_4-tetrahedra and one water group in each of two channels along the c-axis adjacent to the Ca-$Al(OH)_6$ columns. From top to bottom, for one of the channels, there are two sulfate tetrahedra pointing up followed by one pointing down followed by a site containing three water molecules which is only partially occupied. Moore and Taylor[4] made several attempts to examine the possibility of disorder in the SO_4-tetrahedra and to alter the geometry of the channel water groups but were unable to improve their refinement of the X-ray data.

The X-ray structural analysis specifies the location of the heavy ions (Ca, Al, S and O) but is much less specific with respect to the location of the H-ions in the hydroxyl groups or the orientation of the water molecules that are incorporated into the ettringite crystal. This paper reports the results of neutron powder diffraction measurements and Rietveld[5] profile refinement of ettringite specimens where the hydrogen in the hydroxyl groups and water molecules has been replaced with deuterium.

EXPERIMENTAL

Ettringite powder specimens were prepared by precipitation from solutions of reagent grade $Ca(OH)_2$ and Al_2SO_4. Boiled deionized water was used to avoid CO_2 contamination. In this method, first described by Carlson,[6] $Ca(OH)_2$ is dissolved in a sugar solution which increases its solubility. Following the prescription used by Struble[7] we prepared a 10% by weight solution of sugar, adding 0.3 M $Ca(OH)_2$. We found that the $Ca(OH)_2$ would not completely dissolve and were forced to add additional water finally employing 0.15 M $Ca(OH)_2$ in a 10% by weight sugar solution.

A stoichiometric amount of reagent grade $Al_2SO_4 \bullet 18H_2O$ was dissolved in boiled deionized water to make a second solution. The two solutions were mixed, resulting in an immediate cloudy precipitate which was magnetically stirred for several hours. After filtering, the resulting white paste, caked on the filter paper, was removed, remixed with boiled, de-ionized water, stirred for a short time and then re-filtered. This process was repeated several times to remove unreacted $Ca(OH)_2$, Al_2SO_4 and any of the remaining sugar. After filtering and washing, the precipitate was dried over $Zn(Cl)_2$ overnight.[8] Analysis of X-ray diffraction measurements gave the preliminary conclusion that the specimens were ettringite and single phase.

Neutron diffraction structural investigations usually require that the hydrogen atoms in the specimen be replaced with deuterium. The large incoherent scattering cross section and small coherent scattering length of hydrogen leads to diffraction patterns with high, sample dependent backgrounds which are unsuitable for structural analysis. For this experiment, the water molecules and hydrogen atoms in the specimen were replaced with heavy water and deuterium by soaking the specimen in D_2O at room temperature for a period of several weeks. At intervals, the specimen was filtered, dried and a neutron diffraction pattern obtained. The gradual decrease in the specimen background was used as a indicator of the progress of the D-H replacement.

When the D-H replacement had proceeded far enough, the samples were subjected to a final D_2O wash followed by filtration and drying over $ZnCl_2$. Immediately prior to the experiment, the ettringite was removed from the $ZnCl_2$ drying chamber, powdered with a mortar and pestle, placed in a 3 mm diameter vanadium can and sealed in a small aluminum chamber filled with helium gas. Changes in the moisture content of the specimens during the course of the diffraction experiment are not expected to be significant.

Neutron diffraction measurements of the ettringite powder were performed on the PSD-II spectrometer[9][10] at the University of Missouri Research Reactor Center (MURR) with a neutron wavelength of 1.4875 Å. The initial measurements were at room temperature. Examination of the data suggested that the diffraction peaks at high scattering angle were attenuated by the strong Debye-Waller effect from the many D atoms in the specimen which are expected to have significant thermal vibration amplitudes. In order to reduce the impact of this effect, specimens were loaded into a cryo-refrigerator and cooled to low temperature for the structural investigation. It was then easy to obtain data for ettringite over the complete range from 20 K to room temperature.

In Rietveld refinement, a structural model of the crystal is used, along with the characteristics of the diffractometer, to calculate a diffraction pattern that is to be compared to the diffraction data. The structural parameters (atom positions, atom species, site occupancies, thermal vibrations, lattice parameters, ...) and the diffractometer characteristics such as the diffraction peak widths, spectrometer zero and background are refined (varied) using an algorithm to find the minimum least-squares difference between the calculated and measured diffraction patterns. An extensive discussion of the Rietveld method can be found in the recent book by Young.[11] Routine application of this method would not be possible without the effort expended in writing and maintaining the refinement computer codes. *GSAS*[12] was used for the refinements described below.

The starting point for Rietveld analysis of the neutron diffraction data from this experiment was the X-ray single crystal structural solution of Moore and Taylor.[4] In their model, there are nine independent sites for water molecules; eight water sites surrounding the Ca-ions of the central core of the crystal and one "channel" water site as well as four independent sites for the H-atoms forming the $Al(OH)_6$ complex separating the layers of Ca-ions. In total, 22 H-atoms must be located in order to advance a structural model that includes the positions of all the hydrogens in the waters and hydroxides in the ettringite unit cell. In the case of the neutron experiment, to locate the position of each D-atom would require refinement of 3 x 22 deuteron position variables in addition to the scale factor, lattice parameters, spectrometer zero, diffraction peak profile parameters, background parameters, scale factor and at least 5 parameters to represent the thermal vibration amplitudes of the ettringite atomic constituents, that is one parameter for each atomic species (Ca, Al, O, S, D).

Since there are approximately 2000 data points in each diffraction pattern, such an approach - fitting more than 80 parameters - might still be successful. Nevertheless, a considerable reduction in the number of variables in the problem can be obtained by treating the deutero-oxide pairs and heavy waters as "rigid bodies" where the position of the oxygen is known (from the X-ray results) and only the distance between the deuteron and oxygen and the orientation of the heavy waters and deutero-oxides is allowed to vary. In addition, reasonable constraints can be imposed on the thermal vibration parameters of similar atoms in the crystal model. In this fashion, the number of parameters in the refinement can be substantially reduced.

The rigid-body representations of the OD and D_2O objects were defined as O-D separations of 0.97Å for both the deutero-oxides and the heavy water and a D-O-D angle of 109° was prescribed. The position of each deutero-oxide pair was then specified by two angles and one distance and that of each water molecule by three angles and one distance, with the position of the O-atom as the origin of each "rigid body" fixed to the positions given by Moore and Taylor.[4]

It is seldom useful to attempt to refine all of the variables at first and it is customary to allow *GSAS* to vary the background parameters, lattice parameters, spectrometer zero and scale factor in the first few refinement cycles. When the refinement converges to reasonable values for these parameters, as can be confirmed by inspecting the fit-data comparison plots, additional variables are added with the appropriate damping as necessary. A typical refinement of the data, from initial guess to final convergence, might require 120 refinement cycles, each requiring 6-8 seconds on a DEC-Alpha computer.

With so many variables, the multidimensional minimum surface can be expected to have false-minimum "potholes". The program *Atoms*[13] was used to produce drawings of the ettringite unit cell that could be manipulated on the computer as the refinement proceeded. Inspection of intermediate results showed that a few water molecules were "flipped over" with D-atoms toward the Ca-Al column in contrast to the remainder where the D-atoms were on the outside of the column. The orientation of these molecules was altered by hand to point outward, away from the central Ca-Al core. In each case, the quality of the refinement was immediately improved and it could then proceed to lower values of the residual difference.

DISCUSSION

The final results of the refinement of the 20 K ettringite data are shown in Fig. 1. Here, the diffraction data are shown as dots, the calculated diffraction

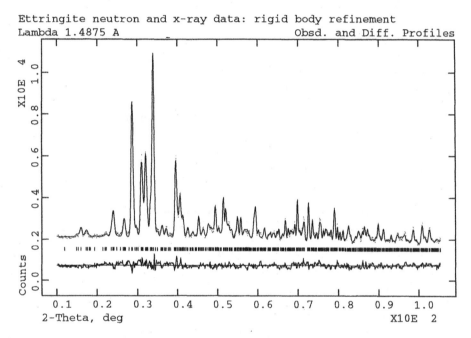

FIGURE 1.   Ettringite neutron diffraction refinement results at T = 20 K. The data is shown by dots in the upper figure while the fit is shown by the solid line. The short vertical lines in the figure mark the position of symmetry allowed reflections while the difference between the data and the fit is shown on the lower curve.

pattern as the solid line and the difference between the data and the fit is the lower curve. The positions of symmetry-allowed diffraction peaks are denoted by the short vertical lines between the data-fit and difference curves. As can be seen, the fit is excellent, with only small differences exhibited between the structural model calculation and the data. There is no evidence of a second phase and the most significant difference between the data and the Rietveld fit is the small residual "hump" in the data around $33°$ 2θ coming from some non-crystalline (or highly disordered) material. The weighted residual[14] for this fit is $R_{wp} = 0.0357$ with the minimum possible residual for this data set; $R_{wp} = 0.0200$.

Table I contains a summary of the structural data for ettringite: the space group, 20 K lattice parameters, atom positions, site symmetries, multiplicities and isotropic thermal factors. In the final refinement, 58 variables were fit. The orientations of the waters in the nine water sites (27 parameters), the orientations of the four deutero-oxide ions (eight parameters), the lattice parameters and spectrometer zero (three parameters), the spectrometer peak profile parameters (three parameters), the background (three parameters), the D-O distance in the deutero-oxide pairs (one parameter), the D-O distance in the heavy waters (one parameter), and the scale factor (one parameter). In addition, 10 parameters were used to account for the thermal motions of the atoms in ettringite. The vibrations of the same species on similar sites were constrained to be the same so that only one parameter each was required for the Ca-, Al- and S-atoms. The four O-atoms on the deutero-oxide sites (O(1) - O(4)) were allocated one parameter. The O-atoms for the column water sites (O(5) - O(12)) were constrained to be the same and the O-atoms associated with the S-tetrahedra (O(13) - O(18)) were allocated one parameter. The thermal parameter for all of the D-atoms on the column water sites was constrained to be the same and required only one parameter. Similarly, the thermal parameter for the D-atoms in the deutero-oxide pairs was constrained to be

Table I Lattice Parameters, space group and symmetry for ettringite at 20K.

Space group P 3 1 c

The lattice is acentric   primitive trigonal   Laue symmetry $3\overline{1}m$

Multiplicity of a general site is 6

The location of the origin is arbitrary in z

The equivalent positions are:

(1)   X   Y   Z   (2) -Y   X-Y   Z   (3) Y-X -X   Z

(4)   Y   X 1/2+Z (5) -X   Y-X 1/2+Z (6) X-Y -Y 1/2+Z

Lattice constants are **a** = 11.1670(5) Å c = 21.3603(13) Å

$$\alpha = 90 \quad \beta = 90 \quad \gamma = 120$$

Cell volume = 2306.8218(0.2070) $Å^3$

Deutero-oxide D-O separation 0.9640(62)

Water D-O separation 0.9408(44) Å

the same. The channel water site was treated independently requiring two thermal parameters: one parameter for the channel water O-atom and one for the two D-atoms. Finally, the occupancy of the channel water site was also included in the refinement.

Table II **Atom positions, thermal parameters and site symmetry, multiplicity and fractional occupancy for ettringite at 20K.**

Name	X	Y	Z	Ui*100	Sym	Mul	Fract
AL(1)	0.000000	0.000000	0.000000	6.1(11)	3	2	1.0000
AL(2)	0.000000	0.000000	0.250000	6.1(11)	3	2	1.0000
CA(1)	0.009000	0.816000	0.875000	1.03(25)	1	6	1.0000
CA(2)	0.994000	0.189000	0.125000	1.03(25)	1	6	1.0000
O(1)	0.994000	0.134000	0.948000	0.40(15)	1	6	1.0000
D(1)	0.9822(38)	0.2043(27)	0.9695(16)	2.42(22)	1	6	1.0000
O(2)	0.996000	0.865000	0.057000	0.40(15)	1	6	1.0000
D(2)	1.0135(37)	0.8056(31)	0.0297(15)	2.42(22)	1	6	1.0000
O(3)	0.004000	0.146000	0.805000	0.40(15)	1	6	1.0000
D(3)	-0.0045(42)	0.2172(29)	0.7825(14)	2.42(22)	1	6	1.0000
O(4)	0.004000	0.876000	0.198000	0.40(15)	1	6	1.0000
D(4)	0.0068(41)	0.7970(25)	0.2156(16)	2.42(22)	1	6	1.0000
O(5)	0.000000	0.348000	0.047000	0.84(10)	1	6	1.0000
D(5a)	0.0666(22)	0.4109(30)	0.0182(13)	4.25(14)	1	6	1.0000
D(5b)	-0.0737(22)	0.3673(33)	0.0452(15)	4.25(14)	1	6	1.0000
O(6)	0.010000	0.663000	0.958000	0.84(10)	1	6	1.0000
D(6a)	-0.0720(20)	0.5775(21)	0.9658(19)	4.25(14)	1	6	1.0000
D(6b)	0.0831(24)	0.6446(35)	0.9656(20)	4.25(14)	1	6	1.0000
O(7)	0.007000	0.345000	0.199000	0.84(10)	1	6	1.0000
D(7a)	0.0816(21)	0.4060(35)	0.2252(13)	4.25(14)	1	6	1.0000
D(7b)	-0.0728(19)	0.3382(39)	0.2183(14)	4.25(14)	1	6	1.0000
O(8)	0.996000	0.655000	0.788000	0.84(10)	1	6	1.0000
D(8a)	0.9244(24)	0.5611(12)	0.7854(19)	4.25(14)	1	6	1.0000
D(8b)	1.0792(19)	0.6517(36)	0.7852(21)	4.25(14)	1	6	1.0000
O(9)	0.263000	0.405000	0.618000	0.84(10)	1	6	1.0000
D(9a)	0.2918(39)	0.4745(21)	0.6493(9)	4.25(14)	1	6	1.0000
D(9b)	0.2930(42)	0.4546(25)	0.5797(7)	4.25(14)	1	6	1.0000
O(10)	0.744000	0.593000	0.374000	0.84(10)	1	6	1.0000
D(10a)	0.7082(43)	0.5246(27)	0.3421(11)	4.25(14)	1	6	1.0000

Table II Atom positions, thermal parameters and site symmetry, multiplicity and fractional occupancy for ettringite at 20K.

Name	X	Y	Z	Ui*100	Sym	Mul	Fract
D(10b)	0.7188(46)	0.5421(31)	0.4121(9)	4.25(14)	1	6	1.0000
O(11)	0.259000	0.406000	0.126000	0.84(10)	1	6	1.0000
D(11a)	0.3304(18)	0.3862(23)	0.1164(13)	4.25(14)	1	6	1.0000
D(11b)	0.2839(25)	0.4894(19)	0.1041(14)	4.25(14)	1	6	1.0000
O(12)	0.768000	0.598000	0.870000	0.84(10)	1	6	1.0000
D(12a)	0.7144(18)	0.6391(22)	0.8814(18)	4.25(14)	1	6	1.0000
D(12b)	0.7024(18)	0.5026(8)	0.8647(17)	4.25(14)	1	6	1.0000
O(13)	0.333330	0.666670	0.420000	3.41(35)	3	2	1.0000
O(14)	0.333330	0.666670	0.814000	3.41(35)	3	2	1.0000
O(15)	0.333330	0.666670	0.070000	3.41(35)	3	2	1.0000
O(16)	0.195000	0.642000	0.518000	3.41(35)	1	6	1.0000
O(17)	0.195000	0.620000	0.723000	3.41(35)	1	6	1.0000
O(18)	0.192000	0.585000	0.982000	3.41(35)	1	6	1.0000
O(19)	0.197000	0.637000	0.243000	-0.86(102)	1	6	0.633(38)
D(19a)	0.2947(9)	0.6806(60)	0.2392(27)	9.47(169)	1	6	0.633(38)
D(19b)	0.1698(52)	0.6778(64)	0.2111(21)	9.47(169)	1	6	0.633(38)
S(1)	0.333330	0.666670	0.491000	-1.7(5)	3	2	1.0000
S(2)	0.333330	0.666670	0.750000	-1.7(5)	3	2	1.0000
S(3)	0.333330	0.666670	0.009000	-1.7(5)	3	2	1.0000

In Fig. 2, the program *Atoms*[13] has been used to draw a 2 x 2 x 2 unit cell representation of the ettringite structure looking down the c-axis. Two of the fundamental structural elements of the crystal are the Ca-Al central cores with one at the center of the figure and the "channels" which occupy the regions between three core columns. In the figure, the bonds between the O- and D-atoms are drawn as sticks and the SO_4-tetrahedra are drawn in as pyramidal solids. Only those atoms actually within the bounds of the 2 x 2 x 2 unit cells are shown so that some of the SO_4-tetrahedra have been "cut-off" with the S-atom, normally hidden within the pyramid, now visible. In a similar manner, some of the O- and D-atom pairs, part of a deutero-oxide ion, are similarly uncompleted, since their mate is located on the other side of the cell boundary. Nevertheless, the fundamental feature of the arrangement of the water molecules is evident - they are clustered on the outside of the Ca-Al core. In addition, the triplet of deutero-oxide pairs separating the Ca-triad and Al-atom can be observed to be oriented essentially symmetrically. What is not so clear in this figure is that the O-D pairs along the central core are tilted so that the

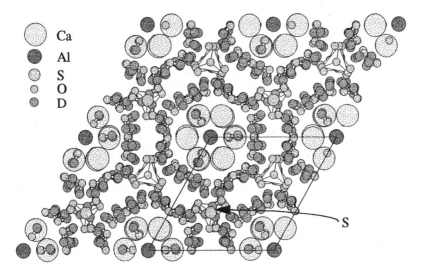

FIGURE 2.   2 x 2 x 2 ettringite unit cells looking down along the c-axis. The key in the upper left of the figure identifies the atom species. The outline of one unit cell is shown in the lower right hand corner. The arrow indicates an exposed S-atom in the channel (see text).

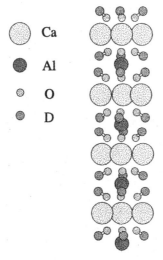

FIGURE 3.   The Ca-Al(OD)$_6$ c-axis column of ettringite with the a-axis out of the paper. The OD ions can be seen to be tilted toward the plane of the Al-atom. The key to the left of the figure identifies the atom species.

OD-ions "cup" the Al-ion at the center of the $Al(OD)_6$ complex. This is illustrated in Fig. 3 which isolates the Ca-Al core of the ettringite unit cell.

The arrangement of the waters in the channel site is also visible in Fig. 2. Three water positions in the channel result from the action of the space-group symmetry elements on the water molecule coordinates. Three D-atoms belonging to the three waters can be seen partially hidden behind the S-atom marked in Fig. 2. The distance between these three is unphysically close (0.92 Å) amplifying similar concerns expressed by Moore and Taylor[4] over the distance between the O-atoms of these same waters in their X-ray structural model. The refinement indicates that this site is only partially occupied (63%)[15] so that only 2 of these sites are likely to be occupied at any one time. Moore and Taylor argue that the partial occupancy of this site would permit cooperative motion among the remaining water molecules that could cause the mean distance between the individual atoms to be larger. There is some support for this hypothesis in the values of the thermal parameters obtained from the refinement as will be described below.

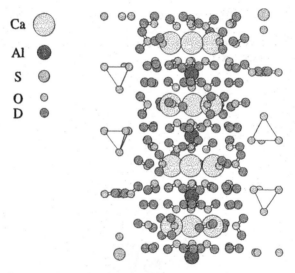

FIGURE 4.   One unit cell of the ettringite crystal with the c-axis vertical in the plane of the paper. Only atoms within the unit cell boundaries are shown so that some O-D bonds and SO_4-tetrahedra are incomplete (see text). Atom species are identified by the key in the upper left of the figure.

Figure 4 shows the ettringite unit cell with the c-axis vertical. Once again, atoms that lie outside of the unit cell boundaries have not been shown so that the SO_4-tetrahedra at the four corners of the figure are incomplete. At the upper right (lower left) the tetrahedron points down (up) and the S-atom and the apex O-atom can be seen. At the lower right (upper left) the basal triplet of O-atoms of the

SO$_4$-tetrahedra are visible. In this case the S-atom and apex O-atom lie over the cell boundary. Otherwise, the SO$_4$-tetrahedra are shown as solids.

In their paper, Moore and Taylor discussed possible orientations for the water molecules and hydroxyl groups on the basis of their data (the Fourier difference maps) and chemical intuition. Each of the Ca-atoms is coordinated by four waters (see Fig. 3 in Ref. 4.). They expected the water molecules that lay above and below the plane of each Ca-triplet to be oriented with their molecular plane roughly horizontal (perpendicular to the c-axis) while those in the plane of the Ca-atoms were expected to be oriented vertically. The calculated diffraction pattern for that orientation of waters differs significantly from the data. Subsequent refinement of the structure suggested that while the water molecules above and below the Ca planes were indeed roughly horizontal, those in the plane appeared to adopt a more complex arrangement, with the water molecule first horizontal and then vertical as one proceeded around the core. Using this observation as a second starting guess, the refinement proceeded smoothly to its minimum.

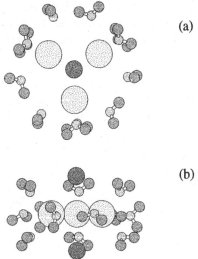

(a)

(b)

FIGURE 5.   a) View along the c-axis of the water coordination of one of the Ca-triads with an Al-atom above and below. The horizontal-vertical alternation of the water molecules can be clearly seen. It is also clear that the water molecules are generally oriented so that the O-atoms are toward the column and the D-atoms face outward. b) View along the a-axis of the same Ca-triad. Note the tilt in the planes of the water molecules above and below the plane of the Ca-atoms.

Figure 5 contains two views of the coordination of water molecules around one of the triads of Ca-atoms in the central core of ettringite. In Fig. 5a, the view is along the c-axis while in Fig. 5b the a-axis points out of the page. The horizontal-vertical alternation of the water molecular plane for those in the plane of the Ca-atoms is seen to be approximately obeyed although none of the molecules is arranged in an exactly regular fashion. In addition, the water molecules above and below the Ca-triad are seen to be tilted toward the position of the Al-atom in a similar manner as the OD-ions located at approximately the same position along the c-axis.

The isotropic thermal factors for atoms in the ettringite cell are shown in Table II. The quantity tabulated is $Uiso*100$ where the thermal correction to the structure factor is written $\exp(-(8\pi^2 Uiso/\lambda^2)\sin^2\Theta)$. The small negative values obtained for the S-atoms and the channel water O-atom are probably a consequence of some residual neutron absorption in the specimen. Any H-atoms remaining (not D-exchanged) would attenuate the beam and masquerade as a negative Debye-Waller effect by reducing the diffraction peaks at low angle more than those at high scattering angle. The other remarkable feature in the thermal parameter results is the large isotropic temperature factor obtained for the two D-atoms of the channel water site. Large amplitude cooperative motion or static disorder would lead to this result and could cause the separation between the D-atoms at this site to be larger than the distance between their mean positions (0.92 Å).

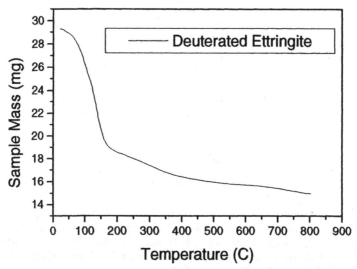

FIGURE 6. TGA data from D-substituted ettringite used for the neutron diffraction measurements.

The separation between the atoms in OD pairs refined to 0.964 Å while the distance between O-atoms and D-atoms in the heavy waters refined to 0.94 Å. Both of these values compare acceptably with the value obtained /for heavy ice (1.01 Å).[16]

The refined value of the occupation of the channel water site can be compared to the water content of the specimen using thermo-gravimetric analysis (TGA). Figure 6 contains the result of TGA analysis on a specimen of the D-substituted ettringite used for the diffraction experiment. If it is assumed that all of the specimen weight loss between room temperature and 600 C is due to the loss of water from the ettringite structure, the ettringite chemical formula can be written as $Ca_6[Al(OD)_6]_2(SO_4)_3 \cdot 27.3D_2O$. This compares favorably with the formula derived from the refinement results (the channel water occupancy): $Ca_6[Al(OD)_6]_2(SO_4)_3 \cdot 25.9D_2O$. The discrepancy is likely due to water not incorporated in the ettringite structure but adsorbed on the powder grain surfaces.

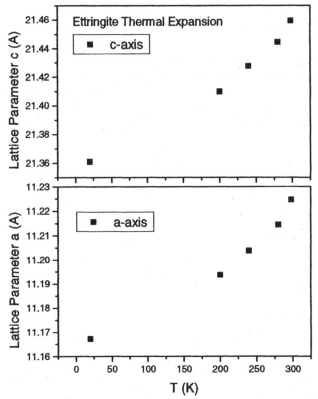

FIGURE 7.   Thermal expansion of the a- and c-axis of ettringite between 20 K and 300 K.

Neutron diffraction spectra were obtained for the same ettringite specimen at T = 300 K, 20 K, 200 K, 240 K, and 270 K. Figure 7 illustrates the change in lattice parameter as a function of temperature over this range. In the span between 20 K and 300 K the a-axis expands 0.51% while the c-axis expands 0.48%. The temperature dependence of the thermal expansion is markedly non-linear. Between 280 K and 300 K, the thermal expansion coefficient for the a-axis is $\alpha = 5*10^{-5}$/K while for the c-axis it is $\alpha = 3.8*10^{-5}$/K. This measurement agrees well with the recent results of Hall and co-workers[17] who measured the thermal expansivity of ettringite in the range 70 C - 115 C (343 K - 388 K). A value of $\alpha = 4.7*10^{-5}$/K can be inferred from their graph of the expansion of the (100) plane d-spacing.

ACKNOWLEDGMENTS

This work was supported by the U. S. Department of Transportation, Federal Highway Administration under the contract DTFH61-97-P-00385. The author wishes to thank Richard Livingston of the Federal Highway Administration for suggesting this experiment and Presbury West of the Construction Technology Laboratories for many helpful discussions on ettringite synthesis. The author is also grateful to Len Barbour of the University of Missouri for performing the ettringite TGA measurement, to Bob Von Dreele for his help with the rigid body refinements and Fred Ross and W. B. Yelon of the Research Reactor Center for many helpful discussions.

REFERENCES

[1] K. L. Scrivener, *The Development of Microstructure During the Hydration of Portland Cement*, Ph.D. dissertation, University of London, (1984).

[2] H. F. W. Taylor, *Cement Chemistry*, Academic Press, London, (1990).

[3] F. A. Bannister, M. Hey and J. D. Bernal, *Ettringite From Scawt Hill, County Antrim [Ireland]*, Miner. Mag. 24, 324, (1936).

[4] A. E. Moore and H. F. W. Taylor, *Crystal Structure of Ettringite*, Acta. Cryst. **B26**, 386-393, (1970).

[5] H. M. Rietveld, *A Profile Refinement Method for Nuclear and Magnetic Structures*, J. Appl. Cryst. **2**, 65-71, (1969).

[6] E. T. Carlson and H. A. Berman, *Some Observations on the Calcium Aluminate Carbonate Hydrates*, J. Res. NBS **64 A**, 333-341, (1960).

[7] L. J. Struble and P. W. Brown, *An Evaluation of Ettringite And Related Compounds for Use In Solar Energy Storage*, NBSIR 84-2942, (1984).

8. A saturated $ZnCl_2$ solution maintains a relative humidity of less than 6.9%. Presbury West, *Saturated Salt Humidity Table*, private communication.

9. R. Berliner, K. McCollough and J. Hilker-Draper, *A Position Sensitive Detector for Neutron Powder Diffraction Studies*, Final Technical Report, University Research Instrumentation Program, Research Agreement DAAL03-86-G-0150.

10. W. B. Yelon, R. Berliner and M. Popovici, *A Perfect Match For High Intensity Neutron Powder Diffraction: Position Sensitive Detection And Focussing Monochromators*, Physica B, in press.

11. R. A. Young, *The Rietveld Method*, Oxford University Press, Oxford, 1993.

12. A. C. Larson and R. B. Von Dreele, *GSAS-General Structural Analysis System*, Los Alamos National Laboratory Report LAUR 86-748 (1994).

13. Eric Dowty, Shape software, 521 Hidden Valley Road, Kingsport, TN 37663.

14. The weighted residual is defined as $R_{wp} = \sqrt{(\sum w(I_o - I_c)^2)/(\sum wI_o^2)}$

15. Moore and Taylor (Ref. 4) obtain f = 0.57.

16. S. W. Peterson and H. Levy, *A Single-Crystal Neutron Diffraction Study Of Heavy Ice*, Acta. Cryst. **10**, 70-76, (1957).

17. Christopher Hall, Paul Barnes, Andrew D. Billimore, Andrew C. Jupe and Xavier Turrillas, *Thermal Decomposition Of Ettringite*, J. Chem. Soc. Faraday Trans. **92**, 2125-2129, (1996).

Modeling ionic interaction mechanisms
in cement-based materials – An overview

J. Marchand[1-2], Y. Maltais[1-2], É. Samson[1-2], V. Johansen[3] and K. Hazrati[1]

(1) Centre de Recherche Interuniversitaire sur le Béton
Université Laval, Sainte-Foy, Canada, G1K 7P4

(2) SIMCO Technologies inc.
1400, boul. du Parc Technologique, Québec, Canada, G1P 4R7

(3) Construction Technology Laboratories
Skokie, Illinois, USA

ABSTRACT

From an engineering point of view, the mechanisms of ion interaction are quite important. Not only do they play an important role in the development of deleterious chemical reactions but they also affect the transport of ions through the materials pore structure. Over the past decade, numerous ways of modeling the interaction of ions with the hydrated cement paste have been proposed. A critical review of these various approaches is presented in this report. Each approach is evaluated on the basis of its ability to account for the influence of parameters such as the nature of the ionic species, the pore solution composition and the solid physical and chemical properties. Problems related to the implementation of these models in a numerical code are also discussed.

Introduction

Over the past decades, it has been clearly established that ions can physically and chemically interact with the hydrated cement paste. Although several aspects of the problem still remain to be clarified, significant progress has been made in understanding the mechanisms of ion interaction. For instance, it is now well known that some ions (such as chloride, sulfate and magnesium) can chemically react with the unhydrated and hydrated phases of the cement paste to form new compounds. With the introduction of new computer codes specifically designed to investigate the mechanisms of chemical equilibrium, multiple-phase diagrams have recently been developed for various systems [1-4]. Globally, these diagrams clearly emphasize the extreme complexity of chemical equilibrium in cement systems which usually comprise numerous different ionic species and solid phases.

It has been clearly established that ions may also physically interact with the hydrated cement paste [5-8]. Although most studies tend to indicate that the mechanisms of physical interaction are dominated by short-range electrical forces (« double-layer » effects), evidence of ion penetration in the layered structure of the C-S-H has been brought forward. For instance, in a pioneer investigation of the influence of $CaCl_2$ on the C_3S hydration mechanisms, Ramachandran [5] distinguished three possible types of interaction. According to the author, chlorides can either be present in a chemisorbed layer on the hydrated calcium silicates, penetrate the C-S-H interlayer spaces, or be intimately bound in the C-S-H lattice. The ability of chlorides to penetrate the interlayer spaces has later been confirmed by Beaudoin et al. [6].

From an engineering point of view, the mechanisms of ion interaction are quite important. Not only do they play an important role in the development of deleterious chemical reactions but they also affect the transport of ions through the material pore structure. In that respect, interaction mechanisms have to be considered in any numerical model aiming at predicting the long-term behavior of concrete exposed to an aggressive environment.

This report attempts to critically review the most recent developments in the field of ion interaction modeling. In the survey, special emphasis is placed on the influence of interaction mechanisms on the mathematical description of transport processes. The report is divided in three parts. A brief overview of the ionic transport mechanisms in cement-based materials is presented in the first part of the report. The second portion of the report is devoted to the interaction of ions in non-equilibrium conditions. The mechanisms of ionic interaction at thermodynamic equilibrium are treated in the third part of the report.

Theoretical considerations pertaining to ion transport in cement-based materials

Concrete degradation mechanisms generally involve the penetration of external ions (such as sulfate, sodium and magnesium) into the material porosity and/or the dissolution of various hydrated and unhydrated phases. The transport of ions in concrete can be associated with the flow of liquid under a capillary potential (or a pressure) gradient. The phenomenon is usually called advection. The penetration of ions can also be associated with an ionic drift originating from a chemical potential gradient. This phenomenon is generally referred to as diffusion. In most practical cases, ions are transported through the concrete pore structure by a combined advection/diffusion process. In some cases, ions can also be transported under the action of an externally applied electrical field. This transport process is referred to as ionic migration.

As previously discussed, in any cement-based materials, ions can be found in three different states:

- they can be free in the pore solution;

- they can be physically bound to the pore wall (or eventually trapped in the interlayer spaces);

- or chemically bound to the hydration products.

In most practical cases, the movement of the ions through the system strictly occurs in the liquid phase, and the ionic flow in the solid and adsorbed phases can be neglected. Hence, the equation describing the total ionic flow (F_{itot}) of an ionic species i through the porous material is given by:

$$\mathbf{F}_{i\,tot} = \theta_w \mathbf{F}_i \qquad (1)$$

where θ is the free water content (expressed in m^3/m^3) and $\mathbf{F}_i$ is the flow of the ion i in free water (expressed in mol/m^2s). The application of the mass balance equation to the material volume leads to :

$$\frac{\partial c_{i\,tot}}{\partial t} = -\,\text{div}\left(\mathbf{F}_{i\,tot}\right) \tag{2}$$

Substituting equation (1) in equation (2), the mass balance equation is now given by:

$$\frac{\partial c_{i\,tot}}{\partial t} = -\,\text{div}\left(\theta_w \mathbf{F}_i\right) \tag{3}$$

It could be demonstrated that the left-hand side of equation (3) can be expanded as follow [9, 10]:

$$\frac{\partial c_{i\,tot}}{\partial t} = \left(1-\phi\right)\frac{\partial c_{is}}{\partial t} + \rho_s\,\alpha\,\kappa\,\frac{\partial c_{ia}}{\partial t} + \frac{\partial\left(\theta_w c_i\right)}{\partial t} \tag{4}$$

where ϕ stands for the total porosity of the material, α (expressed in m^2/kg of solid phase) is the specific area of the material, ρ_s (expressed in kg/m^3 of solid phase) the solid phase density of concrete, κ (m) is the thickness of the adsorbed water layer on the pore wall. In the equation, c_{is}, c_{ia} and c_i are the ion contents in the solid phase, the adsorbed phase and the free solution respectively. c_{itot} stands for the total ion concentration.

Combining, equations (3) and (4), one finally gets :

$$\left(1-\phi\right)\frac{\partial c_{is}}{\partial t} + \rho_s\,\alpha\,\kappa\,\frac{\partial c_{ia}}{\partial t} + \frac{\partial\left(\theta_w c_i\right)}{\partial t} = -\,\text{div}\left(\theta_w\,\mathbf{F}_i\right) \tag{5}$$

As previously emphasized, in most practical cases, concrete is not fully saturated and ions are therefore transported by a combined advection-diffusion process. Accordingly, the free ionic flow can be written as:

$$\mathbf{F}_i = \mathbf{F}_{iu} + \mathbf{F}_{iv} \tag{6}$$

where the subscripts u and v respectively stand for the chemical (diffusion) and capillary (advection) potential applied to the i species. Knowing that the ionic flow can be divided according to equation (6), the right-hand of equation (5) can developed as [9, 10] :

$$\theta_w \frac{\partial c_i}{\partial t} + c_i \frac{\partial \theta_w}{\partial t} + (1-\phi) \frac{\partial c_{is}}{\partial t} + \rho_s \, \alpha \, \kappa \frac{\partial c_{ia}}{\partial t} = \frac{\partial}{\partial x} \left[\theta_w \, D_i \left(\frac{\partial c_i}{\partial x} + \frac{z_i \, F}{RT} c_i \frac{\partial V}{\partial x} \right) \right]$$
$$+ \frac{\partial}{\partial x} \left[\theta_w \, c_i \left(D_w \frac{\partial \theta_w}{\partial x} \right) \right] \tag{7}$$

where F is the Faraday constant, R is the ideal gas constant, T is the temperature (°K), z_i is the ion electronic valence number and V the diffusion potential set up by the drifting ions (in Volt), D_i is the diffusion coefficient of the considered ionic species within the material (m²/s) and D_w is the diffusion coefficient of water in concrete (m²/s).

The first two terms on the right-hand side of equation (7) describe the transport of ions by diffusion. More information on the mechanisms of ionic diffusion can be found in reference [11]. The last term on the right-hand side accounts for the transport of ions by advection. It should be emphasized that the diffusion coefficient of water in concrete (D_w) is a water-content dependent parameter [10, 12, 13].

In the case of a transport of ions in a fully-saturated concrete, equation (7) can be simplified to :

$$\theta_w \frac{\partial c_i}{\partial t} + (1-\phi) \frac{\partial c_{is}}{\partial t} + \rho_s \, \alpha \, \kappa \frac{\partial c_{ia}}{\partial t} = \frac{\partial}{\partial x} \left[\theta_w \, D_i \left(\frac{\partial c_i}{\partial x} + \frac{z_i \, F}{RT} c_i \frac{\partial V}{\partial x} \right) \right] \tag{8}$$

As can be seen, in equations (7) and (8), the chemical and physical interactions of ions with the solid are essentially described by the second and third terms of the left-hand side of the equations. In order to solve these equations, the terms $\partial c_{is}/\partial t$ and $\partial c_{ia}/\partial t$ must be defined for each ionic species present in the system considered. On the one hand, the chemical interaction term ($\partial c_{is}/\partial t$) accounts for the formation or the dissolution of solid phases in the material. On the other hand, the term $\partial c_{ia}/\partial t$ describes the physical interaction of ions with the solid. The transport of ions through the porous system is described by the right-hand side of equations (7) and (8).

It should be emphasized that, in most practical cases, the mechanisms of interaction are strongly affected by the rate of ionic transport through the material pore structure. In some cases, the rate of transport is sufficiently slow to allow the system to locally reach a metastable thermodynamic equilibrium. In some other instances, ions move too rapidly, and the system can not achieve the chemical equilibrium. This is the reason why the mechanisms of ionic interaction in non-equilibrium and equilibrium are further treated in separate sections.

Interaction mechanisms in non-equilibrium conditions

Over the past few years, very little investigations have been specifically devoted to the mechanisms of ionic interaction in non-equilibrium conditions. The main reason behind this lack of interest is most probably related to the fact that, in most practical cases, ions are drifting slowly enough through the concrete pore structure to reach a local thermodynamic equilibrium.

It has been, however, recently established that, in some particular instances, the transport of ions is so fast that local equilibrium cannot be achieved. This is, for instance, the case when ions are transported through the material pore structure under an electro-chemical potential gradient [14]. If the external potential applied on the system is sufficiently important (i.e. over a few V/cm), the second term of the right-hand side of equation (8) dominates the overall transport process, and ions will migrate very quickly. Such a condition is usually met for most migration laboratory tests (such as the Rapid Chloride Penetrability Test – ASTM C 1202) in which high voltages (up to 12 V/cm) are applied on the system.

It has also been observed that the transport of fluids by advection or permeation could also impede the achievement of thermodynamic equilibrium between ions in solution and the surrounding solid. This is particularly the case during a capillary suction experiment for which the concrete sample was initially dry [15]. In that instance, the progression of the liquid front through the material is so fast that it does not allow the system to reach its chemical equilibrium.

In recent years, numerous authors have proposed to treat the mechanisms of ionic interactions in non-equilibrium conditions as a global process [15-18]. According to this approach, no distinction is made between the physical and chemical interaction mechanisms. Consequently, the left-hand side of equation (5) can be simplified in the following way :

$$\frac{\partial c_b}{\partial t} + \frac{\partial(\theta_w c_i)}{\partial t} = - \text{div}\left(\theta_w \ F_i\right) \qquad (9)$$

where c_b stands for the total amount of ions physically and chemically bound to the solid. In that respect, the first term of the left-hand side of equation (9) simply accounts for the rate at which ions are bound by the solid. In most cases, the temporal binding term is simply modeled by linear equations such as [16, 17]:

$$\frac{\partial c_b}{\partial t} = k(c_i - mc_b) \qquad (10)$$

where k and m are material parameters [16]. Despite the relative simplicity of equation (10), the material parameters k and m. are usually difficult to determine experimentally. Most of the time, they are simply fitted on the basis of ion penetration profiles. Given the inherent difficulties of assessing these two parameters, very little values of k and m have been reported in the literature. All published values were obtained on the basis of non steady-state chloride migration experiments.

The effect of k (for m = 0) on the evolution of the chloride ion profile during a migration experiment is illustrated in Figure 1 [15]. Values of k reported by Houdusse [14] for three different 0.45 water/cement ratio neat cement paste mixtures are given in Table 1. More information on the details of the migration test used by the author can be found in reference 15.

Interaction mechanisms in equilibrium conditions

As previously emphasized, in most practical cases, the rate of ionic transport is sufficiently slow to allow the thermodynamic equilibrium to be reached locally. As will be seen in the following paragraphs, the mechanisms of ionic interaction under equilibrium conditions have been the subject of a great deal of interest over the past decade. Numerous approaches have been proposed. In some cases, researchers have chosen to treat the phenomena of chemical and physical interactions separately, others have preferred to use a global approach in which no distinction is made between the two mechanisms.

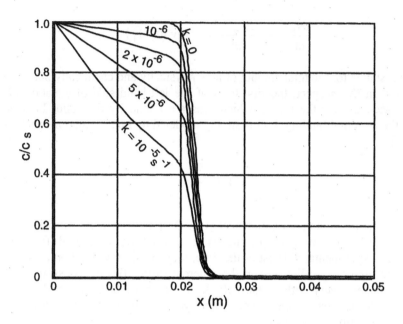

Figure 1 – Influence of the parameter k on the chloride ion profile during an unsteady-state migration experiment (according to ref. 19)

Table 1 – Influence of type of cement on the value of k measured during an unsteady-state chloride ion migration experiment

Type of cement	W/C	Curing (years)	Voltage (V/cm)	k ($\times 10^{-6}$ s^{-1})
Type I	0.45	2	5.3	1
Type III	0.45	2	5.3	2
Type V	0.45	2	5.3	7

Chemical interaction

In equations (7) and (8), the modifications of the pore solution chemistry is considered through the chemical interaction terms ($\partial c_{is}/\partial t$). Accordingly, a chemical interaction term must be defined for each ionic species that is assumed to interact with the solid matrix. If the chemical reactions are assumed to be fast enough with respect to the transport process, the ionic binding mechanism can be treated as an equilibrium problem and the chemical interaction terms can be written:

$$\frac{\partial c_{is}}{\partial t} = \frac{\partial c_{is}}{\partial c_i} \frac{\partial c_i}{\partial t} \tag{11}$$

Equation (11) requires the determination of a new term: $\partial c_{is}/\partial c_i$. This term has to be evaluated for each ionic species. As emphasized by many authors, the phase transformations that may occur in a concrete sample in contact with ionic solutions are quite intricate. In fact, a number of chemical reactions can occur which may lead, for instance, to the formation of new products (such as ettringite, monochloroaluminate, gypsum and brucite) and the dissolution of some hydrated phases (such as portlandite). Upon the alteration process, a number of the concrete hydrated and unhydrated phases may react with the drifting ions.

Given the intrinsic complexity of most cement systems (which comprise numerous solid phases and ionic species), the mechanisms of chemical interaction and phase stability can only be studied with computer-assisted calculations. Over the past few years, numerous models have been developed to study the influence of various ions (such as chlorides, sulfates, ...) on the chemical equilibrium of cement systems [1-3, 19]. Despite some significant differences, all these models have some common features. In all of them, the chemical equilibrium of each solid phase present in the system is verified on the basis of three series of equations :

- an equation accounting for the solubility of each of the solid phase of the material ;
- an equation accounting for the electro-neutrality of the pore solution ;
- and, finally, an equation to calculate the chemical activity of the pore solution.

The main difference between the various models lies in the treatment of the chemical activity of the pore solution. While, in some cases, authors have chosen to assess the activity coefficient on the basis of simple equations (such as the extended Debye-Hückel or the Davies equations), others rely on more sophisticated models (such as the Pitzer equation).

A comprehensive review of the various approaches to calculate the phase stability of cement systems is beyond the scope of this report. However, it should be emphasized that most of these approaches can be used to determine, with a relatively good precision, the pore solution composition of various systems at equilibrium. In fact, the main advantage of these models, from a mass transport point of view, is that any change of the pore solution chemistry is readily taken into account in the calculation.

Given their intricate nature, these models are however extremely difficult to integrate in a ionic transport model. Furthermore, the validity of most of these models is limited to isothermal conditions (25° C in most cases) and temperature effects can hardly be taken into account. More research is warranted, particularly on the numerical aspects of the problem, to efficiently couple the various transport equations with the chemical stability.

Physical interaction

Over the past decade, the mechanisms of physical interaction have received very little attention from researchers involved in the field of cement science. Recent reports tend, however, to underline the paramount importance of these phenomena on some aspects of the durability of concrete [20, 21].

In equations (7) and (8), the physical binding of the ions to the pore surface is described by the following term :

$$\rho_s \alpha \kappa (\partial c_{ia} / \partial t) \tag{12}$$

Under equilibrium conditions, equation (12) can be developed as follow :

$$\frac{\partial c_{ia}}{\partial t} = \frac{\partial c_{ia}}{\partial c_i} \frac{\partial c_i}{\partial t} \tag{13}$$

This new expression for the physical binding term requires the evaluation of the term of the left-hand side of the equation. It has been recently established that this term could be well modeled using a numerical solution of the Gouy-Chapman equation [9]. In cement systems, the double layer (see Figure 2) originates from the dissolution of the cement matrix in contact with the pore water [20]. As a result of this dissolution, a negative charge develops on the surface of the pores [20-23].

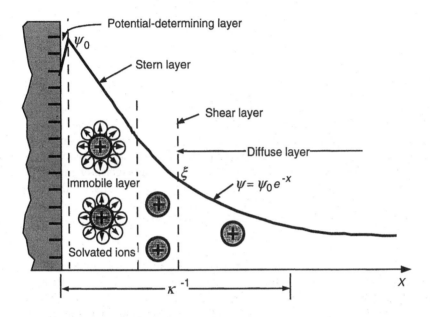

Figure 2 – Formation of the double-layer at the vicinity of the pore wall (according to reference 23)

The presence of these charges at the surface of the pore will force a redistribution of the ions at the vicinity of the pore wall in an attempt to restore the global electroneutrality. This is done by an accumulation of positive charges in a diffuse layer near the interface, thus creating a non-zero electrical potential in this area. The potential measured at the interface between the bound ion layer and the pore solution is called the zeta potential. Reports on the subject indicate that, in cement systems, the value of the zeta potential is relatively constant at -10mV [22].

The electrochemical potential of a given ionic species can be calculated on the basis of the following equation :

$$\mu_i = \mu_i^{\circ} + RT \ln(\gamma_i c_i) + z_i FV(x) \tag{14}$$

The equilibrium of the various ions near the double layer, which corresponds to the equilibrium of the electrochemical potential for each species, leads to the extended Nernst-Planck equation:

$$\frac{\partial}{\partial x}\left(-D_i\left(\frac{\partial c_i}{\partial x} + \frac{z_i F}{RT} c_i \frac{\partial V}{\partial x} + c_i \frac{\partial(\ln \gamma_i)}{\partial x}\right)\right) = 0 \tag{15}$$

Knowing the potential at the surface of the pore or the surface charge density, the distribution of ions in the double-layer can be calculated using a numerical solution of equation (15). As part of a study of the interaction of sulfate ions with well-cured C_3S paste, a series of simulations were carried out using a finite-element code [9]. The simulations were performed in one dimension at ion scale to account for the non-linear evolution of the electro-chemical potential from the pore surface to the bulk solution. The results of these numerical simulations are summarized in Figure 3. As can be seen, these simulations clearly indicate that the physical interaction is a function of the free sulfate concentration in the bulk solution.

The results of these numerical simulations were compared to the amount of bound sulfate ions measured on powdered C_3S samples immersed in sulfate solutions of various concentrations. The amount of bound sulfates was determined on the basis of the method developed by Tang and Nilsson [24]. Since the hydrated C_3S paste samples did not contain any aluminate, the mechanisms of interaction is most probably dominated by the physical binding process. Despite the fact that the simulations were limited to the uni-dimensional case, good correlations were found between the values predicted by the numerical model and the experimental results. This approach (that can be easily implemented in a numerical code) shows promise as a tool to investigate the mechanisms of ionic interaction in porous media. More work is however required to fully assess the reliability of the method for various types of systems (i.e. with and without supplementary cementing minerals, with chemical admixtures...).

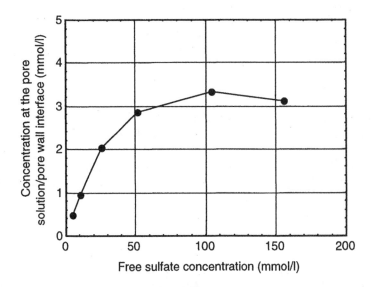

Figure 3 – Numerical simulations of the sulfate concentration at the interface for different bulk solution concentrations

It should be emphasized that one of the main advantages of treating the mechanisms of chemical and physical separately is that this approach is well adapted to the modeling of reversible phenomena. In many practical cases, concrete elements are subjected to variable exposure conditions (wetting and drying cycles, fluctuating ionic concentrations, ...) that may favor, in some instances, the formation of new chemical products, and in some others, the dissolution of various solid phases.

Global approach

Aware of the inherent difficulties of modeling separately the mechanisms of chemical and physical interactions, many authors have chosen to address the problem using a global approach without making any distinction between the two phenomena.

Under equilibrium conditions, the global interaction process can be described by the following equation:

$$\frac{\partial c_b}{\partial t} = \frac{\partial c_b}{\partial c_i} \frac{\partial c_i}{\partial t} \tag{16}$$

Equation (16) requires the determination of the term: $\partial c_b/\partial c_i$. This term can be derived from the so-called interaction isotherm (see Figure 4) that can be easily determined in the laboratory or derived from ionic concentration profiles [15, 24-26]. The experimental technique to obtain an interaction isotherm in the laboratory has been extensively described in numerous recent publications [10, 15, 24-26].

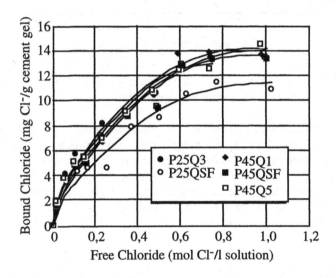

Figure 4 — Chloride binding isotherm expressed on a unit mass of cement gel basis (from reference 25)

It has been demonstrated that such isotherms are non-linear in nature and that they can be well described by a Langmuir isotherm (for low concentrations) and by a Freundlich isotherm (for high concentrations) [15, 24-26] :

$$\text{Langmuir}: \frac{1}{c_b} = \frac{a}{c_i} + b \tag{17}$$

$$\text{Freundlich} : \log c_b = d \log c_i + e \qquad (18)$$

In the coefficients a, b, d and e appearing in both equations are material parameters that have to be determined experimentally. The Freundlich isotherm appears to be better adapted to most practical cases where ionic concentrations are way over the range of validity of the Langmuir isotherm. As can be seen, the obvious advantage of this approach is that the isotherm can be described by a simple equation that can be easily implemented in a numerical code.

The main difficulty in working with this approach is that the shape of the interaction isotherm is influenced by a wide range of parameters. For instance, the isotherm has been found to be strongly affected by the type of binder [10, 15, 24-29]. As can be seen in Figure 4, variations in the shape of the isotherm can be attenuated if the mass of bound ions is expressed per mass of C-S-H gel (instead of mg of bound ions per g of material). However, this approach requires a certain knowledge of the degree of cement hydration. Furthermore, this approach appears to be only valid for OPC mixtures and cannot be applied to concrete made of supplementary cementing materials [25, 28].

As can be seen in Figure 5, the isotherms is also markedly affected by any variation of the surrounding solution. For the standpoint of modeling, this represents a significant problem since, in most cases, the global binding term appearing in equation 16 is solely expressed in terms of a single ionic species. Such a phenomenon limits the global approach to simple practical cases for which the variations in the solution chemistry are limited to a single ionic species.

Another possible drawback of this approach is that interaction isotherms are usually obtained on powdered paste samples. The intricate microstructure of concrete (with its interfacial zones and cracks) is not taking into account during the determination of the interaction curves. There exists very little information on the subject. However, chloride interaction isotherms obtained on well-cured paste and mortar samples tend to indicate that the presence of interfacial transition does not have a predominant influence on the phenomenon of binding (see Figure 6).

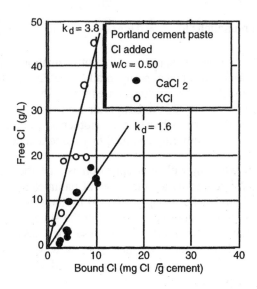

Figure 5 – Influence of solution on the interaction isotherm (from ref. 29)

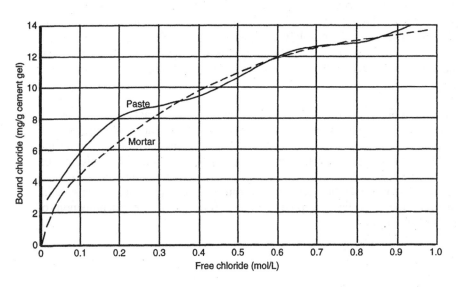

Figure 6 – Influence of interfacial transition zones on chloride binding

Concluding remarks

Over the past decade, the great deal of research specifically devoted to mass transport processes and phase equilibria in cement systems has largely contribute to improve the understanding of the parameters that control the mechanisms of ionic interaction. More work is, however, needed to integrate this knowledge in computer codes.

The approach that consists in separating chemical binding mechanisms from physical interaction phenomena appears to be the most promising avenue. This approach is well-adapted to transport models that account for the coupling between the various ionic fluxes. However, the treatment of phase equilibria has to be simplified in order to be implemented in a numerical code.

More work is finally needed to verify if simpler approaches based on the single-species interaction isotherms can be used to reliably predict the behavior of concrete under certain exposure conditions.

Acknowledgments

The authors are grateful to the Natural Sciences and Engineering Research Council of Canada, for its financial support for this project.

References

[1]Brown, P.W., *Materials Science of Concrete*, American Ceramic Society, **1**, pp.73-94 (1989).
[2]Damidot, D., Glasser, F.P., *Cem. Concr. Res.*, **23**, pp.221-238 (1993).
[3]Damidot, D., Glasser, F.P., 10th Int. Cong. Chem. Cem., Vol. 4, 8 p. (1997).
[4]Diamond, S., *Cem. Concr. Res.*, **11**, pp.383-390 (1981).
[5]Ramachandran, V.S. (1971). *Materials and Structures*, **4**, pp.3-12.
[6]Beaudoin, J.J., Ramachandran, V.S., Feldman, R.F., *Cem. Concr. Res.*, **20**, pp.875-883 (1990).
[7]Ramachandran, S., Seeley, R.C., Polomark, G.M., *Materials and Structures*, **17**, pp.285-289 (1984).
[8]Diamond, S., *Cem. Concr. Aggr.*, **18**, pp.97-102 (1986).
[9]Samson, E., Ph. D. Thesis, Université Laval, Canada. (in preparation).

[10]Maltais, Y., Ph. D. Thesis, Université Laval, Canada. (in preparation).
[11]Marchand, J., Gérard, B., Delagrave, A., Materials Science of Concrete, American Ceramic Society, Vol.5, (in press) (1998).
[12]Hazrati, K., Ph. D. Thesis, Université Laval, Canada. (in preparation).
[13]Pel, L., Ph. D. Thesis, Technishche Universiteit Eindhoven, Netherlands, 127 p. (1995).
[14]Houdusse, O., M.Sc. Dissertation, Université Laval, Canada. (in preparation).
[15]Tang, L., Ph.D. Thesis, Chalmers Univ. of Technology, Sweden, 461 p. (1996).
[16]Johannesson, B.F., Internal Report, Lund Institute of Technology, Sweden, 6 p. (1996).
[17]Johannesson, B.F., *Adv.-Cem. Bas. Mat.*, **6**, pp.71-75. (1997).
[18]Xu, A., Chandra, S., *Cem. Concr. Res.*, **24**, pp.375-379 (1994).
[19]Reardon, E.J., *Cem. Concr. Res.*, **20**, pp.175-192 (1990).
[20]Prezzi, M., Monteiro, P.J.M., Sposito, G., *ACI Mat. J.*, **94**, pp.10-17 (1997).
[21]Prezzi, M., Monteiro, P.J.M., Sposito, G. (1997), *ACI Mat. J.*, **95**, pp.3-10 (1997).
[22]Nägele, E., *Cem. Concr. Res.*, **17**, pp.573-580 (1987).
[23]Zhang, T., Gjørv, O.E., *Cem. Concr. Res.*, **26**, pp.907-917 (1996).
[24]Tang, L., Nilsson, L.O., *Cem. Concr. Res.*, **23**, pp.247-253 (1993).
[25]Delagrave, A., Marchand, J., Ollivier, J.P., Julien, S., Hazrati, K., *Adv.-Cem. Bas. Mat.*, **6**, pp.28-37 (1997).
[26]Arsenault, J., Ph.D. Thesis, Université Laval/INSA de Toulouse, Canada/France, (in preparation).
[27]Pereira, C.J., Hegedus, L.L., 8th Int. Symp. Chem. Reac. Eng., **37**, pp.427-438 (1984).
[28]Nilsson, L.O., Poulsen, E., Sandberg, P., Sørensen, H.E., Klinghoffer, O., Report N° 53, The Road Directorate, Copenhagen, Denmark, 151 p. (1996)
[29]Tuutti, K., Swedish Cement and Concrete Inst., Stockholm, 469 p. (1982)

MECHANICAL PROPERTIES, TESTING, FIBER REINFORCEMENT

FRACTURE BEHAVIOR OF BIAXIALLY CONFINED FIBER REINFORCED HIGH STRENGTH CONCRETE DUE TO IMPACT LOADING

K.-A. Rieder
On leave from the Department of Technical and Applied Physics
Vienna University of Technology
Vienna, Austria A-1040

S. Mindess
Department of Civil Engineering
University of British Columbia
Vancouver, BC, Canada V6T 1Z4

ABSTRACT

An experimental technique for imposing uniaxial or biaxial confinement on specimens subjected to dynamic loading has been developed. A brief description of the experimental technique is given and experimental results obtained on fibre-reinforced high strength concrete (FRHSC) are presented. To quantify the damage of the impacted specimens, the Brazil test was performed on the damaged - but not completely split - specimens under static loading conditions. The influence of different types and quantities of fibres on the biaxially confined impact behaviour is discussed.

INTRODUCTION

Plain unreinforced concrete is a brittle material, with low tensile strength and toughness. One way to enhance the properties of concrete is to add fibres. Their role is to reduce the stress at the crack tip by bridging across the matrix cracks while the concrete is loaded. A number of static and dynamic tests have been carried out in the past to study the effect of fibres on the fracture behaviour of fibre-reinforced concrete (FRC) (1, 2, 3, 4, 5). It was found that at low fibre contents (up to about 1 % by volume) the improvement of the strength of FRC is rather small, whereas an increase in fracture toughness or fracture energy can be achieved. The crack propagation can be slowed down or even stopped when using fibres. If the fibres are strong enough, they can keep the crack widths small and the FRC still can carry stresses in the post cracking stage.

Several different testing techniques have been developed to investigate the impact behaviour of confined concrete (6, 7, 8, 9), but the results of these different tests often cannot be compared. The main reasons for this are the different methods of loading, the different kinds of confinement, and the different ways of analysing the results.

In this paper, a new experimental technique is briefly described to characterise the impact performance of biaxially confined concrete using a modified drop weight impact machine. A detailed description of the new testing technique is given in (10, 11). The description of the instrumented drop-weight impact machine itself can be found in (12, 13). The results obtained with fibre-reinforced high strength concrete (FRHSC) subjected to various pre-stresses are presented and discussed.

EXPERIMENTAL PROCEDURES

A photograph of the 533 kg drop weight impact machine is shown in FIG. 1. The impact loads were applied as line loads along the centre lines of the top and bottom faces of the cubes. This induces tensile stresses perpendicular to the direction of loading ("splitting tension"). The biaxial confinement prevents or at least minimises the movement of the specimens in the splitting direction.

A schematic drawing of the mechanical confinement is shown in FIG. 2. The frame is rigidly connected to the top of the reinforced concrete pedestal. The rigidity of the steel confinement is 10^{-2} mm/kN, as measured under static loading conditions. Four steel rods with a diameter of 38 mm, two of which are equipped with a series of strain gauges to act as load cells, can be moved to produce the selected pre-stress levels.

In order to ensure a homogeneous and well-defined biaxial stress, the specimen is confined by 40 mm thick steel plates. Between the specimen and these steel plates a layer of Teflon is applied to minimise the transverse strains.

After the concrete cube with a side length of 100 mm was placed on the linear support and pre-stressed by tightening the steel rods, the hammer was dropped from a selected height. For this investigation the drop height was 0.225 m, which leads to an impact velocity of approximately 2 m/s. the pre-stress levels in the splitting direction as well as perpendicular to it could be chosen independently of each other. Two voltmeters, directly connected to the amplified output signal of the load cells were used to measure the actual pre-stress. However, the maximum applied pre-stress had to be less than 25 MPa, because of the limited stiffness of the instrumented steel rods. Otherwise yielding of the reduced cross section of the steel rods would damage the strain gauges.

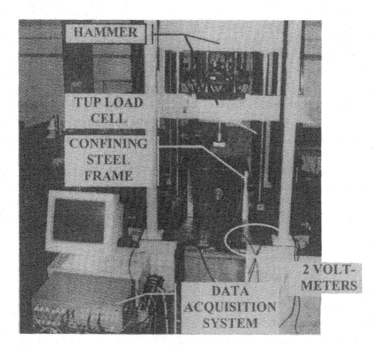

FIG. 1.
Drop weight impact machine with confining frame.

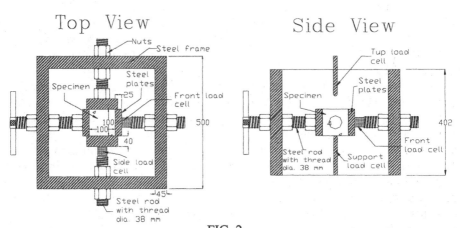

FIG. 2.
Schematic drawing of the test apparatus for the confined impact tests.

The instrumentation consisted of a series of strain gauges mounted within the striking tup of the falling mass (which will be referred to as the "tup load cell") to record the applied load, as well as on the linear support (which will be referred to as the "support load cell") to record the vertical reaction of the confined specimen. Strain gauges were also mounted on one of the steel rods in the splitting direction (which will be referred to as the "front load cell") and on one of the steel rods perpendicular to it (which will be referred to as the "side load cell"). Therefore four load versus time curves at intervals of 8 μs were recorded, using a high-speed PC-based data acquisition system.

The kinetic energy transferred from the falling hammer to the specimen was simply considered as the fracture energy in the impact tests. The kinetic energy lost by the hammer W_{Im} can be obtained from the impulse-momentum relationships (14)

$$W_{Im} = \frac{1}{2} \cdot M_h \cdot \left[\left(\frac{1}{M_h} \cdot \int p(t) \cdot dt - \sqrt{2 \cdot g_c \cdot h} \right)^2 - 2 \cdot g_c \cdot h \right] \qquad (1)$$

where M_h is the mass of the hammer, h is the height of its drop, g_c is the corrected gravitational acceleration, and p(t) is the time dependent load of the hammer during the impact. (A correction factor has to be applied to the acceleration due to gravity g to account for frictional effects between the hammer and the guiding columns, and the air friction: $g_c = 0.91g$). $\int P(t)\, dt$ represents the impulse acting against the hammer during the impact event, which is the area under the load versus time curve.

From the peak value of the "tup load cell" the maximum impact load P_H was calculated. From the peak values of the front load cell, the maximum stresses $\sigma_{x, max}$ were obtained.

Depending on the selected pre-stress and impact height the concrete sample were obviously damaged, but the specimen halves were not always completely separated. The material could still carry some load without breaking apart after it had been impacted. Therefore, the splitting or Brazil test was performed under quasi-static loading conditions on the already impacted cubes. The shapes of the support and the loading device as well as the splitting direction were the same as used for the impact test. The splitting procedure was carried out using a rigid testing machine with a maximum load capacity of 150 kN operating in a displacement controlled mode, with a crosshead speed of 1 mm/min. The residual splitting tensile strength f_t was calculated as

$$f_t = \frac{2 \cdot P_t}{\pi \cdot a^2} \qquad (2)$$

where P_t is the maximum load and "a" the side of the cube. In addition, the vertical displacement δ_V of the loading device was measured. The area under the load versus displacement curve was considered as the "residual splitting energy" G_S:

$$G_S = \int P_t(\delta_V) \cdot d\delta_V \qquad (3)$$

EXPERIMENTAL PROGRAM

The mix design given in Table I, with a water/cement ratio of about 0.35, was used to cast 100 mm cubes in laminated wooden moulds. Companion cylinders with a diameter of 100 mm and a height of 200 mm were cast in plastic moulds. Plain and fibre reinforced HSC with a cylinder compressive strength of 80 ± 5 MPa was obtained with this mix. All specimens were cured in lime-saturated water for 28 days before testing. Four different pre-stress levels (1, 8, 15, and 22 MPa) in the splitting direction, σ_x, were chosen, whereas the pre-stress perpendicular to the splitting direction, σ_y, was kept constant with 1 MPa for this investigation. In addition, unconfined impact tests using the same "splitting tension" test configuration were carried out. For all tests, the impact height was fixed with 0.225 m, which corresponds to an impact velocity of 2 m/s.

Table I.
Material per m^3 of concrete mix.

Cement, ASTM Type I	457 kg
Sand	645 kg
Gravel (size = 7 to 12 mm)	1076 kg
Water	161 l
Silica fume	81 kg
Superplasticizer ASTM C 494 Type A	3.7 – 4.5 l

5 different types of fibres were investigated, which are listed in Table II. For each test series 3 specimens were used.

Table II.Fibres investigated (Note: Fibre Code F0 means no fibres where added).

Fibre Code	Material	Geometry	Cross-section Shape	Length (mm)	Size (mm)
F1	Polyolefin	Straight	Circular	25	0.15 diam.
F2	Polyolefin	Straight	Circular	25	0.38 diam.
F3	Steel	Crimped	Crescent	25	3.0 x 0.8
F4	Steel	Hooked-end	Circular	30	0.5 diam.
F5	Steel	Twin-cone	Circular	35	1.0 diam.

RESULTS

The average values of the impact energy W_{Im}, the maximum load of the tup load cell P_H as well as of the support load cell P_S, and the maximum stress in the splitting direction $\sigma_{x, max}$ at different pre-stresses were calculated and are presented in Table III.

One part of this investigation was to compare the results of the confined impact tests with those of the unconfined drop weight impact test. For this purpose 3 specimens of each series were split with the drop weight impact machine. The drop height of the hammer and therefore the impact velocity were the same as used with the confined impact tests. The results of these tests are presented in Table IV.

The results of the static Brazil test performed with undamaged as well as already impacted (confined impact test), but not completely separated specimens, are presented in Table V and Table VI.

FIG. 3a shows the change in tensile splitting strength f_t of FRHSC, when different types and amounts of fibres are used. In this case the reference value was the tensile splitting strength f_t of HSC, which was 4.3 MPa. These tests were carried out with undamaged (not impacted) specimens under quasi-static loading conditions. FIG. 3b - d show the loss of tensile splitting strength f_t of biaxially confined and impacted FRHSC specimens relative to the f_t value of undamaged FRHSC for four different pre-stress levels. It should be noted that in FIG. 3b - d the influence of the pre-stress during impact on the amount of damage is shown. Though the relative f_t values do indicate the effect of pre-stress on residual strength for each fibre type, they cannot be compared between different fibre types since the values have been normalised to the respective undamaged control specimens.

Table III.

Results of the confined impact tests at different pre-stresses σ_x with $\sigma_y=1$ MPa.

Fibre content	0.5%						1.0%					1.5%				
Fibre Code	F0	F1	F2	F3	F4	F5	F1	F2	F3	F4	F5	F1	F2	F3	F4	F5
σ_x (MPa)	W_{IM} (Nm)															
1	808	902	868	875	863	859	882	870	842	931	857	913	835	871	871	858
8	867	916	867	887	927	848	870	900	877	935	825	962	792	853	878	899
15	798	941	918	880	960	870	903	870	890	903	932	867	949	891	918	854
22	-	884	880	884	937	930	931	850	909	847	873	942	929	868	890	889
	P_H (kN)															
1	245	233	245	249	233	251	225	231	257	250	244	231	255	245	249	245
8	276	264	253	261	260	269	259	259	265	271	275	236	278	273	280	266
15	313	292	263	268	278	278	279	280	285	307	288	288	268	262	267	311
22	-	294	282	269	300	273	283	300	293	322	316	221	296	285	269	271
	P_S (kN)															
1	178	183	189	186	172	180	160	176	187	197	178	172	188	188	208	186
8	189	218	186	198	211	212	195	185	206	225	216	181	209	228	243	212
15	230	223	210	221	222	227	226	217	226	243	243	240	221	212	222	258
22	-	232	211	203	241	219	230	237	231	260	247	165	235	237	208	204
	$\sigma_{x,max}$ (MPa)															
1	20.9	18.6	19.5	18.3	17.9	19.8	18.3	18.8	19.0	17.3	20.0	17.1	19.5	16.9	15.2	19.2
8	24.2	19.2	21.4	18.8	18.3	21.1	19.1	20.6	18.6	17.1	20.1	19.0	21.6	16.3	15.8	17.4
15	27.0	22.9	19.4	20.3	21.1	21.0	19.4	22.1	22.1	20.2	19.7	20.6	19.0	18.8	18.9	20.3
22	-	25.7	25.9	25.0	25.5	26.7	25.0	26.6	25.4	25.8	26.3	24.4	25.6	25.1	25.8	27.0

Table IV.
Results of unconfined impact for different types and quantities of fibre reinforced concrete.

	Fibre content															
	0 %	0.5 %					1.0 %					1.5 %				
	F0	F1	F2	F3	F4	F5	F1	F2	F3	F4	F5	F1	F2	F3	F4	F5
W_{IM} (Nm)	67	74	55	95	127	131	126	91	141	278	164	99	80	137	441	199
P_H (kN)	101	107	82	99	86	120	109	72	98	110	74	96	99	106	130	108
P_S (kN)	99	117	69	99	89	137	108	79	108	111	81	94	100	109	143	117

Table V.

Residual splitting tensile strength f_t of undamaged and impacted specimens.

		Pre-stress σ_x (MPa) during impact				Undamaged
		1	8	15	22	
Fibre code	Content	f_t (MPa) of static Brazil test				
F0	0 %	0.03	0.06	0.17	-	4.3
F1		0.7	0.8	1.2	2.3	3.2
F2		0.5	0.8	1.7	2	3.1
F3	0.5 %	0.6	1.6	2.2	3.6	4.2
F4		1.7	2.3	3.7	3.7	4.1
F5		0.4	1	2.8	3.1	3.8
F1		1.4	2	2.5	2.5	3.7
F2		0.8	0.8	1.4	2.4	3.5
F3	1.0 %	1.5	2.2	2.3	3.1	4.7
F4		3.2	4.1	5.4	5.8	6.3
F5		1.0	2.7	4.6	4.5	5.0
F1		1.6	2.3	3.0	3.0	3.7
F2		1.2	1.5	2.5	3.0	3.8
F3	1.5 %	1.5	5.1	5.1	5.2	5.8
F4		3.8	6.3	6.8	7.4	8.2
F5		1.3	3.1	5.7	6.2	6.2

Table VI.
"Residual splitting energy" G_S of undamaged and impacted specimens.

Fibre code	Content	Pre-stress σ_x (MPa) during impact				Undamaged
		1	8	15	22	
		G_S (Nm) of static Brazil test				
F0	0 %	0.7	0.8	2.8		
F1		45	113	81	146	215
F2		21	37	76	88	70
F3	0.5 %	32	67	73	119	263
F4		113	146	219	199	437
F5		18	26	82	92	206
F1		76	125	170	198	335
F2		33	39	88	96	121
F3	1.0 %	66	103	102	130	269
F4		207	314	325	339	541
F5		23	110	156	164	209
F1		117	150	174	298	312
F2		61	74	170	128	178
F3	1.5 %	70	263	263	232	524
F4		228	412	810	900	943
F5		64	107	225	215	382

DISCUSSION

In general, the peak values of the tup load cell as well as that of the support load cell, P_H and P_S, respectively, increase with increasing confinement. The results of the confined impact tests summarised in Table 3 show no significant change of the tup load and the support load between different materials. All specimens tested have similar maximum loads at the same pre-stress levels. A difference in the impact energy W_{Im} between plain and fibre reinforced HSC can be observed: For plain HSC W_{Im} is slightly lower than for FRHSC.

The maximum stress in the splitting direction $\sigma_{x,m}$ (reading from the front load cell), increases with increasing confinement. The value of this so-called "maximum stress" depends strongly on the experimental set-up, especially on the rigidity of the steel frame: The stiffer the frame the higher this maximum stress will be. The influence of the fibres can be clearly seen on the "maximum stress": $\sigma_{x,m}$ decreases the more fibres are added to the concrete. At a pre-stress of 15 MPa $\sigma_{x,m}$ drops from 27 MPa for plain HSC to 18.9 MPa (a decrease of 30%), if HSC is reinforced with 1.5% of hooked end steel fibres. FIG. 4 shows four different load versus time curves, where the difference of the material response between plain (FIG. 4a and FIG. 4c) and fibre reinforced HSC (FIG. 4b and FIG. 4d) can be seen. Note: The front load cell records the load (stress) in splitting direction during the impact.

Obviously fibres are carrying tensile stresses and therefore they are reducing the maximum dilatation of the specimens during the biaxially confined impact, which leads to the observed lower maximum stresses $\sigma_{x,m}$. If the results for the different types and amounts of fibres are compared, the following conclusions can be drawn:

In terms of $\sigma_{x,m}$ parameter a low value is desired. The lower this value is, the more stresses can be transferred over the fracture surface to reduce the maximum tensile stress at the crack tip. Under static loading conditions mechanically deformed steel fibres have a distinct advantage over other fibres in terms of being very effective in improving pullout resistance (15, 16, 17). This changes for the biaxially confined impact tests: The more pre-stress is applied the less difference in the material response to the impact in terms of the maximum stress $\sigma_{x,m}$ can be observed between the investigated fibres. The influence of the amount of fibres is also very small at high pre-stress levels.

At low pre-stresses up to 8 MPa the amount of fibres as well as the fibre types change the material behaviour. In terms of a reduced $\sigma_{x,m}$ value, fibre type F4 shows the best results at a fibre content of 1.5%. In general, all investigated steel fibres F3, F4, and F5 reduce the $\sigma_{x,m}$ value more effective, if the fibre content is increased to 1.5%. For the Polyolefin fibres F1 and F2 this trend cannot be observed.

As mentioned earlier, the same type of impact test was carried out but without any confinement at all. The results for the impact energy W_{Im} presented in Table 4 show that all steel fibres investigated improve their impact performance, if the content of fibres is increased to 1.5%, while the Polyolefin Fibres reach a maximum at a fibre content of 1%. Among the steel fibres, fibre type F4 showed the best impact performance in terms of an increased impact energy. Nearly all fibres were pulled out, whereas fibre failure was observed for fibre type F4. For fibre type F3 both mechanisms of energy absorption were observed.

It is a well-known fact that fibre failure is less effective than complete pull out of fibres, because in the latter case crack propagation can be stopped or at least minimised. If fibre failure occurs, no stresses can be transferred between the fracture surfaces. Therefore the reinforced materials also tend to fail in a brittle way. The Brazil tests, carried out under static loading conditions with the impacted, but not completely separated specimen halves, confirm the above conclusions. The specimens with contents of fibre type F5 fail brittle, whereas all other FRHSC specimens behave rather ductile.

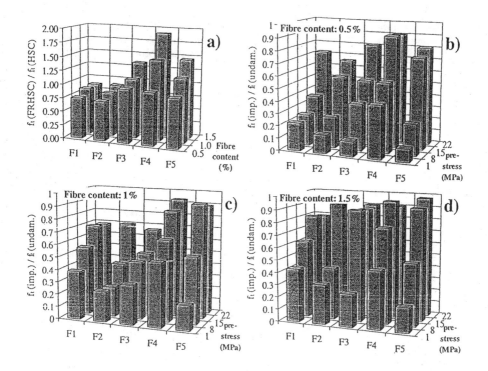

FIG. 3
a): Residual tensile splitting strength f_t of FRHSC compared to that of HSC (static loading),
b), c), d): Residual tensile splitting strength f_t of impacted FRHSC at different pre-stress levels compared to that of undamaged specimens.

FIG. 3a shows the splitting tensile strength f_t of the undamaged FRHSC specimens normalised to the f_t value of HSC, which was 4.3 MPa. Although the compressive strength for all specimens was 80 ± 5 MPa, no matter what type and amount of fibres were used, f_t depends strongly on the amount and type of fibre mixed to the concrete. For the Polyolefin fibre type F1 and F2 the splitting tensile strength f_t drops to 75% of its initial value (plain HSC). An increase of the fibre percentage added does not improve the splitting tensile strength f_t. A completely different behaviour shows fibre type F4. At a fibre content of 0.5%, the f_t value remains the same, but increases 50% at a fibre content of 1%, and 100% at a fibre content of 1.5%. An increase of the f_t value can also be observed for the fibre types F3 and F5, but it is not so pronounced.

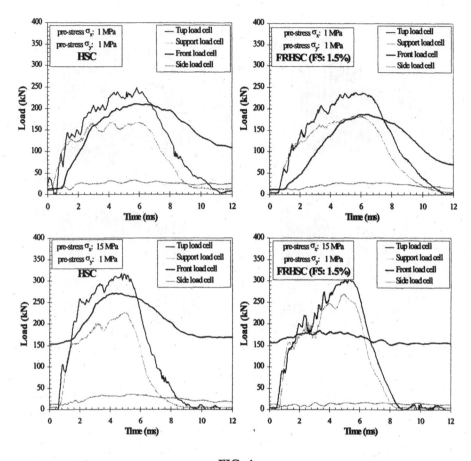

FIG. 4
Various load versus time curves of biaxially confined (a), (c) HSC, and
(b), (d) FRHSC with 1.5% twin-cone steel fibres (fibre code F5).

Another way to quantify the amount of damage due to the confined impact tests of FRHSC is to perform the Brazil test with the impacted specimens under static loading conditions. FIG. 3b – d show the loss of tensile splitting strength f_t of biaxially confined and impacted FRHSC specimens relative to the f_t value of undamaged FRHSC. These results also demonstrate how efficient fibre type F4 is able to minimise the amount of damage. At a pre-stress level of 1 MPa the fibres still can transfer stresses over the cracked surface (40% of the f_t value of the undamaged control specimens), if they are loaded after the impact. The relative loss of the residual tensile splitting strength f_t of different types of fibres becomes less, if the amount of fibres is increased (Note: the absolute f_t values are presented in Table V).

The results for fibre type F5 show very clearly, how much influence the pre-stress has on the amount of damage: At low pre-stress values the specimens with fibre F5 nearly break into two pieces (very low residual tensile strength), when they are impacted. But they are rarely damaged, if the pre-stress level is increased. Therefore it can be concluded that twin-cone steel fibres have the best impact performance, when the maximum displacement is limited (the mechanical confinement defines the maximum possible displacement in the splitting direction).

The results of the residual splitting energy G_S presented in Table VI confirm the above findings. The outstanding performance of fibre type F4 to withstand the tensile stresses during the confined impact nearly without any damage can be observed. The amount of energy remains unchanged, if the pre-stress reaches 22 MPa. It goes without saying that there would be signs of damage, if the impact height were increased.

CONCLUSIONS

The following conclusions can be drawn from the above findings: A fully instrumented drop weight impact machine in conjunction with a mechanical confinement of the specimens was used to investigate the impact behaviour of biaxially confined concrete. There is a significant change in the fracture behaviour of plain and fibre reinforced HSC when the material is confined while being impacted. The more the concrete is confined the more energy it can absorb while impacted. The maximum stress in the splitting direction $\sigma_{x, max}$ drops significantly if fibres are added to the concrete. The hooked ended steel fibres F4 have the lowest $\sigma_{x, max}$ values. The Brazil test performed after the impact test was used to quantify the amount of damage in terms of the residual splitting tensile strength. These results also showed the outstanding performance of this fibre type under these loading conditions. An increase of the fibre content up to 1.5% seems to improve the fracture behaviour of steel fibre reinforced HSC in terms of the tensile splitting strength, the fracture energy, as well as of the impact energy. The same improvement cannot be achieved with Polyolfine fibres F1 and F2 by increasing the content of fibres. Despite the mechanical behaviour, it is very difficult to get an uniform distribution of these fibres at higher volume percentages.

On the whole, it may be said that the resistance of confined FRHSC against impact load is much higher than that of unconfined FRHSC.

ACKNOWLEDGEMENTS

The authors are grateful to the staff of the University of British Columbia, Department of Civil Engineering, Vancouver, Canada where this work was carried out. The help of graduate student Eric Morris, who carried out most of the test, is thankfully acknowledged. One of the authors (Dr. Rieder) received a Schrödinger Fellowship from the Austrian Science Fund FWF, which he gratefully acknowledges. This work was suggested by the Natural Sciences and Engineering Research Council of Canada, through its support for the NCE "Concrete Canada".

REFERENCES

[1]V. S. Gopalaratnam and S. P. Shah, "Properties of Steel Fiber Reinforced Concrete Subjected to Impact Loading," *J. of the Amer. Conc. Inst.,* **83** [8], 117-126 (1985).

[2]S. Mindess, "Fibre-Reinforced Cementitious Materials: A Personal View," *ASCE J. of Materials in Civil Eng.,* **5** [3], 289-292 (1993).

[3]A. Bentur and S. Mindess, *Fibre Reinforced Cementitious Composites.* Elsevier Applied Science, U.K., 1990.

[4]S.P. Shah, Y. Shao and Z. Li, "Characterizing Fracture Behaviour of Fibre Reinforced Composites"; pp. 186-199 in *Advances in Cement and Concrete.* Edited by M.W. Grutzeck, and S.L. Sarkar. American Society of Civil Engineers, 1994.

[5]P. Rossi, T. Sedran, S. Renwez and A. Belloc, "Ultra High Strength Steel Fibre Reinforced Concretes: Mix Design and Mechanical Characterization"; pp. 181-186 in *Fiber Reinforced Concrete: Modern Developments.* Edited by N. Banthia and S. Mindess. 1995.

[6]J. Weerheijm, H. W. Reinhardt, and S. Postma, "Experiments on concrete under lateral compression and tensile impact loading"; pp. 839-848 in *Fracture Process in Concrete, Rock and Ceramics.* Edited by J.G.M. van Mier, J.G. Rots and A. Baker. E&F.N. Spon., London, 1991.

[7]W. Chen and G. Ravichandran, "An Experimental Technique for Imposing Dynamic Multiaxial-compression with Mechanical Confinement," *J. Exp. Mech.* **36** 155-158 (Jan. 1996).

[8]A. J. Zielinski, "Concrete under Biaxial Compressive-impact Tensile Loading"; pp. 479-489 in *Fracture Toughness and Fracture Energy of Concrete.* Edited by F.H. Wittmann. Elsevier Science Publishers, Amsterdam, 1986.

[9]J. Weerheijm, "Concrete under Impact Tensile Loading and Lateral Compression," Ph.D. thesis, TNO Prins Maurits Laboratory (1992).

[10] K.-A. Rieder, and S. Mindess, "New Testing Method to Evaluate the Impact Performance of Biaxially Confined Concrete," submitted to RILEM Materials & Structures, accepted (1998).

[11] S. Mindess and K.-A. Rieder, "Fracture Behavior of Biaxially Confined Concrete Due to Impact Loading," submitted to *Transient Loading and Response of Structures*, Trondheim, Norway, accepted (1998).

[12] N. Banthia, S. Mindess, A. Bentur, and M. Pigeon, "Impact Testing of Concrete Using a Drop Weight Impact Machine." *J. Exp. Mech.* **29** [2] 63-69 (1989).

[13] N.P. Banthia, "Impact Resistance of Concrete," Ph.D. Thesis, Univ. of British Columbia, Vancouver, Canada (1987).

[14] N. Wang, "Resistance of Concrete Railroad Ties to Impact Loading," Ph.D. Thesis, Univ. of British Columbia, Vancouver, Canada (1996).

[15] N.P. Banthia, and J.-F. Trottier, "Concrete Reinforced with Deformed Steel Fibers, Part I: Bond-Slip Mechanisms," *ACI Mat. Journal* **91** [5] 435-446 (1994).

[16] P. Rossi, and G. Chanvillard, "New Geometry of Steel Fiber for Fiber Reinforced Concrete"; pp. 129-139 in *High Performance Fiber Reinforced Cement Composites*. Edited by H.W. Reinhardt and A.E. Naaman. E.&F.N. Spon., London, 1992.

[17] G.H. Tattersall, C.R. Urbanowicz, "Bond Strength in Steel Fiber Reinforced Concrete," *Magazine of Concrete Research* **26** [87] 105-113 (1974).

EFFECT OF FIBER ON AUTOGENOUS SHRINKAGE STRESS OF HIGH-STRENGTH CEMENT MORTAR

Shingo Miyazawa[*], Tokio Kuroi[*] and Hisashi Shimomura[**]
[*] Department of Civil Engineering, Ashikaga Institute of Technology
268-1 Ohmae Ashikaga, 326-8558 Japan
[**] Sub-committee on Steel-Fiber, The Kozai Club
3-2-10 Kayaba-cho Nihonbashi Chuou-ku, 103-0025 Japan

ABSTRACT

The purpose of this study is to investigate the effect of fiber on cracking tendency of high-strength mortar.   Autogenous shrinkage of mortar with water-cement ratio of 0.20 and 0.30 was measured from the time of initial setting to the age of 60 days.   Self stress of mortar was also measured by a ring test, where autogenous shrinkage of mortar was restrained by a steel ring.   Four types of fiber, steel fibers(5 mm and 30 mm in length), vinylon fiber(30 mm in length) and aramid fiber(30mm in length), were used.   It is proved that autogenous shrinkage of high-strength mortar is decreased by addition of these fibers at 2.0 % by volume. The effect is increased with increase in Young's modulus of the fiber. Autogenous shrinkage stress of mortar can be decreased and tensile strength of mortar is increased by addition of fiber.   Among the four type of fibers, steel fiber(30mm) has the greatest effect in improving cracking resistance of high-strength mortar.

INTRODUCTION

Macroscopic volume reduction of hardened concrete caused by cement hydration, which is not caused by external load, moisture movement to or from the surrounding environment, temperature change, nor carbonation, is known as autogenous shrinkage.   It was found by Davis that autogenous shrinkage of ordinary concrete was no more than 100×10^{-6}, which was much less than drying shrinkage[1].   Therefore, it has been thought that autogenous shrinkage could be ignored for the practical purpose.   Recent researches, however, have demonstrated that autogenous shrinkage can be quite large in high-strength concrete[2,3] at early ages.   Therefore, considerably large tensile stress can be generated by restraining autogenous shrinkage, and can be a cause of cracking in high-strength concrete structures[4-6].

There are some researches on reducing autogenous shrinkage and autogenous shrinkage stress of high-strength concrete. It has been reported that autogenous shrinkage can be reduced by using belite-rich cements, expansive admixtures, or chemical admixtures reducing surface tension of capillary water[7]. It has been also reported that the age when a shrinkage cracking occurs in high-strength silica fume concrete can be lengthen by steel fiber[4], and that the tendency of plastic shrinkage cracking of normal strength silica fume concrete exposed to early drying can be eliminated by addition of plastic fibers[8].

EXPERIMENTAL PROCEDURES

Materials and mix proportion

Ordinary Portland cement, polycarboxlic acid ether type superplasticizer and river sand (specific gravity: 2.60, absorption: 1.76%, fineness modulus: 3.15) were used. Two types of steel fiber with different length(5 mm and 30 mm), vinylon fiber(30 mm) and aramid fiber(30 mm), which are commercially available in Japan, were used. Mechanical properties of these fibers are shown in Table I. Water-cement ratio of mortar is 0.20 and 0.30.

Mix proportions of mortar are shown in Table II for Series 1, and Table III for Series 2. Influence of steel fiber length and its content is investigated in Series 1, where steel fibers of 2 mm and 30 mm long were used and the content of fiber was varied as 0%, 0.5%, 1.0% and 2.0% of mortar by volume. Influence of fiber type is investigated in Series 2, where steel fiber, vinylon fiber and aramid fiber were used and the content of fiber was varied as 0% and 2.0%.

Table I. Properties of fibers

	Dimensions (mm)	Specific gravity	Tensile strength (N/mm^2)	Young's modulus (kN/mm^2)
Steel fiber(S30)	$\phi\,0.6\times30$	7.85	$\geq 1000^{*1}$	200
Steel fiber(S5)	$\phi\,0.6\times5$	7.70	588^{*1}	200
Vinylon fiber(V30)	$\phi\,0.66\times30$	1.30	890	29
Aramid fiber(A30)	$\phi\,0.15^{*2}\times30$	1.39^{*2}	3400^{*2}	73^{*2}

($*$1:raw material, $*$2:raw fiber)

Autogenous shrinkage test

Autogenous shrinkage was measured by the method proposed by Japan Concrete Institute[9], except that the dimensions of the specimens was $100\times100\times400$mm. The apparatus for the measurement before demolding is illustrated in Fig.1. Horizontal length change of mortar specimens was measured under sealed

Table II. Mix proportions of mortars (series 1)

Mix	w/c	s/c	Fiber (Vol%)	Unit content(kg/m^3)				SP ×C%	Flow (mm)
				W	C	S	Fiber		
N30-0			0	316	1054	906	0	0.5	230×226
N30-S30-0.5			0.5	315	1050	902	39	0.5	248×246
N30-S30-1.0	0.30	0.86	1.0	313	1043	897	78	0.5	237×235
N30-S30-2.0			2.0	310	1034	889	156	0.5	179×181
N30-S5-2.0			2.0						226×224
N20-0			0	248	1239	867	0	2.2	136×135
N20-S5-2.0	0.20	0.70	2.0	243	1215	851	156	2.2	178×161
N20-S30-2.0			2.0						130×127

SP : superplasticizer

Table III. Mix proportions of mortars (series 2)

Mix	w/c	s/c	Fiber (Vol%)	Unit content(kg/m^3)				SP ×C%	Flow (mm)
				W	C	S	Fiber		
N30-0		0.83		303	1045	873	0	0.5	274×266
N30-S30-2.0	0.30		2.0	299	1012	873	156	0.5	230×223
N30-V30-2.0		0.86		299	1012	873	26	0.5	273×255
N30-A30-2.0				299	1012	873	28	0.5	very stiff
N20-0				223	1239	873	0	2.2	259×255
N20-S30-2.0	0.20	0.70	2.0	217	1221	855	156	2.2	178×172
N20-V30-2.0				217	1221	855	26	2.2	174×168

SP : superplasticizer

condition in a room at 20 ℃. The measurement was started at the time of initial setting of the mortar determined by Test Method JIS A 6204 which is similar to the ASTM method. The initial setting time of mortars with fiber was assumed to be the same as that of mortar without fiber. For the first 24 hours after casting, the measurements were done with mixtures in the mold by dial gauges. In order to eliminate the restraint by the mold, a polytetrafluoroethylene (Teflon) sheet (1 mm in thickness) was put on the bottom of the steel mold. During the first 24 hours, temperature of the specimens rose due to cement hydration. Therefore, the thermal expansion strain was excluded from the measured strain on the assumption that the thermal expansion coefficient of the mortars was $15 \times 10^{-6}/℃$. After demolded at the age of 24 hours, the specimen was sealed with aluminum tape in order to prevent from evaporation, and the length change was measured with a contact gauge for up to 2 months.

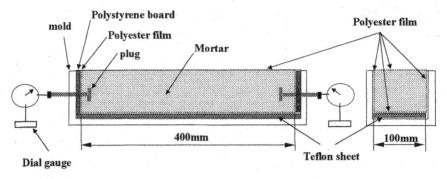

Fig. 1 Measurement of autogenous shrinkage of mortar (before demolding)

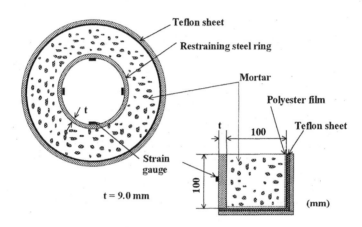

Fig. 2 Measurement of restraining stress of mortar

Self-stress test

This test was arranged for investigating tensile stress of mortar generated under external restraint. Mortar was cast in the steel ring mold shown in Fig.2. The mortar was isolated from the exterior ring and the bottom of the mold by a polytetrafluoroethylene (Teflon) sheet (1 mm in thickness) and polyethylene sheet (0.1 mm in thickness). Shrinkage of mortar was partially restrained by the inner steel ring, thickness of which (t) was 9.0mm. Self stress in mortar was obtained from the strain of the inner steel ring, which was measured by four electric wire strain gauges(gauge length: 5 mm). The specimens were stored in the room at 20

°C and 80%R.H., and the surface of the mortar was subjected to two different conditions, sealed or wet conditions. For the specimens under wet condition, the mortar surface was covered with sponge (50 mm thick), and 500 cc of water was supplied to the sponge every day.

RESULTS AND DISCUSSIONS

It was proved from autogenous shrinkage test that expansion was not observed in the tested mortars. Therefore it can be said that compressive self stress due to the restraint by the exterior ring was not generated in the self stress test. It was also found by measurements that stress in perpendicular direction due to restraint by the inner steel ring was relatively low.

When mortar is assumed to be a elastic body during the test, tensile self stress of mortar in the tangent direction due to the restraint by the interior steel ring can be calculated from the next equation. Notations in the equation are presented in Fig. 3.

$$\sigma(r) = \varepsilon_s E_s t\, r_1 \left(r_2^2 / r^2 + 1 \right) / \left(r_2^2 - r_1^2 \right) \tag{1}$$

where,

$\sigma(r)$: normal tensile stress of mortar in the tangent direction
r : radial coordinate
ε_s : compressive strain of interior steel ring
E_s : Young' modulus of interior steel ring
t : thickness of interior steel ring (9 mm)
r_1 : exterior radius of interior steel ring (150 mm)
r_2 : interior radius of exterior steel ring (250 mm)

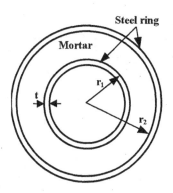

Fig.3 Notation in equation (1)

In the cross section of the mortar specimen, $\sigma(r)$ is dependent on r and is maximized when r is equal to r_1. In the following figures, $\sigma(r)$ is presented as the mean tensile stress over a cross section of the specimen.

Effect of fiber content (Series 1)

Test results of autogenous shrinkage for 0.3 w/c mortars with different fiber content is shown in Fig. 4. It can been seen that autogenous shrinkage of mortar containing 2.0% of steel fiber is lower than that of plain mortar, while autogenous shrinkage is not decreased by fiber addition of 1.0% or less. As shown in Fig.5, test results show that self stress can reach to a considerably large value in a few days after casting even if no evaporation is permitted. The tensile stress is decreased with increase in steel fiber content, within the range form 0% to 2.0%.

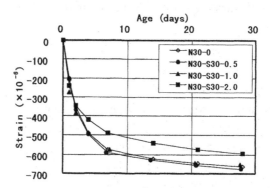

Fig. 4 Effect of fiber content on autogenous shrinkage (steel fiber, 30 mm)

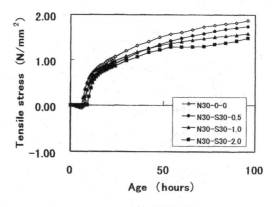

Fig 5 Effect of fiber content on autogenous shrinkage stress (steel fiber, 30 mm)

Effect of fiber length (Series 1)

   Autogenous shrinkage of mortar containing fibers with different length is shown in Fig.6. For both water-cement ratio of 0.2 and 0.3, autogenous shrinkage of mortar with 30 mm steel fiber is less than that with 5 mm one. Influence of fiber length on autogenous shrinkage is more significant in mortar with 0.2 w/c mortar than 0.3 w/c. For the mortar with 0.20 w/c, effect of fiber length is significant, especially for the first 24 hours. Tensile stress of mortar with 30 mm steel fiber is less than that with 5 mm one in the both water-cement ratio, as show in Fig.7.

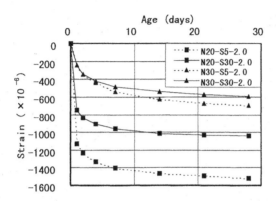

Fig. 6 Effect of fiber length on autogenous shrinkage (steel fiber, 2.0%)

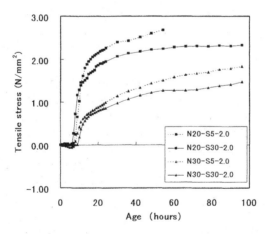

Fig. 7 Effect of fiber length on autogenous shrinkage stress (steel fiber, 2.0%)

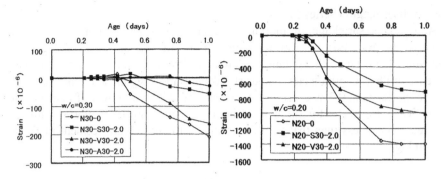

Fig. 8 Autogenous shrinkage of mortar at early ages (30 mm fiber, 0% or 2.0%)

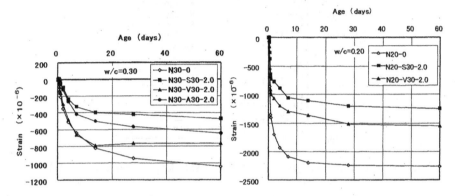

Fig. 9 Autogenous shrinkage of mortar at later ages (30 mm fiber, 0% or 2.0%)

Effect of fiber type (Series 2)

The tests results for autogenous shrinkage in the mortars with different types of 30 mm fiber are shown in Fig.8 for the first 24 hours and in Fig.9 for the later ages. Autogenous shrinkage is decreased by addition of each type of fiber at 2 % by volume. Autogenous shrinkage is initiated at about the age of 12 hours after casting for 0.30 w/c mortar and about 8 hours for 0.2 w/c mortar.

When the aramid fiber was used, the mortar with 0.30 w/c was very stiff, and the mortar with 0.20 w/c was too stiff to be mixed, since the raw fibers got untied at the time of mixing.

At the age of 24 hours, the ratio of autogenous shrinkage strain of the fiber mortars to that of the plain mortar ($\varepsilon_f / \varepsilon_p$) is 50% or less for the steel fiber and

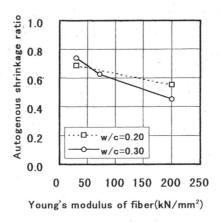

Fig. 10 Relation between Young's modulus of fiber and autogenous shrinkage of mortar (60 days)

70~80% for the vinylon fiber. The effect is larger at early ages than at later ages. The ratio $\varepsilon_f/\varepsilon_p$ at 60 days versus Young's modulus of the fibers is shown in Fig. 10. It can be said that the effect of fiber increases with increasing Young's modulus of the fiber.

Self stress of the sealed mortar with different types of fiber is shown in Fig.11. Compared with the results shown in Fig.8, it can be seen that the age when tensile stress begins to increase is about the same as the age when autogenous shrinkage begins to occur. In mortar with 0.20 w/c, tensile stress grows rapidly during the first 24 hours, and its development after 24 hours is relatively little. On the other hand, tensile stress grows gradually in mortar with 0.30 w/c. These trend is similar to those observed in autogenous shrinkage generation. For the mortars with 0.3 w/c, tensile stress is decreased by addition of any kind of fiber under both sealed and wet conditions (Fig.11 and Fig.12). For the mortars with 0.2 w/c, self stress is decreased by addition of the steel fiber under sealed condition although the effect is not significant for this mix proportion, and little effect is observed with the vinylon fiber.

Effect of curing (Series 2)

Self stress of mortar under wet condition, where water is continuously supplied to the mortar surface from the time of casting is shown in Fig.12. Tensile stress of the mortars under wet condition is less than that of the sealed mortars (Fig.11). But considerably large tensile stress is still observed under wet condition. It can be said that curing water only permeates into the surface layer of the specimens due to the low permeability of the mortar with low water-cement ratio, while the inside of the specimen is subjected to self-desiccation[10]. These experimental results suggest that autogenous shrinkage has a great influence on plastic shrinkage cracking which has been thought to be caused by evaporation from the surface.

Although the test methods for tensile strength of fiber reinforced mortar and concrete have not been established, it is reported that initial cracking strength obtained from splitting tensile strength test can be used for estimating their tensile strength[11]. The test results for splitting tensile strength and bending strength of

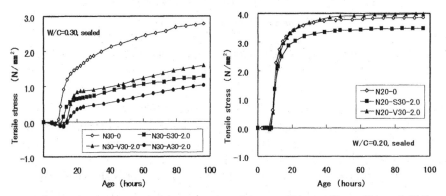

Fig.11 Autogenous shrinkage stress of mortar (30 mm fiber, 0% or 2.0%)

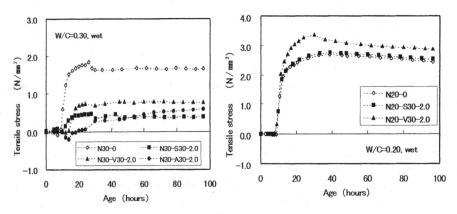

Fig.12 Self stress of mortar under wet condition (30 mm fiber, 0% or 2.0%)

TABLE IV. Tensile stress of mortar corresponding to initial cracking load (Age: 1day)

Mix	w/c (%)	Splitting test (N/mm^2)	Bending test (N/mm^2)
N30-0		2.05	4.64
N30-S30-2.0	0.30	3.44	9.02
N30-V30-2.0		2.76	4.85
N20-0		3.69	6.51
N20-S30-2.0	0.20	4.35	8.48
N20-V30-2.0		3.37	6.46

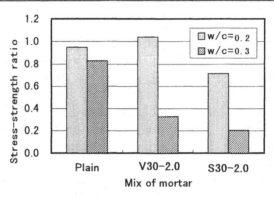

Fig.13 Stress-strength ratio of mortar (sealed, Age: 24 hours)

the mortars at the age of 24 hours are shown in Table IV. Tensile strength of the mortars is increased by addition of the steel fiber. The ratio of tensile self stress at the age of 24 hours to the initial cracking strength at the same age (stress-strength ratio) is shown in Fig.13. The values for the plain mortars and the 0.2 w/c mortar with the vinylon fiber are nearly equal to 1.0. This means that cracking probability is very high in these specimens, although no cracking was observed in the tests. Stress-strength ratio for 0.3 w/c mortar is considerably decreased by addition of vinylon fiber and steel fiber. Stress-strength ratio for 0.20 w/c mortar is considerably decreased by addition of steel fiber. Therefore, steel fiber is more effective than the other type of fibers for crack control in high-strength mortar, when compared under the same mix proportion.

CONCLUDING REMARKS

It is proved that tensile self stress due to restraint of autogenous shrinkage can be so large in high-strength mortar that it can be a cause of cracking at early ages.

In order to improve cracking resistance of high-strength mortar, steel fiber and vinylon fiber can be effectively used. Especially, mortar containing 30 mm steel fiber at 2.0% by volume has lower autogenous shrinkage strain and higher tensile strength, which means that the possibility of cracking at early ages can be significantly decreased by steel fiber.

REFERENCES

[1]H. E. Davis, "Autogenous volume change of concrete," *Proceeding of ASTM*, 403, 1103 (1940).

[2]F. de Larrard, "Creep and shrinkage of high-strength field concretes," *SP121-28*, ACI, 577 (1990).

[3]E. Tazawa and S. Miyazawa, "Autogenous shrinkage of cement paste caused by self desiccation in cementitious material," *Ninth International Congress on Chemistry of Cement*, New Delhi, India, 712 (1992).

[4]A. M. Paillere, M. Buil and J.J. Serrano, "Effect of fiber addition on the autogenous shrinkage of silica fume concrete," *ACI Material Journal*, 86, No.2, 139 (1989).

[5]E. Tazawa, Y. Matsuoka, S. Miyazawa and S. Okamoto, "Effect of autogenous shrinkage on self stress in hardening concrete," *RELEM Int. Symposium on Thermal Cracking in Concrete at Early Age*, Munchen, 221 (1994).

[6]E. Tazawa and S. Miyazawa, "Influence of autogenous shrinkage on cracking in high-strength concrete," *Proceedings of the 4th International Symposium on Utilization of High-strength/High-performance Concrete*, Paris, .321 (1996).

[7]E. Tazawa and S. Miyazawa, "Influence of constituents and composition on autogenous shrinkage of cementitious materials," *Magazine of Concrete Research*, Vol.49, No.178, 15 (1997).

[8]Per Arne Dahl, "Cracking tendency of Krenit fiber reinforced concrete exposed to plastic shrinkage," *SINTEF Report*, Trondheim, Norway (1988).

[9]Committee Report, Japan Concrete Institute, *Autogenous Shrinkage of Concrete*, edited by E. Tazawa, E & FN SPON (1998, to be published).

[10]S. Miyazawa and P. J. M. Monteiro, "Volume change of high-strength concrete in moist conditions," *Cement and Concrete Research*, Vol.26, No.4, 567 (1996) .

[11]K. Kobayashi, et.al., "Estimation of tensile strength and bending strength of steel fiber reinforced concrete by splitting test," *Proceedings of the 34th Annual Conference of the Japan Society of Civil Engineers*, 5, 135 (1979) (in Japanese).

EFFECT OF HIGHER STRENGTH ON PAVEMENT PERFORMANCE

Will Hansen, Ashraf R. Mohamed, Christopher R. Byrum, and Elin Jensen
Department of Civil and Environmental Engineering
University of Michigan
2340 G. G. Brown Building
2350 Hayward
Ann Arbor, MI 48109-2125

ABSTRACT

A comprehensive study is underway at the University of Michigan as part of a Federal Highway Administration (FHWA) project on selected concrete pavements. Some of these pavements have exhibited excellent long-term performance especially near the joints and free edges. The major objective of this study is to determine the Portland Cement Concrete (PCC) properties associated with excellent long-term pavement performance in the context of increasing strength. This paper presents detailed information about PCC mixtures and properties used in pavement applications across the United States. Regional trends in mix design parameters and PCC properties are discussed. A thorough laboratory study of concrete from various regions is underway. The strength data observed cover a wide range; split tensile of 2.1 to 4.83 MPa (300 to 700 psi), compressive strength of 31 to 76 MPa (4500 to 11,000 psi), and elastic modulus of 21 to 48 GPa (3.0E+06 to 7.0E+06 psi). This wide range in PCC mechanical properties associated with excellent pavement performance suggests that prediction of PCC performance cannot be based only on mechanical properties. A thorough study involving all other aspects of pavement performance is currently in progress to identify the major contributors to good pavement performance.

INTRODUCTION

In concrete applications other than pavements (e.g. bridges, columns for high rise buildings, off shore structures, etc.) there is a trend toward using higher strength. Design strengths of 55 MPa (8,000 psi) or higher have been utilized with success. The weight-strength-cost trade-off has favored the use of higher and higher strength mixes in structural applications[1]. However, the same general trend has not been observed in paving applications, where design strengths remain in the 24 to 34.5 MPa (3,500 to 5,000 psi) range. Fortuitous circumstances, such as favorable long-term environmental conditions and strong aggregate[2], have resulted in concrete pavements in Washington State which have long-term strengths greater than 69 MPa (10,000 psi) in compression and 4.0 MPa (600 psi) in split tension. These pavements have exceeded their original design lives with respect to 80 KN (18-kip) equivalent single axle loads.

Pavements are exposed to daily and seasonal varying harsh environmental conditions and millions of load repetitions. There are many factors which influence long-term performance of PCC pavement systems. Good performers are typically characterized by freedom of structural distress such as faulting and spalling near joints and transverse cracking. Spalling is defined as the deterioration and breaking up of concrete joints and cracks. This concrete distress type is the final manifestation of many forms of material deterioration mechanisms including Freeze-Thaw and Alkali-Aggregate Reaction (ASR) in addition to increased slab deflection during loading. Transverse cracking is a distress associated with restrained thermal slab contraction and drying shrinkage. This phenomenon is more pronounced in slabs longer than 4.5 m (15 ft.). Faulting of joints and cracks is the permanent offset in slab elevation on either side of the joint or crack. It is caused by the loss of load transfer. These distress types develop with repeated truck loading and slab movement due to daily and seasonal temperature and moisture variations.

The primary focus of the on-going FHWA study is to determine the concrete properties, in the context of increased strength and associated properties such as permeability, which are responsible for the good performance of jointed concrete pavements. More than 15 concrete pavements were selected throughout three different climatic zones within the US for field and laboratory evaluation including distress condition survey, coring of samples for detailed laboratory investigation of mechanical properties, permeability and petrography.

This paper reports some of the aspects associated with typical regional trends in mix proportions and corresponding mechanical and physical properties available in the LTPP database. Further, the preliminary findings of the laboratory investigation will be discussed in relation to the PCC mechanical properties,

REGIONAL TRENDS IN PAVEMENT PCC FROM THE LTPP DATA

The FHWA Long Term Pavement Performance (LTPP) program has sampled concrete from over 200 concrete pavement sites across the US and Canada. Figure 1 shows a map of the North American continent, which shows the Geographic regions used by LTPP. Also shown on the figure are symbols identifying the test sections studied in detail as part of the FHWA High Strength

Concrete (HSC) study. The LTPP sections were reviewed in detail prior to beginning the field testing for the HSC project. Some of the LTPP sections were selected for more detailed review as part of the HSC study. Sections were selected from LTPP which had shown primarily excellent long-term performance (more than 20 years), or which had high or low PCC strength.

For each SHRP site, six-inch diameter cores were obtained and tested as part of the LTPP program for various PCC properties. Along with a detailed concrete sampling and testing program, mix design information, where available, was provided by the various state departments of transportation. There are a wide variety of PCC mixture types being used across the country. Tables 1 and 2 show typical variation in PCC properties and mixture parameters for pavement concrete in the LTPP database.

There are both general regional trends in PCC mixture parameters and properties and clear State to State trends. The general trends are likely related to the evolution of PCC mixtures within regions of different paving conditions. Regions which are generally hot, such as the South West and South East have evolved mixes which use higher w/c ratio. This general trend may have evolved to ensure adequate workability in such hot conditions. There is also common use of type II low heat of hydration cement in the warmer regions of the south west and along the west coast in general. Regions which have considerable freeze thaw, such as the upper Mid-West appear to have evolved mixes which have more paste volume with low w/c and higher air and cement contents. Regions, which are generally cooler, but non-freeze and have good aggregates, such as Washington State have evolved into the highest strength mixes. These mixtures have This combination typically results in both high strengths and high stiffness. low paste volume, low w/c, high strength aggregate type mixes, and low air content. Similar to the Washington State mixes are the California mixes. These mixes have low paste volume, low air, and high strength aggregate. However, they use high mix water contents and w/c ratio. These mixes often have properties of higher split tensile strength and lower stiffness than would be expected for average concrete. Mixes used in Georgia and Florida have properties of average paste volume, high w/c ratio, and low to medium aggregate strength. The result is a PCC with lower strength and stiffness. Wisconsin PCC uses very high paste volume with low w/c and high air content and lower aggregate strength. This results in higher stiffness and compressive strength but not necessarily higher split tensile strength. State to State variations are likely based on variations in State DOT mixture specifications and design methodology, variations in primary aggregate types within each state, and variations in placement and curing specifications.

Figure 2 shows bar charts for different pavement sections in five regions or states within the USA; California, Georgia/Florida, Minnesota, Wisconsin, and

Washington State. Each bar represents the average of one or two tests per parameter per site. Bars not shown are missing information. The highest strength mixes are found in Washington State.

Preliminary Discussion of LTPP Database Trends

A preliminary review of the effects of PCC strength on pavement performance levels in LTPP indicates that PCC strength plays a secondary but not insignificant role in the durability of pavement systems. The primary pavement deterioration mechanisms are hidden within the foundation layer properties and climate/traffic parameters. Water, deicing salts, and freeze thaw cycles are the enemy of any pavement system. The data does show surprisingly that pavements which use PCC with the properties of high stiffness and low w/c ratio are generally rougher than high w/c ratio low stiffness PCC pavements. This rather significant trend may indicate that these higher strength mixes may be less workable mixes resulting in rougher pavement. This trend may also point to high strength mixes being more susceptible to developing significant slab warping problems. This phenomenon is related to heat of hydration and relaxation properties of the PCC. Higher strength mixes generate higher heat and would relax less. These properties would likely lead to greater warping susceptibility. The data also indicates that higher strength PCC pavements generally have less cracking and deterioration of cracks. This observation is likely related to fatigue properties. Higher strength materials will generally sustain many more load repetitions at a given stress level.

In general, higher strength concrete may be more sensitive for use in pavement applications. Extra care must be taken when using higher strength concrete in pavement applications to insure that a "flat slab" condition is obtained. It appears that if the initial sensitivity can be overcome by stricter temperature control and curing specifications, pavement deterioration will be slower and less severe for higher strength concrete.

DETAILED FIELD AND LABRATORY STUDY

In order to investigate further the preliminary findings and trends found in the literature and LTPP database, a detailed field study has been conducted. The main focus of this investigation is to identify the concrete properties, in the context of increasing strength, which resulted in good pavement performance. The locations of the test sections were selected according to the criteria mentioned before, and are shown in Figure 1. Cores were taken from these locations for laboratory investigation regarding mechanical and physical properties of various concretes. In the following section, typical mechanical properties of these cores will be reported.

Experimental Evaluation of Cores' Mechanical Properties

Mechanical testing was performed using an MTS closed loop servo-hydraulic testing machine with maximum load capacity of 4,500 KN (1000 Kip). Tests include compression strength (ASTM C 39), split tensile testing (ASTM C 496), and static elastic modulus (ASTM C 469). Cores were prepared for testing according to ASTM C 42.

Cored cylinders were instrumented for compression testing with three extensometers in the axial direction (gage length of 100 mm), and a circumferential gauge attached to a chain for post-peak deformation control as shown in Fig. 3. The specimen diameter was typically 149.2 mm. The specimen length varied from 205.7 to 271.8 mm., determined by the slab thickness for each core location. Compressive strength results were corrected for varying height to diameter ratio according to the ASTM C42.

MECHANICAL TEST RESULTS AND DISCUSSION

Table 3 lists the mechanical properties (corrected compressive strength for H/D ratio, split tensile strength, elastic modulus, and Poisson's ratio) obtained on the cores. Also included in this table are the corresponding data available on some of these sections from the LTPP database, the concrete age at testing, the state and SHRP identification number by state and section ID.

In general good agreement was found between the HSC test results and values in the LTPP database except for significantly higher split tensile strength values reported in the LTPP database for California and Washington State.

As shown in Figure 4, the compressive strength of different cores ranged from 31 to 75.8 MPa (4500 to 11000 psi), confirming the same trend observed in the database. The relation between the compressive strength and elastic modulus shown in this figure matches that predicted by the ACI equation (up to about 41.4 MPa). Similar observations hold for the ACI equation for predicting split tensile strength from compressive strength results as shown in Figure 5. The expected trend between compressive strength and water-cement ratio, as predicted by an Abram's type equation, is found to hold despite the wide variety of different concrete. This is seen from Figure 6. It should be noted that the water-cement ratio values are those reported in the LTPP database. This type of information is not available for all the sections studied, and is currently being investigated using petrography analysis. It is clear that the quality of the mechanical properties data is good. These results will be used in combination with several other properties such as permeability, in determining correlation between material characteristics, PCC properties, and pavement performance.

Typical complete compressive stress-strain curves as obtained for four different concretes are shown in Figure 7. The four types of concrete were associated with excellent pavement performance in the field, yet their mechanical properties are widely different. As compressive failure has seldom been observed in pavement structures, the associated lateral deformation of these four types of concrete will be studied further. As shown in Figure 8, higher strength concrete is usually accompanied by less lateral deformation (which is related to tensile strain capacity of the concrete). And the lateral deformation response is linear up to higher stress-strain ratio than for normal strength concrete This signifies in general that there is less microcracking associated with higher strength concrete, due to improved matrix strength and enhanced interfacial properties (especially the bond strength) between the aggregate and the matrix. This trend has been found for all test specimens as shown in Figure 9.

PRELIMINARY CONCLUSIONS

1. The PCC mechanical properties obtained through the detailed laboratory investigation compared overall well with most of the trends found in the LTPP database.

2. A wide range in strength (tensile and compressive) has been found in concrete pavements which have shown excellent long term pavement performance.

3. ACI prediction equations for compressive strength vs. elastic modulus, and compressive strength vs. the split tensile strength have shown good agreement with the data obtained in the FHWA project for widely different concretes.

4. The lateral strain at ultimate load in compression was found to decrease with increasing strength.

5. Additional work is currently being conducted at the University of Michigan to correlate the mix characteristics and concrete mechanical and physical properties to pavement performance.

ACKNOWLEDGMENTS

This study was sponsored by the Federal Highway Administration (FHWA) as part of a study entitled, "Effect of Higher Strength and Associated Concrete Properties on Pavement Performance". The contents of this paper reflects the views of the authors and do not necessarily reflect the views or policy of the FHWA.

REFERENCES

[1]Diamond, S., (1984), "Very High Strength Cement Based Materials - A Prospective," Very High Strength Cement Based Materials, MRS Proceedings, Vol. 42, pp. 233-243.

[2]Wood, S. L., (1992), "Evaluation of the Long-Term Properties of Concrete," PCA R&D, Serial No. 1895.

[3]Perera, R., et al., "Study to Investigate the Development of Roughness in Pavement", in press, FHWA LTPP.

[4]Byrum, C., Hansen, W., and Kohn, S., "The Effect of PCC Strength and Other Parameters on the Performance of PCC Pavements," Sixth International Purdue Conference on Concrete Pavement Design and Materials for High Performance (1997).

[4]ACI Committee 363 (1984), "State-of-The-Art Report on High Strength Concrete," ACI Journal, Vol. 81, No. 4, pp. 364-411.

[5] Aïtcin P. C., Sakar, S.L., and Diatta, Y., (1987), "Microstructural Study of Different Types of Very High Strength Concretes," Mat. Res. Soc. Symp., Vol. 85, pp. 261-272.

[6]Carrasquillo, R. L., Nilson, A. H., and Slate, F. O., (1981), " Properties of High Strength concrete Subject to Short Term Load," ACI Journal, pp. 171-178.

Table I. Variation in PCC Properties in the FHWA LTPP Database.

Parameter	Core Age year as of 1/1/98	Compressive Strength MPa	Split Tensile Strength MPa	Elastic Modulus GPa	Poisson's Ratio	Unit Weight Kg/m³
Average	19.7	52	4.44	31.4	0.18	2338
Min.	5.7	27.3	2.9	13.8	0.09	2146
Max.	40.3	78	7.0	47.0	0.29	2498
Std. Dev.	6.7	8.5	.76	5.9	0.04	64

Table II. Variation in PCC Mix Parameters in the FHWA LTPP Database.

Parameter	w/c ratio	Air Content, %	Coarse Agg., kg/m³	Fine Agg., kg/m³	Cement kg/m³	Water kg/m³	Slump mm.	Paste Vol., m³/m³
Average	0.47	5.1	1098	737	325	150	45.0	0.307
Min.	0.27	0.0	345	523	241	96	12.7	0.235
Max.	0.72	9.8	1406	1381	462	233	153.0	0.368
Std. Dev.	0.08	1.4	179	145	33	24	18.0	0.028

Table III Concrete Strength Results for Field Cores, From This Study and LTPP Database.

	Section ID	LTPP State	State	PCC age at testing LTPP Years	PCC age at testing UM Years	Comp. Str. LTPP MPa	Comp. Str. UM MPa	Sp. Tens. Str. LTPP MPa	Sp. Tens. Str. UM MPa	E-Mod. LTPP GPa	E-Mod. UM GPa
Dry Region Group	3017	6	CA	18.60	19.50	43.83	36.83	5.00	3.68	28.28	26.83
	3021	6	CA	17.60	23.70	44.45	32.47	5.11	3.46	21.03	25.14
	7456	6	CA	20.10	23.90	50.41		5.65		29.14	
	CS1	6	CA		26.00		44.79		3.48		32.62
	CS3	6	CA		26.00		52.79		4.18		33.72
	I-10	6	CA		31.50	N/A	37.62		4.18		28.34
	3019	53	WA	5.70	11.20	63.90	55.61	6.21	4.18	34.14	40.55
Wet Freeze Group	3006	19	IW	18.80	21.80	58.48	45.81	3.40	3.51	31.55	31.56
	3055	19	IW		28.10	58.93	41.67	3.96	2.82	23.97	25.76
	4054	27	MN	19.4	24.9	57.31	54.39	3.95	3.69	38.28	39.06
	3801	39	OH	11.40	13.40	59.21	45.43	3.25	3.73	25.86	25.24
	3008	55	WI	19.00	21.70	72.17	61.51	4.71	4.50	46.90	44.85
Wet No Freeze Group	3011	53	WA	15.00	20.40		59.64	7.00	3.84	36.38	41.76
	3812	53	WA	27.30	32.60		75.04	6.15	4.40	45.69	46.93
	GA1-5	13	GA		26.10		39.43	N/A	2.86		24.97
	GA1-6	13	GA		26.10			N/A	3.27		

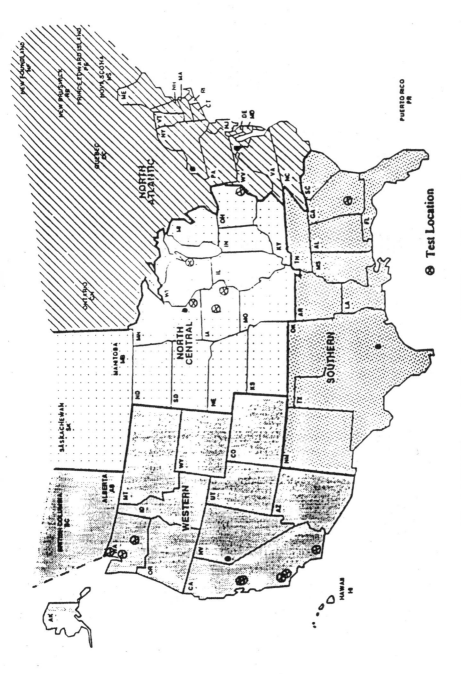

⊗ **Test Location**

Fig. 1 LTPP Geographic Regions and HSC Project Test Locations

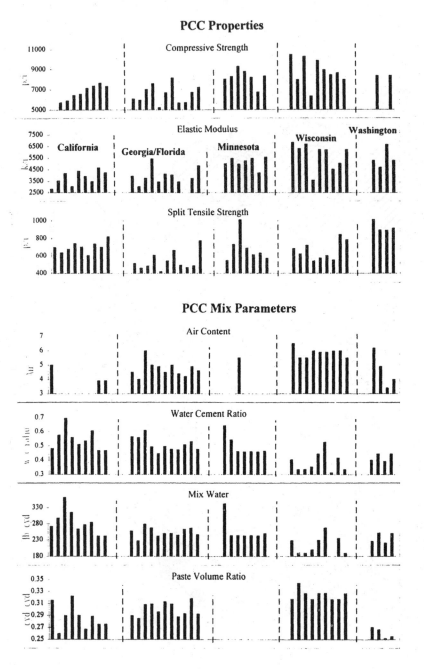

Fig. 2 Variation of PCC Properties and PCC Mix Parameters in the LTPP Database for Various States.

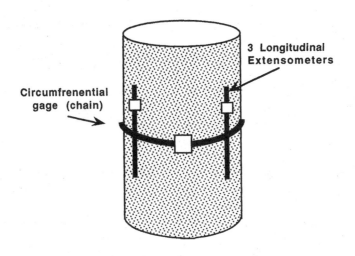

Fig. 3 Set-up for Compression Testing in Closed-loop Testing

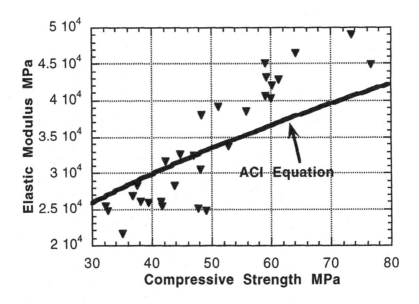

Fig. 4 Relation Between Compressive Strength and Elastic Modulus as Obtained in the HSC Laboratory Study.

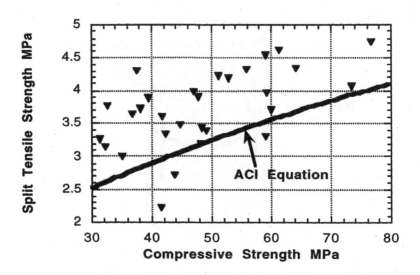

Fig. 5 Relation between Compressive Strength and Split Tensile Strength as Obtained In the HSC Laboratory Study

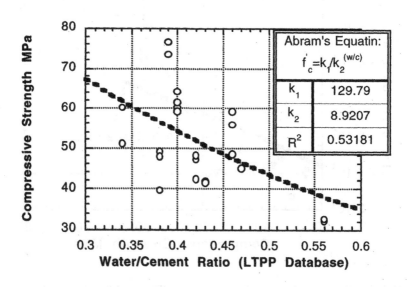

Fig. 6 Relation between Compressive Strength from the HSC Laboratory Study and Water/Cement Ratio from LTPP Database

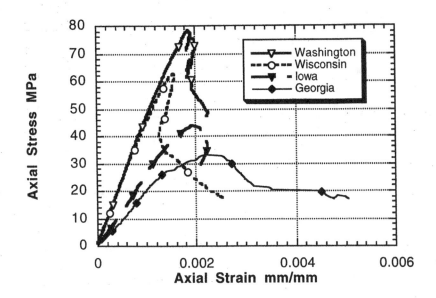

Fig. 7 Typical Stress-Strain Curves for Four Different Concretes Investigated in this Study.

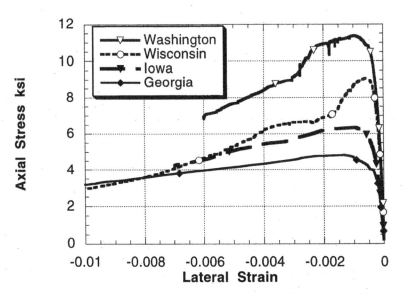

Fig. 8 Stress-Lateral Stain Curves for the Four Concretes in Figure 7

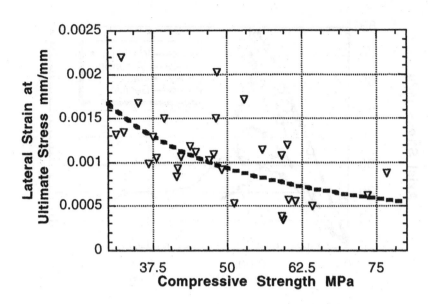

Fig. 9 Relation between Compressive Strength and Lateral Strain at Ultimate Load For all Concretes Investigated In this Study.

A NEW TENSILE TESTER FOR CEMENTITIOUS MATERIALS

Tahar El-Korchi
Department of Civil and Environmental Engineering
Worcester Polytechnic Institute
Worcester, MA 01609

Shuji Tsubota
Kyowa Kako Co.,Ltd.
Miyagi, Japan

Houssam Toutanji
University of Alabama
Huntsville, AL 35801

ABSTRACT

A novel technique for testing cement materials in tension is presented. This test uses the ASCERA hydraulic tensile tester. The advantages of this technique include; minimizing misalignment and stress concentration from gripping the specimen, larger specimen volume under tensile stress, and determination of fracture initiation. The test uses cylindrical specimens that are fractured in a hydraulic chamber which allows specimen self-alignment to occur. Tensile results using this technique can be 20-30% higher when compared to the traditional uniaxial tensile test. Data is presented on cement mortar and concrete. Due to the applied hydraulic pressure, this test applies a tensile stress in the axial direction and compressive stress in the lateral direction ($\sigma_2 = \sigma_3$), and a small degree of triaxial effect is induced. Results will be presented with respect to a tensile failure criterion developed for cementitious materials under tension-compression loading. The effect of the lateral pressure was clearly defined by the stress concentration factor. The failure line agrees well with the analytical results.

INTRODUCTION

An accurate determination of the tensile strength of cement and concrete composites is important. Most cracks in cement based materials are associated with some form of tensile stress loading. Tensile strength is usually the weakest intrinsic strength and may be one of the initial causes of structural failure.

Numerous tests have been used to determine the direct tensile strength [1-12] of cementitious materials and a thorough review is presented by Hannant [2] and later by Toutanji et al. [13]. However, there are major difficulties in tensile

---

testing of brittle materials which arise from gripping and misalignment of the specimen during testing. Any deviation from pure uniaxial tension will result in a complex stress state in the specimen and fracture will again occur at loads less than the intrinsic tensile strength of the material.

Previous workers have sought to eliminate the former problem with various non-standardgripping systems with varying degrees of success. Some examples include (a) adhesively bonded end plates [2], (b) a tapered section gripped by wedge action in a tapered end plate [3], (c) a friction grip with the lateral force supplied by two stiff side plates by means of a scissor action [4,5]. These efforts generally loaded the specimen through a universal joint or a ball and socket joint in an attempt to minimize the loading eccentricity.

A technique which minimizes the above mentioned problems could accurately measure the tensile strength of cementitious materials. Such a technique would be extremely desirable. This paper reports on the development of a simple testing technique based on a commercially available hydraulic tensile testing machine.

THE HYDRAULIC TENSILE TECHNIQUE

Concern over the problems associated with gripping and alignment in tensile testing of advanced structural ceramics led Baratta and Driscoll [14,15] to the development of the hydraulic tensile tester. This design was later modified and commercialized by ASCERA® [16].

The tensile test is a self-aligning hydraulic technique that utilizes a hydraulic tension tester manufactured by ASCERA®. The technique was initially developed for determining the tensile characteristics of high-strength ceramics. The tensile specimens are cylindrical bars measuring 16 mm in diameter and 120 mm in length. Forty mm on each end of the cylindrical bar is inserted into a steel piston and adhesively bonded in place with high-strength epoxy. The specimen-piston assembly is inserted into a high pressure chamber. The pressure is applied and increased until the specimen is broken apart by the hydraulic pressure acting against the pistons. The pressure chamber, with specimens inserted, is shown schematically in Fig. 1. The internal pressurization of the specimen suspended between the O-ring seals minimizes bending stresses. The nominal tensile fracture stress may be calculated by using the hydraulic pressure at failure and geometric parameters of the specimen-piston assembly [16].

The hydraulic pressure causes a triaxial stress state with minor effects on the uniaxial tensile strength which will be discussed later in the paper. In addition, to eliminate the effects of stress intensification at the specimen to piston bond transition, data from specimens that fractured within one half the radius of the specimen from the piston glue line are not used. Fractures occurring in these locations are considered invalid tests [15].

The stress is related to the hydraulic pressure and test geometry by the following expression:

$$\sigma = \sigma_{nom} = \left(\frac{A - A_S}{A_S} \right) x \, P$$

(1)

Where σ_{nom} is the nominal fracture stress, A is the cross-sectional area of the piston, A_s is the cross sectional area of the specimen, P is the pressure at failure [4,15,16].

This ASCERA hydraulic tensile technique has been adopted for cement composites and has been called the *cementitious composites axial tensile test* (*CCATT*). The CCATT has been used to evaluate the tensile strength of a number of structural materials, such as: cement-based composites [17,18], silica-fume cementitious composites [13,14], carbon fiber-reinforced cementitious composites [13,14], glass fiber-reinforced cementitious composites [13,14], cement composites wrapped with FRP sheets [23] and graphite and high-strength ceramics [18,19]. Test results showed that the uniaxial tensile strength of cementitious composites tested using the ASCERA hydraulic tensile tester is higher than those obtained using the traditional uniaxial tensile test. Furthermore, the tensile strength of cementitious composites, obtained using the ASCERA hydraulic tensile tester, is about 12% of the compressive strength [19] as compared to less than 10% using the traditional tensile test [22,23]. This is basically due to the minimization of load eccentricity and gripping effect inherent in traditional uniaxial tensile tests. The use of the hydraulic tensile technique provides a number of advantages [11,18], including: (a) minimization of bending stresses along the cross-section of the specimen due to the uniformity of the stress distribution, (b) minimization of gripping effects that may cause premature failure of the specimen at gripping, (c) enables testing of relatively large volumes even though small size specimens are tested, (d) possible identification of the origin of fracture for studying the fracture mechanics of failure.

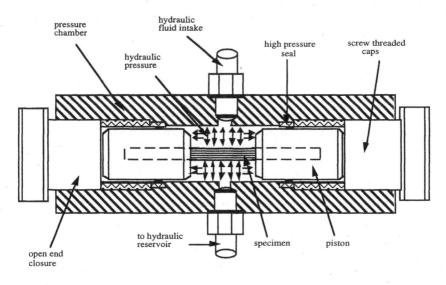

FIG. 1.
Schematic of the ASCERA pressure chamber with specimens inserted.

WEIBULL STATISTICS

The statistical nature of fracture in brittle materials such as ceramics is well known. The variability or scatter in the strength of these materials is the result of both the large range of flaw sizes and stress intensity factors of these flaws (location and orientation effects) present in most brittle materials. Mathematical descriptions of brittle fracture statistics are generally based on the weakest link concept: A body will fail when the stress at any flaw is sufficient to initiate fracture at that flaw. Although there is no weakest-link distribution function which describes brittle fracture that is universally accepted, the Weibull distribution [24, 25] is widely used because of its mathematical simplicity and the general agreement of this distribution with actual fracture data.

The Weibull model views the body under load as a chain of elements with each element having a probability of failure. The observed strength of the specimen is then simply the strength of the weakest link. Mathematically this probability of failure of a given specimen can be represented by the two parameter Weibull function as:

$$P_f = 1 - e^{-\int_v \left(\frac{\sigma}{\sigma_0}\right)^m dv}$$

(2)

where

P_f - is the probability of failure

σ - is the applied stress

σ_0 - is a normalizing constant, the characteristic strength, related to the mean strength of the distribution

m - is a second parameter, the Weibull modulus, a measure of the dispersion of the distribution

v - is the volume under tensile stress

The Weibull modulus, m, is a measure of the scatter or dispersion of the distribution (similar to the standard deviation of a normal distribution). The scatter is inversely related to m, a small Weibull modulus is indicative of a large dispersion in strength values. Typical ceramics have reported Weibull moduli in the range 5-15. Metals and alloys, which fail in a ductile manner, have reported values in the range 30-50. The scaling constant is an indirect measure of the average strength. If m, specimen size, and loading configuration are held constant, the mean strength is then directly related to characteristic strength, σ_0.

The Weibull modulus, m, and σ_0 are usually estimated by the use of ranking statistics. Fracture strengths are measured on a number of specimens. The strengths are ranked from weakest to strongest. The probability of failure, P_f, is then assigned each specimen based on its ranking, i, by:

$$P_f = i / (N+1) \tag{3}$$

where, N is the total number of specimens tested. The data is then plotted as in the form $\ln \ln [1 / (1-P_f)]$ versus $\ln \sigma$. The basic probability distribution can be manipulated to obtain Eq (4). The relationship in Eq (4) is fitted by linear regression to the data and values for m and σ_0 are obtained.

$$\ln \ln [1 / (1-P_f)] = m \ln \sigma - m \ln \sigma_0 \tag{4}$$

Although Weibull statistics are generally not applied to the strength of portland cement based materials due to the large number of specimens required to achieve reliable estimates of m and σ_0, it is interesting to note that Weibull used portland cement materials in developing his statistical model [24].

This paper will present tensile strength data on mortar and concrete specimens. A comparison of the hydraulic tensile tester *CCATT* and the traditional uniaxial tensile tester (UTT) will be made on mortar specimens. Tensile strength of concrete specimens using the CCATT will be presented. Fractography of fractured surfaces will be discussed in terms of apparent strength and the effect of confining lateral pressure on tensile strength will be presented.

EXPERIMENTAL PROCEDURE

Tensile strength tests using the hydraulic tensile tester were conducted on mortar specimens, and two different concretes with crushed rock and natural gravel aggregate. The mortar mix consisted of cement:sand:water of 1:1:0.41. The concrete mix consisted of 50 percent aggregate, a 10 cm (4 inch) slump, and 272 Kg (600 lbs) of cement. The concrete was over 1 year old and cored from large blocks. The mortar was cast as final cylinder specimens.

RESULTS AND DISCUSSION

The tensile strength results of the mortar mix using the CCATT is shown in Fig. 2. Tensile strength using a traditional uniaxial tensile test-UTT (using an MTS-Sintech® electro-mechanical testing machine with computerized control and data acquisition) is used for comparison on the same Weibull plot. The Weibull plot in Fig. 2(a) shows that the characteristic strength (CS) obtained using the hydraulic tensile tester is larger than the that using the conventional UTT method. In addition the Weibull modulus (WM), a measure of scatter of the data, is much larger for the CCATT than the UTT method indicating lower scatter. In Fig. 2(b), a Weibull plot shows the two concretes to have similar characteristic strength, however, the gravel appears to have a larger value for Weibull modulus indicating less scatter in the concrete with the gravel aggregate. Both concretes have a much lower characteristic strength and lower reliability than the corresponding mortar mix. This is not surprising given the

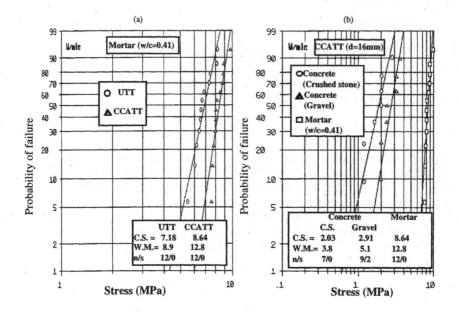

FIG. 2.
Weibull plots summarizing tensile strength of mortar and concrete specimens

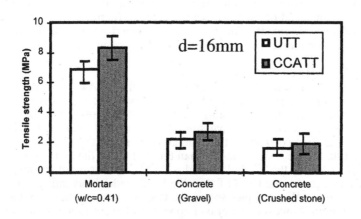

FIG. 3.
A comparison between UTT and CCATT tensile strength results.

influence of the transition zone and the aggregate size vis-a-vis the specimen size. Fig. 3 shows that for all mixes, the average tensile strength obtained using the CCATT is greater than the UTT.

Fractography analysis was conducted to assist in evaluating the cause in variation associated with the different mixtures, specifically the concrete samples. Fig. 4 shows typical fractured surfaces in the concrete and mortar samples. Image analysis of the fractured surfaces shows that the tensile strength is highly sensitive to the location of the fracture plane across the specimen. The fracture plane may traverse through the aggregate or around the aggregate through the aggregate-paste interface. In addition, other features such as air voids, tortuosity of the path and aggregate cleavage planes have a large impact on the apparent tensile strength.

TRIAXIAL EFFECT ON TENSILE STRENGTH

As mentioned earlier, the hydraulic tensile tester produces a confining pressure which produces a multiaxial loading condition on the specimen. It is well known that as the lateral pressure is increased, the tensile strength decreases. Tsubota and El-Korchi [26] developed a tensile failure criteria based on the hydraulic tensile test (σ_1 in tension, $\sigma_2 = \sigma_3$ in compression). A finite element analysis was conducted to assess the effect of lateral confining stress on uniaxial tensile strength. FEM axisymmetric models, BC's and applied pressure are shown in Fig. 5. The principal-stress ratio ($\sigma_1 / |\sigma_2|$) is varied between 0.092 and 11.76 by changing the specimen diameter. Effect of multiaxial loading is analyzed by a combination of the classical theory of stress concentration and Weibull theory. Using the stress concentration factor K_t, ($K_t = \sigma_{max} / \sigma_{nom}$) and the tensile stress equation for the CCATT ($\sigma_{nom} = \sigma P$), the maximum stress is written by $\sigma_{max} = K_t \sigma P = K_t (\sigma_1 / |\sigma_2|) P$. The stress distributions and contour lines along the x-axis are shown in Fig. 5. The tensile stress is uniformly distributed near the center of the test zone. The stress concentration is remarkably increased when the pricipal stress ratio is increased. In Fig. 5, as the confining pressure is increased from 10.34 kgf/cm^2 to 108.6 kgf/cm^2 the stress concentration ratio increases by 33 percent (σ_{nom} increases from 39.1 to 51.6 kgf/cm^2).

CONCLUSIONS

This study shows that the hydraulic tensile tester is a viable method for tensile testing of cement and concrete materials. Tensile strength values obtained using the CCATT are substantially higher than traditional tensile testing methods due to a decrease in testing artifacts associated with eccentricity and bending in the UTT methods. Reliability is also increased with the CCATT method. The triaxial effect associated with hydraulic confining pressure could be significant at large diameters above 25 mm. However, the effect is negligible with the current specimen sizes (16 mm) used for this study.

Concrete with Crushed
Rock

Concrete with Natural
Gravel

Mortar

FIG. 4.
Fractography images showing typical fracture surfaces from the CCATT test
specimens.

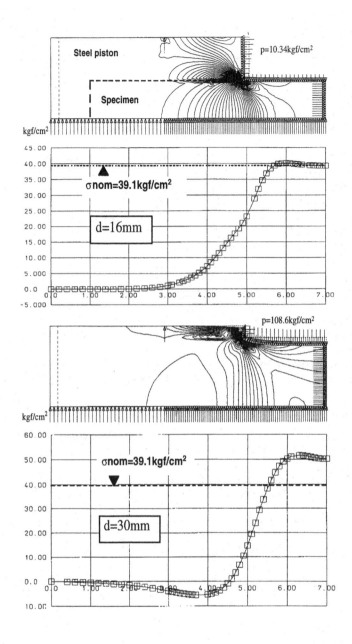

FIG. 5.
Stress contour lines and principal stresses along X-axis of tensile specimen.

REFERENCES

[1]B. W. Christ and S. R. Swanson, *Journal of Testing and Evaluation*, **4** [6] 405-417 (1976).

[2]D. J. Hannant, *The Structural Engineer*, **50** [7] 253-258 (1972).

[3]RILEM Bulletin "Direct Tensile Test of Concrete," 20, pp.84-89, September 1963.

[4]C. Avram, I. Facaoaru, O. Mirsu, I. Filimon, I. Tertea, "Concrete Strength and Strains," *Elsevier Scientific* (1981).

[5]J. D. Todd, "The Determination of Tensile Stress-Strain Curve for Concrete," Proc., Inst. of Civil Engineers, **4** [2] 202-221 (1954).

[6]R. Humphries, "Direct Tensile Strength of Concrete," *Civil Engineering and Public Work Review*, **52** [614] 882-883 (1957).

[7]D. P. O'Cleary, and J. G. Bryne, "Testing Concrete and Mortar in Tension," *Engineering*, 18, pp.384-385 (1960).

[8]C. D. Johnston and E. H. Sidwell, "Testing Concrete in Tension and Compression," *Magazine of Concrete Research*, **20** [65] 221-228 (1968).

[9]M. A. Ward and D. J. Cook, "The Development of a Uniaxial Tension Test for Concrete and Similar Brittle Materials," *Materials Res. and Standards*, MTRSA, **9** [5] 16-20 (1969).

[10]R. P. Johnson, *Structural Engineer*, **42** [4] 135-141 (1964).

[11]B. P. Hughes, and G. P. Chapman, "Direct Tensile Test for Concrete Using Modern Adhesive," RILEM Bulletin, 26, pp.77-80 (1965).

[12]R. H. Elvery and W. Haroun, "A Direct Tensile Test for Concrete Under Long Or Short-Term Loading," *Magazine of Concrete Research*, **20** [63] 111-116 (1968).

[13]H. A. Toutanji, T. El-Korchi, *Journal of Testing and Evaluation*, **22** [3] 226-232 (1994).

[14]F. I. Baratta and G. W. Driscoll, "A New Axial Tension Tester for Brittle Materials," Army Matls. & Mechanics Res. Center, AMMRC TR 69-02, Watertown, MA, 1968.

[15]G. W. Driscoll and F. I. Baratta, "Modification to an Axial Tension Tester for Brittle Materials," Army Materials. & Mechanics Res. Center, AMMRC TR 71-3, Watertown, MA 1971.

[16]ASEA Cerama AB, "Instruction and Operation Manual for ASCERA Hydraulic Tensile Testing," ASEA CERAMA AB, Robertsfors, Sweden, 1988.

[17]H. A. Toutanji, T. El-Korchi, ASTM Journal of Cement, Concrete, & Aggregates, **18** [2] 78-84 (1996).

[18]T. El-Korchi, H. A. Toutanji, R. N. Katz, G. L. Leatherman, H. Lucas, and C. Demers, "Tensile Testing of Fiber Reinforced Cement Composites," *Proc. Mater. Res. Soc.* 211 (1991).

[19]H. A.Toutanji, T. El-Korchi, R.N. Katz, G.L. Leatherman, Cement Concrete Res., **23** [3] 618-626 (1993).

[20]H.A. Toutanji, T. El-Korchi, G.L. Leatherman,R.N. Katz, "Tensile Strength of Carbon Fiber-Reinforced Cement Composites," *Proc. Mater. Res. Soc.*, 245, pp. 359-364., (1992).

[21]R.N. Katz, H. Toutanji, D. Friel, G.L. Leatherman, T. El-Korchi, W. Rafaniello, "Room Temperature Tensile Strength of AlN," *Proc. of Ceramic Eng. and Sci.*,14, 718, (1993).

[22]H.A. Toutanji, D. Freil,T. El-Korchi, R.N. Katz, G. Wechsler, W. Rafaniello, "Room Temperature Tensile and Flexural Strength of Ceramics in the AlN-SiC System ", *Journal of European Ceramic Soc.*, 15, 5, pp. 425-434, (1995).

[23]H.A. Toutanji, T. El-Korchi, S. Tsubota,"New Approach to Tensile Strength Testing of Cement Based FRP Composites Wrapped Specimens" Proc. ICCI 98, Tucson Arizona, Jan 1998.

[24]W. Weibull, "A Statistical Distribution Function of Wide Applicability", *Journal of Applied Mechanics*, p. 293-297, Sept. 1951.

[25]W.D. Kingery, H.L. Bowen, D.R. Uhlmann, *Introduction to Ceramics*, 787, John Wiley and Son, New York, 1976.

[26]S. Tsubota, T. El-Korchi, 'Tensile Strength of Cementitious Materials Under Triaxial Loading," Proc. Architectural Institute of Japan, paper No 23398, pp. 795 (1997).

PROCESSED CELLULOSE FIBER REINFORCED PORTLAND CEMENT CONCRETE

Neeraj Buch
Pavement Research Center of Excellence
Department of Civil and Environmental Engineering
Michigan State University, East Lansing, MI 48824
U.S.A.

Mohammad Owais Rehman
Pavement Research Center of Excellence
Department of Civil and Environmental Engineering
Michigan State University, East Lansing, MI 48824
U.S.A.

ABSTRACT

A comprehensive laboratory and field test program (over 300 specimens were molded and tested) was carried out to develop information about cellulose fiber reinforced concrete. The concrete compressive strengths were in the range of 40-60 MPa.. The objective of the study was to investigate mechanical properties concrete mixtures with varying amounts of cellulose fibers. These mixtures were then in turn compared to a standard control concrete mixture design. The paper will discuss results based on compressive strength, flexural strength, impact properties, fracture toughness, and stress intensity factors, as a function of time and curing regimes. Due to the crack arresting properties of cellulose fibers it is hypothesized that there will be an improvement in long term durability properties of these mixture designs. The potential of dimensional stability and long-term durability will also be discussed in the paper. Preliminary results have shown substantial improvement in impact, fracture, and flexural strength properties of the concrete over standard concrete mixtures.

INTRODUCTION

An unreinforced cement matrix is brittle and susceptible to premature tensile failure. The primary reasons for incorporating fibers into portland cement concrete is to improve impact resistance, toughness and tensile strength. Fiber reinforced concrete (FRC) is portland cement concrete reinforced with more or less randomly dispersed fibers, both man-made and natural. Over the past thirty years a significant amount of research has been performed in the field of FRC (1). The objective of this paper is to study impact of processed cellulose fibers on the mechanical properties of portland cement concrete.

Cellulose fibers are classified as natural fibers, derived from the processing of wood. The "kraft" (details of which can be found elsewhere) (1) involves the cooking of wood chips in a solution of sodium hydroxide, sodium carbonate and sodium sulfite. Different grades of fibers can be obtained containing more or less the three main constituents: cellulose, hemicellulose and lignin, by bleaching. Bleaching reduces the amount of cellulose fiber but practically eliminates the presence of lignin. The processed cellulose fibers have relatively good mechanical properties when compared to many manmade fibers like polypropylene, polyester, acrylic, etc (1). Table I summarizes mechanical and physical properties for processed cellulose fibers, polypropylene and nylon.

Processed cellulose fibers provide a desirable balance between mechanical, physical and durability characteristics when placed in a cement matrix. Figure 1 illustrates the three step process of fiber action. Fibers with a high elastic modulus as compared to polypropylene and nylon and if placed closely (2,000,000 fibers per gram) prove effective in *micro-crack supression* (2,3) . The close fiber spacing, increased surface area and bond strength are effective in *micro-crack stabilization* (2,3). Closely spaced fibers with a higher length-diameter ratio are efficient in stabilization of cracks. The geometric features of processed cellulose fibers are illustrated in Figure 2 (4). The close spacing of the cellulose fibers compared with the coarser synthetic fibers ensures presence of fibers at crack tips, resulting in effective suppression and stabilization. The small diameter and hydrophillic surfaces lend themselves to a close packed structure resulting in a dense and uniform microstructure.

Microcrack Suppression   Microcrack Stabilization   Crack Stabilization

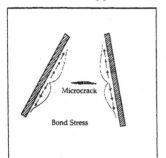 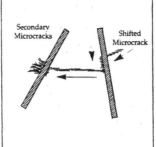 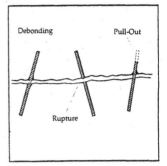

FIG. 1
Mechanisms of Fiber Action in Concrete (2,3).

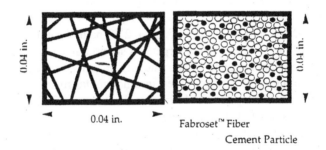

FIG 2.
Geometric Features of Cellulose Fibers (4).

TABLE I. Comparison of Properties between Cellulose, Polypropylene and Nylon Fibers (After 1).

Property	Preferred	Cellulose	Polypropylene	Nylon
Elastic Modulus, Gpa.	High	60	4	4
Bond Strength, MPa.	High	1.5	0.4	NA
Tensile Strength, MPa.	High	500	600	700
Effective Dia., mm.	Low	0.015	0.1	0.05
# of Fibers/gm	High	2,000,000	12,000	45,000
L/D Ratio	High	200	120	200
Surface Characteristics	Hydrophilic	Hydrophilic	Hydrophobic	Hydrophilic
Density, gms/cc	Medium	1.5	0.9	1.1
Alkali Resistance	High	High	High	High
Fiber Spacing, mm.	Low	0.53	2.8	1.7
Fiber Count, 1/cc	High	90	0.6	3.3
Specific Surface	High	0.13	0.033	0.052

EXPERIMENTAL PROGRAM

The mechanical properties of four concrete mixtures; control, control + 1kg/m^3 cellulose fibers (CF1), control + 2kg/m^3 cellulose fibers (CF2), control + 3kg/m^3 cellulose fibers (CF3) were investigated. The basic mix ingredients were Type I portland cement, 6AA (Coarse Agg.-MDOT specification) crushed limestone (nominal maximum size 25mm.), 2NS natural sand, and air entraining admixture. All the concrete mixes targeted a slump range of 25-75 mm. (without fibers) and air content range of 4-7% . The mix proportions are presented in Table 2. The mixture ingredients were mixed in a horizontal rotary drum mixer with a capacity of 0.1m^3. The cylinders (100mm in diameter and 200mm in height) were molded and consolidated in accordance with ASTM C 31-90a, and the flexural beams (100mm×100mm×350mm) were molded and consolidated in accordance with ASTM C 31-90a. For the concrete mixture designs identified in Table 2, the following testing plan was developed to characterize the mechanical properties of concrete mixtures.

- Consistency Tests
  initial and final set times,
  slump and slump loss over time tests, and
  air content determination.

- Strength Tests
    - compressive strength tests over time,
    - flexural strength tests over time, and
    - fracture
    - impact

Five replicate samples were prepared for each test type to verify repeatability and confidence level. The molded specimens were laboratory cured for the first 24-hours, demolded and then transferred to an environmental chamber (73 F and R.H.=100%) till testing age was reached. In the subsequent sections of this paper results from the various tests listed above will be presented.

TABLE II.   Mixture Designs for the Experimental Program

Water	Cement	Fine Agg.	Coarse Agg.	AEA	CaCl$_2$	Fibers
182	415	610	1060	415 ml/m^3	13	0,1,2,3

Note: All weights are in kg/m^3, unless otherwise stated.

DISCUSSION OF RESULTS

Figure 3 summarizes the consistency properties of the four mixtures under investigation.

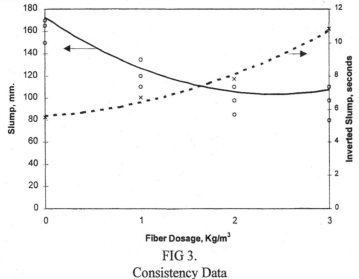

FIG 3.
Consistency Data

Figure 4 illustrates the effect of processed cellulose fibers on concrete compressive strength. In general, concrete mixtures with cellulose fibers exhibited a higher compressive strength when compared to the control mixture design. There is approximately a 3% increase with CF1, a 5% increase with CF2 and a 15% increase with CF3 in the 28-day compressive strength over the control mixture. Statistically, at a $\alpha=5\%$ (95% confidence) CF3 exhibits the most significant gain in 28-day compressive strength over the control mixture, whereas, all mixtures (CF1, CF2, and CF3) exhibit a significant improvement in 1-day compressive strength over the control mixture. The desirable performance of the processed cellulose fibers can be attributed to their high elastic modulus, close spacing and strong bonding to the cement matrix. It should be pointed out that the real effectiveness of fibers in general is realized in flexure, fracture toughness and impact resistance. Any improvement in compressive strength is an added advantage.

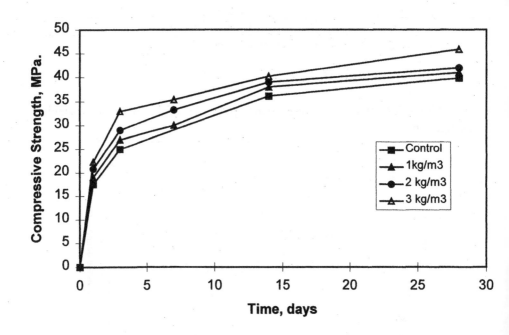

FIG 4.
Average Compressive Strength Data

The brittleness of plain cement concrete is a drawback in concrete pavements and other structures subjected to impact loading. In this study impact load tests were performed to investigate the influence of fibers on the behavior of plain cement concrete. The impact test consists of subjecting a disc of concrete (150mm in diameter and 64 mm. thick) to repeated impact load by dropping a 4.5 kg hammer from a height of 460 mm. The load is transferred from the hammer to the specimen via a steel ball 64 mm in diameter. The number of blows that cause the first visible crack and then failure of the concrete disc specimen are recorded. There is a large degree of variability associated with the test procedure, however, Figure 5 summarizes (# of blows to failure) the effect of fibers on impact resistance. Concrete with processed cellulose fibers exhibits a 50% improvement in impact resistance over plain concrete.

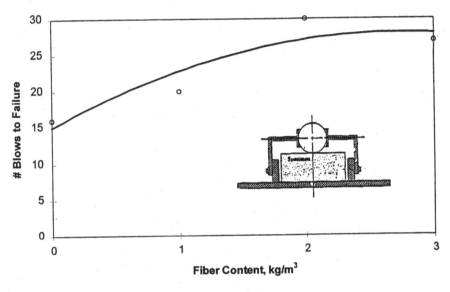

FIG 5.
Average 28-day Impact Resistance

Both the early age and later age flexural strength results are summarized in Figure 6. Inspection of the data confirms the effectiveness of processed cellulose fibers, especially in enhancing early age strength characteristics of the concrete mixtures. There is approximately 20% increase in the 28-day flexural strength over the control mixture. Statistically, at a α=5% (95% confidence) CF1, CF2 and CF3 exhibit significant gain in 1- and 28-day compressive strength over the control mixture. It was observed that the flexural strength peaked at a fiber content range of 2-2.3 kg/m³.

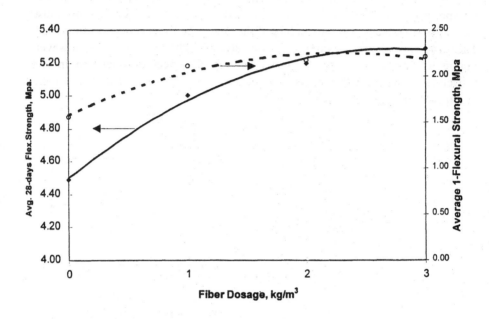

FIG 6.
Flexural Strength Test Data

Fracture toughness represents energy absorption capacity of flawed concrete (flaws due to shrinkage micro-cracks for example). Specialty cellulose fibers increase the fracture toughness through crack stabilization. .Concrete mixtures with processed cellulose fibers exhibit improved fracture toughness properties and fracture strength as illustrated in Figures 7 and 8. Toughness may be represented by the area under the load-deflection curve (ASTM C 1018), however, there are some accuracy concerns associated with this test procedure.

Alternatively, toughness may be expressed as the area under the load-crack mouth opening displacement curve in a notched flexure test. The results suggest that cellulose fibers enhance fracture toughness of the concrete, toughness was observed to peak at 2-2.3 kg/m^3 fiber content.

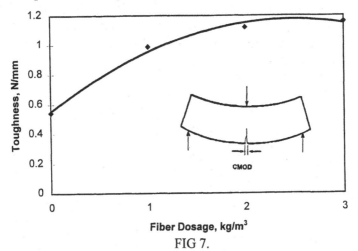

FIG 7.
Fracture Toughness as a Function of Fiber Dosage

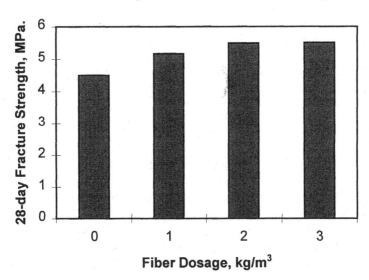

FIG 8.
Average 28-day Fracture Strength

The three cellulose fiber concrete mixtures were cast against the control mixture (no special surface treatment was used, other than cleaning with a wire brush) with a contact surface area of 10,332 mm^2. After 28-days of moist curing, the bond strength of the interface under pure shear was measured. The 28-day bond strength test results (Figure 9) suggest that processed cellulose fibers improve the bond strength. The improved bond characteristics can be taken advantage of in full-depth concrete patches, where a fiber reinforced concrete patch can be placed against existing concrete.

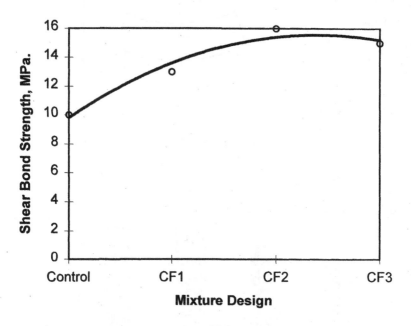

FIG 9
Average 28-day Shear Bond Strength

It has been hypothesized that suppression and stabilization of micro-cracks are fundamental to the superior performance of cellulose fibers. Electron scanning microscopy of the fractured surfaces was used to study the action of cellulose fibers with cracks. The micro-crack suppression and stabilization action of cellulose fibers that mitigate crack propagation is illustrated in Figure 10a. Once the cellulose fiber crosses a micro-crack, the high bond strength and surface area of the fibers provide a pull-out resistance which controls the widening of the crack. Figures 10b and c illustrate the close fiber spacing and the intricate grid formation of the cellulose fibers in the concrete matrix.

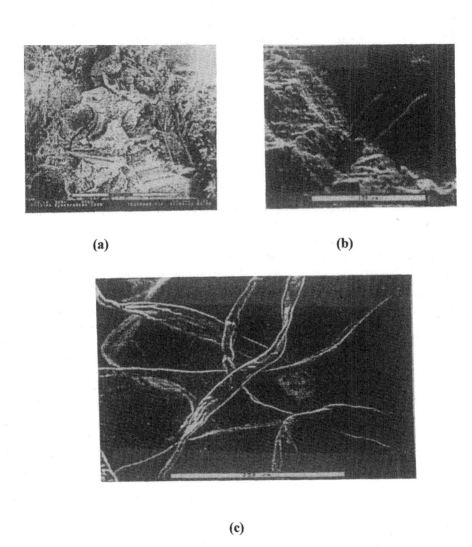

(a)   (b)

(c)

FIG 10
Scanning Electron Microscope Images

CONCLUSIONS

The cellulose fibers provide a balance between geometric and mechanical properties for the reinforcement of concrete. The laboratory processing of cellulose concrete has shown the ease with which cellulose fibers can be dispersed. The improved reinforcing properties of processed cellulose fibers can be attributed to the suppression and stabilization characteristics of the cracks. The high fiber count, large surface area and high elastic modulus of the processed cellulose fibers are ideal for reinforcement efficiency. As the data has indicated that these fibers have a significant potential to improve compressive and flexural strength of concrete mixtures, concrete mixtures with fibers have vastly improved impact and fracture toughness properties.

REFERENCES

1. *Fiber Reinforced Concrete*, Edited by J.I. Daniel, Portland Cement Association, SP039.01T, Skokie, Illinois, 1993.
2. Balaguru, P.N. and Shah, S.P., *Fiber Reinforced Cement Composite*, McGraw Hill, Inc., 1992, 531 pp.
3. Bentur, A., and Mindess, S., *Fiber Reinforced Cementitious Composites*, Elsevier Science Publishers, Ltd., 1990, 449 pp.
4. Soroushian, P., Secondary Reinforcement-Adding Cellulose Fibers, *Concrete International*, American Concrete Institute, June, 1997.
5. Nagi, M. and Whiting, D. "Strength and Durability of Rapid Highway Repair Concrete" *Concrete International*, Sept. 1994, pp. 36-41.

RHEOLOGY

CONTROL OF CONCRETE RHEOLOGY

Leslie Struble, Xihuang Ji, and Guy Salinas
Department of Civil Engineering
University of Illinois
205 N. Mathews
Urbana IL 61801–2352

Richard Szecsy
Pioneer Concrete of Texas, Inc.
8505 Freeport Parkway, Suite 200
Irving, TX 75063

ABSTRACT

Measurements of rheological parameters (yield stress and plastic viscosity) provide information about the flow behavior of concrete that may be especially useful in designing concrete for specific processing technologies. The rheological parameters suitable for pumping and for highly flowable concretes are discussed. Ways to manipulate concrete to achieve target rheological parameters are also discussed. Both yield stress and plastic viscosity are increased by increasing the aggregate volume fraction and by packing the aggregate less densely. Yield stress is increased by the flocculation of fine particles, and therefore is reduced by addition of superplasticizer. An example is discussed in which these parameters were manipulated so as to produce concrete with a higher yield stress but a lower plastic viscosity for use as a pavement overlay.

INTRODUCTION

Rheology is the study of flow behavior. When studying flow behavior, the parameters measured are typically viscosity (apparent viscosity is defined as stress divided by strain rate) and, if the material is plastic or pseudoplastic, yield stress, the stress at which a measurable strain rate is obtained. The flow behavior of fresh concrete controls how the concrete may be processed, how easy it is to mix, place, and consolidate the concrete. Rheology also provides important information about the microstructure of the material; for example a suspension usually shows a yield stress because particles are flocculated. The focus of this paper, however, is on concrete technology. The objective is to demonstrate that concrete rheology is important and how it can be used to improve performance. Experimental results from our laboratory are shown in order to illustrate certain points and to describe the rheological behavior of fresh concrete.

Concrete is known to show plastic (Bingham) flow behavior (1), as illustrated in Figure 1, so its flow behavior may be described using yield stress and plastic viscosity (defined in the figure). The apparent viscosity varies with strain rate; in order to describe concrete flow behavior, it is necessary to measure stress at two or more levels of strain rate. Although the yield stress and plastic viscosity tend to be well correlated, they are not directly related (1, p. 141–142). As discussed later in this paper, there are ways to increase yield stress without causing much increase in plastic viscosity, and vice versa.

Slump is widely used as an measure of workability (indeed, in the US slump is generally the only such measure), and has been shown (1, p. 185) to correlate well with yield stress. As yield stress increases, slump decreases. If yield stress provided a full description of concrete rheology, slump would probably be a suitable rheological

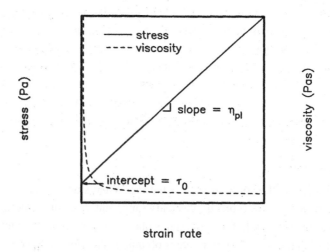

Figure 1. Shear stress and apparent viscosity versus strain rate for a plastic material, showing yield stress and plastic viscosity.

test, especially for use under field conditions. But yield stress does not provide a full description, and it is necessary to also measure viscosity.

There is much to be gained by using rheological parameters (yield stress and plastic viscosity) to describe concrete workability. The rheological parameters have a scientific basis, whereas slump is an empirical test. Tattersall and Banfill have already criticized the empirical workability tests (1, pp. 6–9). We would only add that empirical tests do not allow us to take advantage of the broad knowledge that has been developed concerning suspension rheology.

It is clear from practical experience that concrete rheology depends on paste rheology. One can alter concrete rheology by changing any number of paste parameters – for example, by changing the water–to–cement ratio, by adding a superplasticizer to prevent flocculation, and by allowing cement particles to hydrate. We have found it useful to consider concrete as a suspension of aggregates (coarse and fine), with cement paste as the fluid phase. This convenient simplification allows us to analyze concrete viscosity based on the paste viscosity and the effects of aggregate (volume fraction, packing, and shape).

The rheology of cement paste is very sensitive to shear history, and this sensitivity is likely to be important in concrete. Therefore it is necessary to consider shear history as well as strain rate when designing rheological experiments. For example, in order to measure flow behavior of cement paste to predict the behavior of concrete, it is necessary to shear the paste at the same stress to which it is subjected in concrete. It has been shown by researchers at Construction Technology Laboratories that a rather high shear treatment is necessary to reproduce the shear history of the paste in concrete (2).

Another important aspect of the flow behavior of cement paste concerns how the measurement affects the yield stress. To study changes in yield stress as a function of time, the typical technique is the static flow curve, in which one measures stress at various strain rates, then extrapolates to zero strain rate to estimate yield stress. However, the flow required breaks down the delicate hydrated microstructure, thereby altering the subsequent measured yield stress. In our research on cement paste, we have successfully applied dynamic flow techniques (creep/recovery and low amplitude oscillatory shear) to measure changes due to hydration. This work has been the subject of several papers, including a recent review (3). There appears to be no similar dynamic technique for concrete.

CONCRETE PROCESSING

There are several methods used to process concrete in which rheology is likely to provide a major technological contribution. Of particular interest to us has been pumping and highly flowable concrete, and these are discussed here. This section is rather subjective, although some (concrete) examples are given in the section on experimental studies.

Pumping

In pumping concrete, the primary parameter controlled in the field is the aggregate volume fraction. If the volume fraction is too great, the mix is often deemed "unpumpable". Aggregate size and shape are also important, but only relative to the size of the pipe. When the pipe diameter gets down to about 50 mm, not only is the aggregate important (volume fraction, size and shape), but also is the cement content. If a mix is difficult to pump, it is entirely possible that the the pump operator will add water to facilitate pumping, making the mix difficult to place because of segregation.

Highly Flowable Concrete

Highly flowable concrete is a new type of concrete which has excellent flowability and high resistance to segregation and bleeding, designed for use with heavily reinforced formwork or sophisticated mold shapes. The category includes self–levelling concrete, designed for producing slabs with little or no vibration. The mixture is designed for high flowability, presumably corresponding to low yield stress, and resistance to segregation, provided by high viscosity. It typically contains a water reducing agent and/or superplasticizer to provide low yield stress (high slump), often contains some type of thickener (viscosity agent) to increase the viscosity, and may also contain other fine particles (blast furnace slag, fly ash, limestone powder) to increase the viscosity. The original material (4) utilized polysaccharide as a thickener. Careful attention is often paid to aggregate grading to provide optimum packing for low yield stress.

Rheological Parameters

Because concrete viscosity varies with strain rate, it is necessary to consider what strain rate is appropriate for predicting behavior during a specific processing operation. Some guidance is provided by Barnes et al. (5, p. 13); they listed typical strain rates for various manufacturing processes, including the following:

1. flow through a pipe is a high strain–rate process, about 10^0 to 10^3 s^{-1}.

2. settling of particles is a low strain–rate process, perhaps as low as 10^{-4} s^{-1}

Based on that analysis, we expect the pumpability of concrete to correspond to viscosity at high strain rate. De Larrard et al. (6) used the LCPC concrete rheometer in the field to evaluate pumping performance. They concluded that there is a direct relationship between pumping resistance and the concrete plastic viscosity. They also showed a relationship between pumping flow rate and rheological properties (both yield stress and plastic viscosity). However, it is apparent viscosity at high strain rate that one expects to relate to pumpability. Indeed, as is shown in the experimental studies, existing concrete rheometers do not even allow the measurement of viscosity at the higher strain rates expected for flow through a pipe. However, at increasing strain rate apparent viscosity approaches plastic viscosity, so the relationship reported by de Larrard et al. is reasonable.

The analysis of Barnes et al. suggests that aggregate settling would correspond to viscosity at low strain rate, which in a plastic material is proportional to the yield stress. Therefore it is especially interesting to note that the literature on highly flowable concrete (discussed above) clearly considers that settling occurs at low viscosity, not low yield stress. However, no specific strain rate values were considered.

Settling of a single small particle in a fluid is usually described using Stokes Law,

$$U_0 = \frac{2a^2(\Delta\rho)g}{9\eta} \tag{1}$$

where U_0 is the settling velocity, $\Delta\rho$ is the difference in density between the particle and the surrounding fluid, a is the diameter of the particle, and η is the viscosity of the fluid. The effects of interactions between particles have been analyzed in considerable detail because the problem of sedimentation has considerable technical significance. The velocity is reduced considerably by other non attracting particles (due simply to steric interactions); the effect of other particles (at a volume fraction ϕ) has been characterized using $U/U_0 = (1 - \phi)^5$ (7). Applying Stokes Law to large particles such as coarse aggregate in concrete requires consideration of strain rate; if the fluid is not Newtonian, it is probably suitable to use the limiting viscosity at low strain rate.

EXPERIMENTAL STUDIES

This section presents experimental studies of concrete flow that demonstrate some of the general behaviors already discussed in this paper.

Procedure

The rheometer used in these studies was based on the BTRHEOM (8). It utilizes a parallel plate geometry, with the plates open and vaned to prevent slip, and an outside wall. The container is shown in Figure 2. A version of the rheometer with an inner wall, like the BTRHEOM, was used in our early studies (9). Without the inner wall, individual values of stress and strain rate may be computed from measured values of torque and angular velocity, using equations we have described elsewhere (10). With the inner wall, only the yield stress and plastic viscosity values may be computed, using equations developed by Hu and de Larrard (8).

The general experimental procedure is to mix the concrete in a pan mixer, transfer it to the rheometer, then measure torque at a series of rotational velocity values starting at the high end. The highest strain rate we can apply is 23 s^{-1} (though such a high strain rate produces errors in flow behavior, such as segregation, for some concretes).

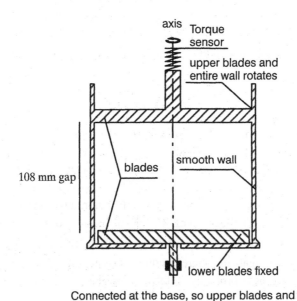

Figure 2. Schematic of concrete rheometer container.

Results

Ordinary concrete, as noted previously, shows more–or–less plastic flow behavior, with a yield stress of several hundred Pa (perhaps in excess of 1000 Pa) and a
plastic viscosity of a few hundred Pa·s. The flow curve in Figure 3 is typical. The
concrete was made using a total aggregate volume fraction of 0.60, the coarse aggregate was a crushed stone with a nominal maximum diameter of 25 mm, the sand content was 35%, and the w/c was 0.40.

The effect of adding aggregate is to increase both yield stress and plastic viscosity. Both parameters were seen to increase following a power–law relationship
(Figure 4, similar to the concrete in Figure 3 except that the w/c was 0.45).

The effect of aggregate grading was studied by varying the relative amounts
of coarse and fine aggregate (expressed using percent sand), while keeping the total
amount of aggregate constant. Results are shown in Figure 5. The coarse aggregate
was a river gravel with a nominal–maximum diameter of 25 mm, the total aggregate
volume fraction was 0.60, and the w/c was 0.45. Both yield stress and viscosity had
minimum values at an intermediate sand content. We attribute this effect to changes
in packing of the combined coarse and fine aggregate. The aggregate showed a maximum volume at about the same sand content (40%) at which the minimum in yield

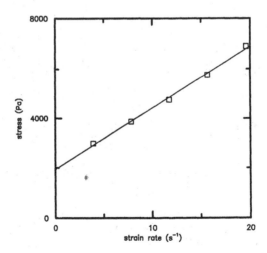

Figure 3. Concrete flow curve.

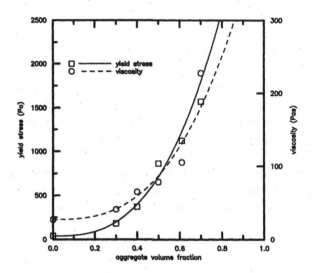

Figure 4. Yield stress and plastic viscosity as a function of aggregate volume fraction.

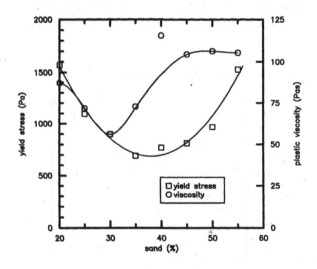

Figure 5. Yield stress and plastic viscosity as a function of sand content (reprinted from ref. 11, with permission).

stress was observed. It was interesting and unexpected that the minimum in viscosity was observed at a somewhat lower sand content (30%).

Increasing the paste w/c reduces both yield stress and plastic viscosity. These effects are shown in Figure 6 (using the same materials as Figure 5).

Addition of superplasticizer reduces both yield stress and plastic viscosity, but its main effect is on yield stress. Addition of 0.8% superplasticizer[1] (expressed as % solid superplasticizer by weight of cement) to the concrete shown in Figure 3 reduced the yield stress from 1930 Pa to 60 Pa and reduced the plastic viscosity from 250 Pa·s to 150 Pa·s.

Addition of fly ash was seen to have little effect on concrete flow, probably because the particle size distribution of the ash is generally similar to that of the cement it replaces. Replacement of 15% (by weight) cement with Class C fly ash[2] in a concrete similar to that shown in Figure 3 (except that w/c was 0.45 instead of 0.40) reduced the yield stress from 1120 Pa to 940 Pa and left the plastic viscosity basically unchanged (105 Pa·s to 108 Pa·s).

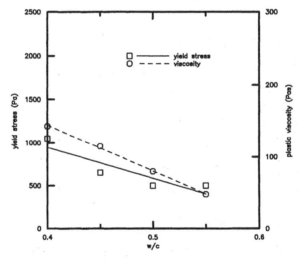

Figure 6. Yield stress and plastic viscosity as a function of w/c (reprinted from ref. 11, with permission).

1. Daracem 100 from W.R. Grace.
2. from American Fly Ash.

When silica fume is added without a superplasticizer, the apparently floccu-lated particles cause an increase in both yield stress and plastic viscosity. Replace-ment of 10% cement with silica fume[3] in a concrete similar to that shown in Figure 3 (same base concrete as used for fly ash, above, w/c was 0.45 instead of 0.40) in-creased the yield stress from 1120 Pa to 2310 Pa and increased the plastic viscosity from 105 Pa·s to 200 Pa·s.

Many of these (and other) effects were combined in a study of a high perfor-mance concrete (HPC) designed for study as an overlay for an airport pavement. We designed an ordinary concrete (OPC) for comparison. Both are low slump concretes as required for use with a paving machine. The mix proportions and properties of these concretes are given in Table I. The HPC had a lower w/s (0.25 for the HPC compared to 0.42 for the OPC) and contained both silica fume (replacing 10% of the cement by weight) and a superplasticizer[4]. The coarse aggregate was a crushed stone, nominal maximum diameter of 10 mm for the HPC and 25 mm for the OPC. The sand content, selected to provide a minimum volume, was 49% for the HPC and 39% for the OPC. The total aggregate volume fraction was the same in the two concretes. The flow behavior of these concretes is shown in Figure 7. The HPC had a higher yield stress but a lower plastic viscosity than the ordinary concrete. We believe that this

Table I. Mix proportions and properties of pavement concretes

	OPC	HPC
water (% by weight)	7.2	5.4
cement (% by weight)	17.1	19.4
silica fume (% by weight)	0	2.2
fine aggregate (% by weight)	29.8	36.0
coarse aggregete[a] (% by weight)	45.8	37.1
superplasticizer (% by weight of cementitious material)	0	0.3
slump (mm)	75	100
air content (% by volume)	6.7	5.0

[a]Coarse aggregate nominal maximum diameter 25 mm for OPC and 10 mm for HPC.

3. EMS965 from Elkem.
4. AdvaFlow from W.R. Grace.

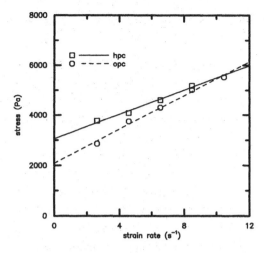

Figure 7. Flow curves of high performance and ordinary pavement concretes.

pavement will be easier to process using a paving machine – its lower plastic viscosity will provide for greater fluidity during vibration, but its rather high yield stress, which probably reflects the stickiness due to the silica fume, will provide better dimensional stability after the paving machine has moved on. What is more important, however, is the example provided by these two concretes of how rheological behavior may be altered through control of the materials and their proportions. In particular, we were able to increase the yield stress while lowering the plastic viscosity.

DISCUSSION

The experimental results are consistent with the view of concrete as a suspension of coarse and fine aggregate in a continuous phase of cement paste. The various parameters affecting concrete viscosity may be analyzed using the Krieger–Dougherty equation. This equation is widely used to describe the viscosity of dispersed suspensions, and we have used it previously to describe cement paste (12). The equation is

$$\eta = \eta_c \left(1 - \frac{\phi}{\phi_m} \right)^{-[\eta]\phi_m} \tag{2}$$

where ϕ_m is the maximum possible volume fraction for the particular assemblage of particles, which has a value of 65% for randomly close–packed spheres, and $[\eta]$ is intrinsic viscosity, which equals 2.5 for spherical particles and has a higher value for non–equant shapes.

It is clear from our concrete studies that the yield stress of a suspension depends on the yield stress of its continuous phase, but we do not yet know how to describe this behavior quantitatively. Indeed, we have found no publication dealing with yield stress of suspensions whose continuous phase is non–Newtonian. That we observe a direct relationship between paste and concrete yield stress is not surprising. It is widely recognized from everyday experience that addition of a superplasticizer reduces the yield stress of both paste and concrete. Similarly, it is widely recognized that hydration of cement particles increases the yield stress of both cement paste and concrete; although, as noted earlier in the paper, measuring changes in yield stress as a function of hydration time is difficult for cement paste and not possible for concrete. However, these effects are not understood quantitatively, so it is not possible to model the effects of concrete mix parameters on yield stress.

This equation has been used to evaluate the effects on concrete viscosity of various changes in the constituents and their proportions. This evaluation is not meant to be a quantitative analysis, but rather simply to estimate the approximate impact of each change. Reasonable values were obtained for aggregate volume fraction and maximum volume fraction using data reported by Szecsy (9), and for paste viscosity using data reported by Struble and Ji (11). Results of the analysis are presented in Table II. It appears from this analysis that the main effect on concrete viscosity is obtained through modifications in the paste, either changing the w/c using a superplasticizer. Changing the aggregate grading, shape, and volume fraction have considerable effects, but not as great as the changes in paste.

OBTAINING SPECIFIC RHEOLOGICAL PROPERTIES

Equation 2 shows that the viscosity of a suspension is affected by the viscosity of its continuous phase. In concrete we consider the continuous phase to be the paste, and concrete viscosity may be changed by manipulating the paste viscosity. The same effect may be used to increase paste viscosity by using thickeners (e.g., gums). Yield stress appears to show a similar effect, but there is no theoretical model.

Equation 2 also shows that viscosity depends on the concentration of particles and on how densely the particles are packed. Again, yield stress appears to show a similar effect. In concrete, the aggregate amount and grading affect viscosity and yield stress. Viscosity is increased as the volume fraction of particles is increased, and at a given volume fraction, the viscosity is reduced by packing particles more dense-

Table II. Effects of concrete parameters on viscosity

Parameter	Change in Parameter	Change in Apparent Viscosity (Pa·s)
paste w/c (η_c)	0.23 to 0.32	310 to 5
paste dispersion (η_c)	fully flocculated (no superplasticizer) to fully dispersed	150 to 5
aggregate volume fraction (ϕ)	0.6 to 0.5	125 to 25
aggregate grading (ϕ_m)	0.6 to 0.8	75 to 25
aggregate shape [η]	elongated (6) to equant (2.5)	110 to 25

ly. In paste, viscosity and yield stress are affected by the volume fraction and packing of cement and other fine particles. We can increase or reduce the paste viscosity by reducing or increasing the water–to–cement ratio (w/c), and at a given w/c, we can reduce paste viscosity by packing particles more densely (by adding fine particles, as in DSP).

Flow behavior also depends on flocculation and hydration. While these affect both yield stress and viscosity, they are especially important to the yield stress. In cement paste, yield stress and viscosity are high if the cement particles are flocculated (i.e., agglomerated due to attractive surface forces). Therefore they are decreased when a water reducer or superplasticizer is added that disperses the particles, and increased when cement is replaced by finer particles, more likely to flocculate. The increase in yield stress observed with silica fume could be manipulated more effectively by using particles more prone to flocculation and hence to thixotropic flow behavior, such as clay particles. Yield stress and viscosity are also increased when cement hydration products bind together the individual cement particles. However, the effects of flocculation and hydration are very sensitive to strain rate and become progressively less important as strain rate is increased. Therefore flocculation and hydration have a greater effect on yield stress than on viscosity, especially viscosity at high strain rate. However, it should be noted (as discussed in the experimental studies) that concrete viscosity is typically measured at rather low strain rates, up to about 20 s^{-1}; at this level flocculation and hydration are still quite important.

CONCLUSIONS

Rheology provides important information about the flow behavior of concrete. The flow behavior may be especially useful in designing concrete for specific processing methods. Pumping and highly flowable concrete are technologies where rheology offers particular potential for improvement.

Concrete constituents and proportions can be adjusted to provide targeted values of yield stress and plastic viscosity. Both rheological parameters are affected by paste and aggregate volume fraction and packing density. Yield stress more than viscosity is sensitive to the flocculation of fine particles, which is reduced by addition of superplasticizer and increased by replacing cement with finer particles. Analysis suggests that viscosity may be controlled by changing either paste or aggregate properties, with the greater effects coming from changes in paste w/c or dispersion.

ACKNOWLEDGEMENTS

The experimental research was funded by the National Science Foundation (through an NYI award to LJS), by the Portland Cement Association, and by the Federal Aviation Authority Center of Excellence at the University of Illinois.

REFERENCES

1. G.H. Tattersall and P.F.G. Banfill, *The Rheology of Fresh Concrete*, Pitman Advanced Publishing Program, London, 1983.

2. F.J. Tang and S. Bhattacharja, "Development of an Early Stiffening Test", RP346, Portland Cement Association, Skokie IL, 1997.

3. L.J. Struble, W.–G. Lei, G.–K. Sun, and R.S. Szecsy, , accepted for publication in *Cem. Concr. Aggregates*, 1998.

4. M. Hayakawa, Y. Matsuoka and T. Shindoh, in *Special Concretes: Workability and Mixing*, pp. 183–190, E & FN Spon, London, 1993.

5. H.A. Barnes, J.F. Hutton, and K. Walters, *An Introduction to Rheology*, Elsevier, Oxford, 1989.

6. F. de Larrard, C. Hu, T. Sedran, J.C. Szitkar, M. Joly, F. Claux, and F. Derkx, *ACI Mater. J.* **94** 234–243 (1997).

7. W.B. Russel, D.A. Saville, and W.R. Schowalter, *Colloidal Dispersions*, Cambridge University Press, Cambridge, 1989.

8. C. Hu and F. de Larrard, "The Rheology of Fresh High—Performance Concrete", *Cem. Concr. Res.* **26** 283–294 (1996).

9. R.S. Szecsy, "Concrete Rheology", PhD Thesis, Department of Civil Engineering, University of Illinois, Urbana IL, 1997.

10. X. Ji, L.J. Struble, and R.S. Szecsy, "Concrete Rheometry", submitted to *Mater. Struct.*, 1997.

11. L.J. Struble and X. Ji, "Rheology", chapter in *Handbook of Analytical Techniques in Concrete Science and Technology*, edited by Ramachandran and Beaudoin, Noyes Publication, Park Ridge NJ (in press).

12. L.J. Struble and G.K. Sun, "Viscosity of Portland Cement Paste as a Function of Concentration", *Adv. Cem. Based Mater.* **2** 62–69 (1995).

MIX-DESIGN METHOD OF CONCRETE, WITH IMPACT ON RHEOLOGICAL REQUIREMENTS

T. Roshavelov
Institute of Construction Engineering "L. Karavelov"
Sofia 1373, Bulgaria

Ö. Petersson, P. Billberg
Swedish Cement and Concrete Research Institute (CBI),
Stockholm 100 44, Sweden

ABSTRACT

The everlasting topic of concrete mix design is critically reviewed. The new challenges to concrete composition are summarised and a descriptions of the Solid Suspension Model is presented. Analysis of the theoretical background of the model, as well as some of the advantages and existing uncertainties in the theory are discussed. Experimental verification of the applicability of the model on compositions for self-compacting concrete mixes is given.

INTRODUCTION

The application of packing studies for the optimisation of granular mixtures dates back to the end of the nineteenth century. The long line of proposed models contains some quite successful equations that provide reasonable agreement between theoretical and experimental results of the packing density. The practical aim of these theories is to estimate the water requirement of concrete mixtures on the basis of its theoretically or empirically predicted porosity.

During the last few years the theory of the Solid Suspension Model developed by T. Sedran and F. de Larrard (1) has been successfully applied in a number of constructional applications as a tool for mix design and optimization of concrete mixture proportions (2, 3). As the authors point out, this method gives a remarkable fit between experimental and theoretical results. This has promoted extensive additional research in order to estimate the applicability of the method and to establish the range of applications that the mix design approach affords.

PACKING THEORIES AND CONCRETE MIXTURE PROPORTIONS

In this section we will concentrate only on some of the most commonly used models that have recently been under extensive study. The theories of the packing models will not be given, as they are described repeatedly in the specialised litterature. Here, only a summary of the models will be presented, as special attention will be given to the group of theoretical models, rather than to the purely empirical ones.

Toufar's model

Goltermann and Johansen (4, 5) discussed on the use of different models that give best fit of the theoretical to the experimental packing densities. They showed that the model proposed by Toufar, Klose and Born was best for larger particle diameter ratios (above 0.22, as initially found by Petersen) and the model described by Aim and Goff better describes the packing densities in the range of lower diameter ratios (below 0.22). The authors of (5) proposed a modification of Toufar's model that improves the correlation between the theoretical and experimental results. In this modified model an assumption is made that fine and coarse aggregates are of different sizes, each characterised by a parameter called "characteristic diameter". This leads to problems when two aggregates have overlapping fractions while their characteristic diameters are fairly different. The model gives an overestimation of the packing degree for small amounts of fine aggregate in the binary mix, but does not influence the determination of the optimal packing.

Linear Packing Density Model (LPDM)

De Larrard et al. (6) developed the theory of LPDM for multi-component mixtures. It takes continuous particle-size distributions (PSD) into account, but differing from Toufar's model, calculation of the dry packing of the mixtures involves integration of the PSD functions for the components present in the mixture. The model gives a favourable fit between experiments and theory for dry mixtures, excluding the considerable overestimation of the packing density for compositions of maximum packing. In (7) it has been shown through experiments, that LPDM models satisfactorily the packing density of systems for which small particles dictate packing, i.e. the coarse grains "flow" in a matrix of finer grains, but cannot model a particle system in which large particles dictate packing. One of the possible realistic explanations of this phenomenon is as follows:

When aggregate particles move with respect to each other, both the shape and size of individual aggregate inter-space voids change. For this to happen, the paste must move from one inter-space void to another. If the micro-mortar is not fully packed, water can separate from it and move between the inter-spaces, which allows the particles not to change location. In such mixes the experimental results are in good agreement with LPDM. When approaching the optimal composition none of the classes are fully packed, and so LPDM cannot be applied to these mixes. Increasing the fine particles content results in a decrease of water content. Here, the critical minimum distance between coarser particles limits the movement of the particles of the fully packed size class. With a further increase of fine particle content, the micro-mortar volume fraction increases, as do the distances between larger particles. The fully packed micro-mortar can then move between the inter-spaces as a single phase and no freely moving water is needed. In such mixes the experimental and theoretical values once again approach each other. The fully packed size class forms a continuous particle skeleton in the system and dictates the water requirement. Smaller particles move freely in inter-spaces of the skeleton while larger particles are discrete and are surrounded by the phase comprising the skeleton, e.g. the fully packed size class, smaller particles and water.

Solid Suspension Model (SSM)

The authors of LPDM realised that packing models can adequately predict the dry packing density of compact granular mixes more or less successfully, but can not provide a solution to the basic question: how to optimize the granular skeleton with respect to flow properties of fresh concrete? To perform movement, the solid particles need space in order to rotate and translate. If such is not available, the confined compact system does not move (blocks) or in the case of non-restricting conditions it moves (flows) but the flow is accompanied by considerable dilatancy. With this problem in mind, a next generation model - SSM, has been developed. It combines two powerful scientific tools as a means of concrete mix design: the very old concept of optimization of dry material granulometry and the relatively new idea of characterization of fresh concrete behaviour by its rheological parameters. Determination of the water demand of a granular mixture is carried out by particle packing studies, i.e. the space not occupied by the solids, and the assumption that the water requirement determines the rheology of the whole system. Though SSM predicts the dry packing density of solid constituents with precision higher than 99% (1), its second aim, i.e. characterization of concrete flow properties through modification of Mooney's concentration equation is rather unfavourable. Some of the possible sources for the latter consideration are summarized below and are based on the premises of the SSM theory.

The modification of Mooney's equation: Mooney's theory successfully extends Einstein's equation for the viscosity of a dilute suspension of rigid spheres in a viscous liquid to suspensions of higher concentrations. The original form of Mooney's equation (8) is:

$$\eta_r = \exp\left(\frac{2.5\phi}{1-k\phi}\right) \tag{1}$$

where:

η_r -   relative viscosity;

ϕ -   volume fraction of the suspended spheres;

k -   coefficient, interpreted by Mooney as a self-crowding factor, whose value range is predicted approximately by the theory, but is left for experimental determination. In polydisperse fractional solutions, k becomes a variable factor, λ_{ij} (r_i / r_j), which is a function of their radius ratios. Consequently, in the case of polydisperse suspensions, equation (1) becomes:

$$\eta_r = \exp\sum_{i=1}^{n}\left(\frac{2.5\phi_i}{1-\sum_{j=1}^{n}\lambda_{ji}\phi_j}\right) \tag{2}$$

Based on Mooney's theory, the authors of (1) derive the basic equation of SSM in the form:

$$\eta_r = \exp\left(\sum_{i=1}^{n}\frac{2.5y_i}{\dfrac{1}{c}-\dfrac{1}{\gamma_i}}\right), y_i \neq 0 \tag{3}$$

where:

η_r -   relative reference viscosity;

y_i -   volume fraction of the suspended particles;

c -   packing density of the mixture;

γ_i -   virtual packing density of class i particles, in the presence of n-1 fractions. Its value is calculated on the basis of the theory developed for the LPDM, for which the crowding factor has been substituted by mathematical fitting rather than by a theoretically based approach.

As the mathematical transformations given in the theory of the model seem not to be quite convincing to the authors, in the present paper only an analysis of the equation (3) is presented. However, it must be noticed that original Mooney's equation (1) and SSM's equation (3) are obviously of a different mathematical and physical nature.

If we substitute $1/\gamma_i = a_i$, $2.5y_i = b_i$ $(0 < y_i \leq 1)$ and $\ln(\eta_r) = C > 0$, equation (3) can be rewritten as:

$$C = \sum_{i=1}^{n} \frac{b_i}{\left(\dfrac{1}{x} - a_i\right)} = f(x); \text{ and } \frac{d f(x)}{dx} = \left(-\sum_{i=1}^{n} \frac{1}{\left(\dfrac{1}{x} - a_i\right)^2}\right) \bullet \left(-\frac{1}{x^2}\right) > 0 \qquad (4)$$

From equation (3) and (4) the following important conclusions can be drawn: The relative reference viscosity is the product of n positive and increasing functions, each one tending to infinity. Equation (3) does not have a single solution, but n roots. Each of these roots lies on an interval limited by a_{i-1} and a_i, i.e. the number of the roots depend on the number of the components in the mixture composition. The physical meaning of the values of a_i less than unity is not clear, i.e. the virtual single class packing density can take values less or equal to unity by definition.

The particles are classified into two groups - round and crushed: One of the great advantages of SSM is the possibility for each granular class to enter the model with its own shape factor that describes the nature of the grains. When PC and the binders start to react, the volume of the gel system changes with time. This later results in a change of the shape and the size of the powder particles, and as a consequence, a change in the packing density with time, with different rates depending on the nature of the reacting system. On the other hand, aggregate shape and texture should be classified much more completely, as spherical particles are a myth and crushing may result in a variety of shapes, characterization of which as "crushed" is not simply a matter of poor vocabulary.

The surface forces do not change packing density of the solid system: The fact that surface forces do hinder packing does not need comment. Therefore, the model should take into account the surface forces between the smaller particles that cause aggregation. Moreover, experimental packing density has meaningful values only in the case in which full de-flocculation of the solid phase is reached and that fact is indirectly recognized in the procedure for estimation of the packing densities of powder materials proposed in (3). The use of a plasticizer/superplasticizer to reach deflocculation of fine particle systems and to allow the particles to fill the spaces of their own size class is an imperative condition. An analytical model can describe only system in which this condition is fulfilled. In non-plasticized mixes, inert fine particles generally increase the water requirement, as they are in a flocculated state and cannot fill spaces of their own size class.

Partly, the problem has been overcome by the water demand test introduced for the estimation of powder materials packing density (3). However, the procedure is not physically well founded, as it uses descriptive terms as: " ... weight of water *just necessary* to make the mix change from a wet-like soil state (with formation of *balls*) to a smooth homogeneous paste is found... ". The problem is, that at high speeds of rotation of the impeller, formation of balls continues till much higher water quantities than required for the system to transform into a homogeneous paste. Therefore, periodical terminations of the mixing are needed in order to estimate the paste continuity. On the other hand, visual assessment of homogeneity of the paste is rather uncertain, as well as any descriptive technique.

The air in the two-phase system is not treated as a third component: The air volume can not be considered either as part of the solid or liquid phase. Air voids are far too deformable to be considered as solids and far too discrete to be considered as part of the continuos fluid phase. As this seems to be one of the very important questions that the model must clarify, the experimental part of the current research has been performed with the aim of understanding where the air content of the concrete mix must refer.

As can be seen, some of the premises of SSM are different from the one developed for the original LPDM and Mooney's equation. The above given considerations determined the need of further study of the applicability of the model. Additional research has been done at CBI in order to establish the validation of SSM that is presented in the next section.

EXPERIMENTAL

Several mixes have been designed by use of the procedures described in (9) and the results are critically analyzed. A software version for the mix design has been developed for the purpose, as well.

Materials

The materials for the trial mixes are typical for the Swedish construction industry. Two types of natural aggregates are used: AB Underås (FA) and Vendels (CA, partly crushed), with PSD curves in the range 0-8 mm and 8-16 mm, respectively. The grain density of Underås (FA) is 2,65 kg/dm³ and of Vendels (CA) 2,687 kg/dm³. The cement (CEM I 42.5) is ordinary Portland cement, with low C_3A content (2%), that is supposed to minimize the water demand and improve the sulphate resistance of concrete for engineering applications. Limestone filler (KÖ 500) is incorporated to improve the fluidity of the mixes and as a partial replacement of the cement. The superplastisizer (Glenium 51) is a low slump loss poly-carboxylate ether and cross-linked polymer with 35% solid content. The saturation quantity (by solid content) is estimated to be 1.2% of the binders' content.

The PSD for the aggregates is determined by standard sieve series (ISO 565: 1990 (E)) and for the cement and the limestone powder by a Laser analyzer.

Aggregate dry mixes and concrete compositions

To verify the model, the dry packing densities of ten binary mixtures of aggregates were tested and compared with the theoretical predictions.

Compositions of four self-compacting concrete mixtures have been projected with use of the software. Cement replacement levels have been kept constant at 30 and 35%, while binders-to-total solids and coarse-to-fine aggregate ratios were optimized for a minimal viscosity of the concrete mixture, at porosity equal to 0.170. The dosage of the superplastisizer was 1/2 of the saturation dose, i.e. 0.6%. As no air-entraining agent was used and the concrete was expected to be self-compacting, the air content was set at 2%. The relative reference viscosity was 5×10^4, as proposed by LCPC (2). Mixture proportions were designed for a confinement equal to 50 mm. Four mixtures were prepared and tested for slump flow, air content and compressive strength of the hardened specimens. Each test was performed on three specimens and the results were averaged. The mixture proportions are presented in Table I.

Table I. SCC Compositions

No	$\dfrac{CA^1}{FA}$	$\dfrac{LP^2}{C+LP}$	$\dfrac{C+LP^3}{Solids}$	$\dfrac{W^4}{C}$	Cement (C), kg	Limestone filler (LP), kg	Underås (FA), kg	Vendels (CA), kg	Water (W), dm³	Super plast., kg
1	1.0	0.35	0.25	0.395	380.0	204.6	877.0	877.0	147.9	3.51
2	0.95	0.30	0.26	0.362	425.1	182.2	886.4	842.1	151.6	3.64
3	0.95	0.35	0.22	0.443	334.1	179.9	934.4	887.7	145.9	3.08
4	1.05	0.30	0.26	0.361	425.4	182.3	843.7	885.8	151.3	3.65

[1]CA/FA – Coarse-to-fine aggregate;
[2]LP/(C+LP) – Limestone filler-to-total binders;
[3](C+LP)/Solids – Binders-to-total solids;
[4]W/C – Water-to-cement.

RESULTS

The results from dry packing experiments are presented in Table II. The experimental dry packing density is the ratio between the bulk (compacted) density, estimated through the procedure of de Larrard (10), and the grain density. The deviations between experimental and theoretical values were less than 1 % and agreed fully with the results shown by other authors (see for example 1, 2, and 10).

TABLE II. Dry packing results

$\dfrac{FA}{(FA+CA)}$ [1]	0,00	0,10	0,20	0,30	0,40	0,50	0,60	0,70	0,80	0,90	1,00
EV [2]	0,632	0,676	0,720	0,781	0,802	0,814	0,815	0,796	0,792	0,777	0,764
TV [3]	0,627	0,672	0,721	0,771	0,808	0,819	0,813	0,801	0,787	0,773	0,762

[1] $FA/(FA+CA)$ – Fine-to-total aggregate;
[2] EV - Experimental values;
[3] TV – Theoretical values from the SSM;

It must be noticed, that some authors (5) use an approach based on preliminary optimization of aggregate mixtures that is supposed to provide best performance when applied in a concrete composition. Such an approach is rather uncertain, because the optimal values for binary, ternary and even for combination between optimized mixtures of aggregates are not necessarily the ones in the suspension.

The mixtures # 1 and 2 contain constant cement replacement levels, while the levels of the paste and aggregate ratios were optimized with the software for each of the compositions. The mixtures # 3 and 4 were projected with ratios different from the optimal packing degrees in order to verify the precision of the model. The optimal values of the coarse-to-fine aggregate ratios are lower than usually reported. This evidence may be related to the granulometry of the fine aggregate. AB Underås contains 12% grains with diameter higher than 4 mm and therefore, its fraction 4-8 mm could be considered as coarse aggregate.

In the first two mixes, the measured slump flow was 750 and 740 mm, respectively, which is about 15% higher than required for SCC. The air content was 2.5% in the both specimens, i.e. 20% higher than set in the software. The fresh concrete mixture #3 showed slump flow of 490 mm and air-content 4.5%, while mixture #4 showed slump flow of 560 mm and air-content 3.2%. The four mixes showed stable behaviour during the flow test (rather higher viscosity) that was evaluated visually.

DISCUSSION

The experimental results show some deviations from the expectations. Moreover, the further the dry mixture from the optimal one, the larger the lack of coincidence. Therefore, two findings have to be discussed.

The higher values of the slump flow for the optimized compositions (#1 and 2), suggest higher water content than required, because the superplastisizer dose was just 1/2 of the saturation dose and can not be considered as an overdose. Consequently, the theoretical porosity is higher than the actual. With the same formalism, the higher measured air content in the mix suggests exactly the opposite. The following questions arise: Is the theoretical porosity of the mix higher than the actual? Can the air content not be considered as part of the fluid phase? Is the value of the relative viscosity too high for SCC?

In mixes # 3 and 4, one or more than one classes of grains are fully packed and determine the overall packing of the mix. In this case the deviation in the theoretical packing density dependence of the relative viscosity is even more pronounced. If we suppose the mixes had the same viscosity, then it could be concluded that the slump will be reached by increasing the dose of the superplastisizer. This however, will not result in a considerable decrease of the air content and again we have the case of overestimation of the actual porosity. All of the above, may result in overestimation of the real porosity, compared with that experimentally observed. This consideration however, should be studied more deeply.

Along this line of reasoning, one more question arises: the SSM provides a value called "relative reference viscosity" of the concrete mixture. It is not very clear however, how the value of this parameter correlates with concrete workability or any of the well known rheological parameters. The reason for this is that the relative viscosity is the ratio between apparent viscosity (of the suspension and disperse media, respectively), and as such, it can not describe the flow curve of a Bingham material. Any assumptions that concrete has Newtonian behaviour in the presence of chemical admixtures seem unreasonable. The primary aim of the SSM was to correlate the value of the relative viscosity to a rheological parameter(s) (3). It is suggested that the value of relative reference viscosity correlate with the plastic viscosity of concrete. Two pieces of evidence confirm this suggestion.

1. The relative reference viscosity is not dependent on the amount of superplasticizer. The only condition is that the quantity of the superplasticizer be high enough in order for the granular mixture to be in a disperse state, i.e. not flocculated. The same dependence has been found for the value of the plastic viscosity.
2. This parameter depends on the porosity or the volume concentration of the mixture, as well as plastic viscosity does.

However, a number of observations suggest the opposite:

1. If the relative reference viscosity of SCC is set to be 50 000 and the viscosity of the water is known to be 0.001 Pas, than the plastic viscosity of SCC should take values around 50 Pas. This value is rather low for concrete. Moreover, de Larrard suggests values for the plastic viscosity of this type of concrete in the range between 100 and 200 Pas.

2. There is no clear explanation at the moment, why at different packing densities (lower than 1) we can have the same plastic viscosity. If this could be explained with the anomaly of the viscosity, described by Chang & Powell (11), then we have to correlate the relative reference viscosity to the apparent viscosity, rather than to the plastic viscosity.

As authors of the model point out (3), one of the great advantages of the model is the possibility for the confining effects exerted by the form and the reinforcement to be taken into account. Calculations have been performed with and without confining, with fixed paste volume. Our experiments with SCC confirmed this feature of the model.

CONCLUSIONS

As it can be seen from the results, the model works quite well in general. Deviation of about 0.5% in the optimal compositions are considered very good in concrete mix design. However, the discussed deviations of the results need further study. It is the authors' opinion, that only when the theoretical uncertainties of the model are overcame, can it be considered as a basis of adequate physical theory for fresh concrete composition-properties, and the construction industry will then be equipped with a powerful and effective tool for concrete mixture proportioning.

ACKNOWLEDGEMENT

The authors would like to thank The Swedish Institute for the financial support given to T. Roshavelov, which helped him to take part in the research.

Note added in proof. In the course of preparation of this manuscript, as a result of some correspondence with Dr. de Larrard (12), the authors wish to add a remark concerning the discussed issues in the paper. A new version of a model, relating the packing density of a suspension and its rheological characteristic is under preparation and soon will be published in a monograph. As its author states, in the formalism of the next generation model, called Compressible Packing Model (CPM), many of the above discussed imperfections of SSM are overcome.

REFERENCES

1. Sedran T., F. de Larrard, D. Angot, Prévision de la compacité des melangs granulaires par le modéle de suspension solide. I – Fondements théoriques et étalonnage du modéle, Bull. liaison Labo. P et Ch., 194, nov-déc., 1994, p.59

2. de Larrard F., T.Sedran, Optimization of ultra-high-performance concrete by the use of a packing model, Cem. Concr. Res., Vol. 24, No.6, 1994, p. 997.

3. Sedran T., F. de Larrad, , RENÉ - LCPC: a software to optimize the mix design High-Performance Concrete, BHP'96, 4th International Symposium on the utilization of High-Strength/ High-Performance Concrete, ed. F. de Larrad and R. Lacroix, Paris, May, 1996, p. 169.

4. Johansen V., P.J. Andersen, Particle Packing and Concrete Properties, In: Materials Science of Concrete II, Ed. by J. Skalny and S. Mindess, Vol. II, 1991, p. 111.

5. Goltermann P., Johansen V., Palbol L., Packing of Aggregates: An Alternative Tool to Determine the Optimal Aggregate Mix, ACI Mater. J., sept.-oct., 1997, p. 435.

6. de Larrad F., M. Buil, Granularité et Compacité dans les matériaux de Génie Civil, Matériaux et Construction, RILEM, Vol. 20, 1987, p. 117.

7. Kronlöf Anna, Filler effect of inert mineral powder in concrete, PhD Thesis, Technical research center of Finland, VVT, 1997.

8. Mooney M., The Viscosity of Concentrated Suspension of Spherical Particles, Journal of Colloid Science, 6, 1951, p.162.

9. Sedran T., F. de Larrard, F. Hourst, Contamines C., Mix design of self-compacting concrete (SCC), Production Methods and Workability of Concrete, Ed. by P.J. Bartos, D.L. Marrs and D.J. Cleland, E&FN Spon, London, 1996, p. 439.

10. Sedran T., F. de Larrard, D. Angot, Prévision de la compacité des melangs granulaires par le modéle de suspension solide. II - Validation Cas des mélanges confinés, Bull. liaison Labo. P et Ch., 194, nov-déc., 1994, p.71.

11. Chang S., Powell R.L., Effect of particle size distribution on the rheology of concentrated bimodal suspensions, J. of Rheology, Vol.38, 1994, pp. 85-98.

12. de Larrard F., Private correspondence, January, 1998.

DURABILITY

COMPOSITION OF ASR GELS AND EXPANSION OF MORTARS

Mitsunori Kawamura, Noriyuki Arano and Tsutomu Terashima
Department of Civil Engineering, Kanazawa University, Kanazawa, Ishikawa 920-8667, Japan

ABSTRACT
The purpose of this study is to relate the composition of ASR gels to the expansion characteristics of mortars. Mortars made by the use of Blue Circle calcined flint and a relatively low alkali cement were immersed in $0.6N$ NaOH solution with and without $Ca(OH)_2$ after pre-curing in water. The massive ASR gel produced within relatively wide cracks in reactive aggregate grains in mortars were analyzed by BSE-EDS analysis. Measurements of the length changes of mortars were made along with BSE-EDS analyses for the wide bands of gel formed in the mortars. The intrusion of NaOH solution into reactive aggregate grains without production of the expansive ASR gel in mortars with 10% silica fume and 20% fly ash supported the validity of a concept that the presence of $Ca(OH)_2$ is the prerequisite for the formation of the expansive ASR gel within reactive aggregate grains in mortars. The increase in the alkali content in ASR gel in mortars decreased their expansions. However, the calcium content in ASR gel did not so explicitly influence the expansivity of the gel as the alkali content in it.

INTRODUCTION
Since the overall expansion of ASR affected concretes mainly depends on the amount and the composition of ASR gels produced, many workers have been concerned about the relationship between the composition of ASR gels and expansion of concrete. Idorn (1) analyzed gels from deteriorated concrete structures, dividing the gels into three groups with different CaO contents. SEM-EDS analyses for gels within cracks in ASR deteriorated concretes showed that the CaO content in gels increased with distance from reaction sites (2). The presence of gels with relatively high calcium contents in concrete was regarded as the result of transport of sol through the cement paste matrix, picking up additional calcium (2,3,4). However, the roles of calcium in ASR expansion in concrete have not completely been understood. Powers and Steinouer hypothesized that an expansive ASR gel with a low calcium content or a non-expansive gel with a high calcium content must be formed depending on the migration rate of Ca^{2+} and alkalis ions through the gel layer formed and their concentrations surrounding reacting aggregate grains (5,6). The behavior of alkalis and calcium in the process of gel formation has been pursued by the use of SEM-EDS analysis (7). The significance

of the presence of calcium hydroxide in the formation of expansive gels was demonstrated by Chatterji et al. (8), Diamond (9), Struble (10) and Thomas et al. (11). At present, further examinations on the chemical composition of ASR gel formed under various conditions are needed for elucidating the roles of calcium in the expansion of concrete due to ASR. Relating the expansion of mortars to the composition of ASR gels formed within the mortars may lead to the establishment of general relationships between the composition of ASR gel and its expansivity.

The result that there were considerable variations in the compositions of the ASR gels formed in a model opal-cement paste composite, was obtained by Scrivener and Monteiro (12). Laing et al. (13) showed that the composition of gel changed with age and depended on the location in a model concrete system. However, the swelling properties of the gels were not referred in these studies.

The major purpose of this study is to relate the chemical composition of ASR gels to their expansion behavior. The gels formed in mortars made by the use of Blue Circle calcined flint were analyzed by BSE-EDS measuring expansions of the mortars. In order to produce relatively large amounts of ASR gel to be analyzed within mortars containing the reactive aggregate, unlimited amounts of 0.6N NaOH solution were supplied into the mortar specimens which were made by the use of a relatively low alkali cement. Namely, the mortar specimens showing no expansions during the pre-curing period of 56 days and 28 days in water and a moist environment, respectively, were immersed in 0.6N NaOH solution and 0.6N NaOH solution saturated with $Ca(OH)_2$ for producing ASR gels with different compositions. In a series of mortars, silica fume and fly ash were incorporated at the replacement of 10% and 20% for cement to reduce amounts of $Ca(OH)_2$ in mortars, respectively.

EXPERIMENTAL OUTLINE

MATERIALS
The reactive aggregate with a size fraction of 2.5 mm to 0.6 mm was a calcined flint produced in Blue Circle Ltd.. Its dissolved silica (S_c) and the reduction in alkalinity (R_c) in ASTM chemical test were 1063 and 70 mmol/l, respectively. The Japanese Toyoura standard sand was used as a non-reactive aggregate. The cement used was an ordinary portland cement with an equivalent percentage of Na_2O of 0.67. Its chemical compositions are given in Table I..

Table I. Chemical composition of cement (%)

CaO	SiO$_2$	Al$_2$O$_3$	Fe$_2$O$_3$	SO$_3$	MgO	Na$_2$O	K$_2$O	Ig.Loss
63.6	22.0	5.2	2.7	1.9	1.5	0.40	0.41	1.3

PRODUCTION OF MORTAR SPECIMENS FOR EXPANSION TEST
Two series of mortars were produced. A series of mortars were prepared with an aggregate : cement ratio of 2.25 and a water : cement ratio of 0.6. The 10 % of aggregate by mass was replaced by the reactive aggregate. In the other series of mortar, the 10% and 20 % of cement was replaced by silica fume and fly ash, respectively, to make mortars with reduced amounts of $Ca(OH)_2$ before immersion

in NaOH solution. Considerable decreases in the amount of $Ca(OH)_2$ in mortars were confirmed by DSC analyses.

A group of mortar bars, 25.3 by 25.3 by 285.5 mm, were immersed in 0.6N NaOH solution without $Ca(OH)_2$ after curing in water for 56 days at 20°C; the other embedded in the solid calcium hydroxide powder bed in 0.6N NaOH solution. Thus, in the former solution, only Na^+ and OH^- ions intruded into mortar bars, and in the latter, there were a great amount of solid calcium hydroxide as a source providing ASR reaction sites with Ca^{2+} ions. The 0.6N NaOH solution corresponds to the pore solution in cement pastes with a water : cement ratio of 0.5 which is produced by the use of a cement with an equivalent percentage of Na_2O of 0.83 according to an equation between OH^- ion concentration and the alkali content of cement established by Diamond (9). In order to reveal the relations between compositions of gels formed at different temperatures and their expansion behavior, a series of mortar specimens were immersed in the same solution after curing in sealed vinyl sacks at 38°C in a moist environment for 28 days.

PROCEDURE OF BSE-EDS ANALYSIS

As described later, mortar bars in NaOH solution with solid calcium hydroxide showed almost the same expansion behavior as those in NaOH solution without calcium hydroxide during early periods of immersion. After a certain immersion time, expansion curves for the former deviated from those of the latter. At 20°C, at about two months after the initiation of divergence in expansion curves (at the immersion time of 182 days), mortar slices for BSE-EDS analysis were cut from the middle portion of mortar bars with and without a mineral admixture, and then dried using ethanol replacement, followed by vacuum drying at a room temperature overnight. The dried slices were impregnated with ultra-low viscosity epoxy resin under vacuum, and then polished using silicon carbide abrasive papers. In mortars immersed in the solutions at 38°C, mortar samples for BSE-EDS analyses were made at the immersion time of 21, 63 and 133 days. The final polished surface was then sputter coated with a 30 nm thick layer of gold-palladium alloy.

The polished mortar samples were examined using an Hitachi S-2250N SEM equipped with a back scatter detector and Horiba EMAX-5770W energy dispersive X-ray analyzer. The SEM was operated at 25 KeV. A standard magnification of 500 was used in EDS analyses at 10 to 20 spots selected within gel areas, 10 to 20 μm wide and about 150 μm long, within cracks in reactive grains in mortars. The widths of gel areas for EDS analyses are large enough for the purposes of this study.

RESULTS AND DISCUSSION

ALKALI-SILICA REACTION PROCESS IN ADMIXTURE-FREE MORTARS AT 20°C.

It was found from SEM examinations on the polished surfaces of mortars with reactive aggregate that cracks with 10 to 20 μm wide went through central parts in many reactive grains in mineral admixture-free mortars in NaOH solution with and without $Ca(OH)_2$. A SEM micrograph showing such a grain is given in Fig.1.

Smaller grains found in Fig.1 are non-reactive standard sand. No signs of the ASR was found in non-reactive sand grains. Fig.2 shows an enlarged BSE image for a part of wide cracks in Fig.1. The wide cracks are found to be impregnated with the massive ASR gel. Fig.3 shows X-ray dot maps for Si, Na, K and Ca in the same areas as shown in Fig.2. The dot maps indicate that the gel contained less silica than the surrounding unreacted portions and considerable amounts of calcium existed in the gel. It was also found from EDS spot analyses at 14 spots on the gel areas that the compositions of the gel varies from spot to spot to some extent. The fact that gels existed within other fine cracks in the reactive grain, was also confirmed by the dot maps for Ca and Si (Fig.3). However, it was impossible to confirm the network of fine cracks in the reactive grain in SEM micrographs (Fig.2). Little ASR gels were found from careful SEM examinations along the periphery of all the reactive grains which appeared on the polished surfaces of samples. These results suggest that the pore solution intruded into fissures within reactive grains rather than reached to the cement paste-aggregate interfaces. The presence of many fissures in the original reactive gains (14) indicates that potentially reactive surface areas within reactive grains are far greater than those along their apparent periphery.

It was found from some preliminary examinations that considerable amounts of calcium were taken into ASR gels produced within reactive grains during polishing by the use of water in the production of mortar samples for BSE-EDS analyses. The results of BSE-EDS analyses presented in Figs. 1 to 3 were obtained with mortar samples which were intentionally treated with water so as to make ASR gel formation visually more conspicuous through X-ray dot maps for Ca. Naturally, the quantitative BSE-EDS analyses, the results of which will be presented in the following, were made on mortar samples polished with the use of kerosene.

Lumley (14) reported that there were many fissures in the cristobalite aggregate gains used in this study. The fissures formed networks within grains as a whole. Lengths across areas surrounded by fissures in the networks were approximately 50 μm. An attention should be paid to the fact that average sizes of the areas surrounded by thin concentrated striped zones in the dot maps for Ca (Fig.3) approximately coincided with the sizes of areas surrounded by fissures in the original reactive aggregate grains.

From the results of SEM examinations and EDS analyses described above, the following supposition may be proposed for the process of ASR reaction in mineral admixture-free mortars with the calcined flint in NaOH solution with and without $Ca(OH)_2$. Not so active ASR reaction must have occurred in mortars during the first 56 days curing in water at 20°C and 28 days curing in a moist environment at 38° C. As pore solutions intruded into reactive grains through fissures, ASR reaction must have taken place at silica surfaces along fissures. Naturally, alkaline pore solutions might intrude into fissures in the reactive aggregate grains during the pre-curing periods. As the result, ASR gels were produced within the fissures. The expansion of gels widened fissures, some of which might grow up to considerable wide cracks (Fig.1). Widening of fissures within reactive grains may lead to the overall expansion of mortar bars. As shown in Fig. 4, the extension of a wide crack in a reactive grain into the cement paste matrix appears to support the validity of the supposition on the expansion process in the mortars .

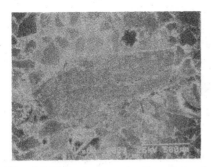

Fig. 1 SEM micrograph of a calcined
flint grain in mortar in NaOH
solution at 20°C.

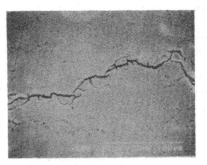

Fig.2 BSE micrograph relatively
wide strips of ASR gel
in mortar.

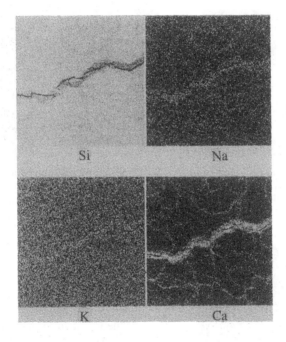

Fig. 3 X-ray dot maps for Si, Na, K and Ca in the areas around
a wide ASR gel strip.

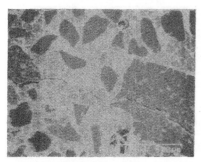

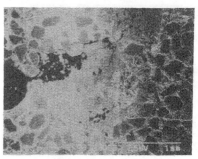

Fig. 4 Cracks caused by expansion
of ASR gel formed in a
reactive aggregate grain.

Fig. 5 Evacuation of portions
in a reactive aggregate
grain in mortar.

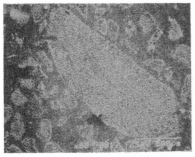

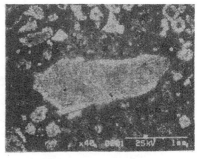

Fig. 6 SEM micrograph of a reactive
aggregate grain in with 20%
fly ash in NaOH solution.

Fig. 7 SEM micrograph of a reactive
aggregate grain in mortar
with 10% silica fume.

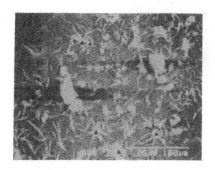

Fig. 8 EDS spot analysis for
relatively large flakes on
the Polished Surfaces
of Mortars.

ALKALI-SILICA REACTION PROCESS IN ADMIXTURE-FREE MORTARS AT 38°C.

As shown in Fig.5, several cracks with 10 to 20 μm wide within many reactive aggregate grains existed. These cracks ran across the grains in the same manner as in mortars stored at 20°C. It is found from Fig.5 that parts of reactive grains were impregnated with the epoxy resin. Thus, a conspicuous contrast between the features in reactive grains in mortars immersed in the NaOH solutions at 20°C and 38°C for 182 days and 133 days, respectively, was that the evacuation of large parts in reactive grain was found in many grains only in mortars immersed in the NaOH solutions at 38°C. The evacuation of parts of reactive grains indicates that the ASR sol formed within the grains permeated into the cement paste matrix through cracks. These features of reactive grains in mortars show that the ASR reaction followed by the absorption of water by gel far more rapidly progressed at 38°C than did at 20°C.

ALKALI-SILICA REACTION IN SILICA FUME- AND FLY ASH-CONTAINING MORTARS AT 20°C.

SEM examinations for each reactive aggregate grain found in the polished surfaces of mortars with a replacement of 10% silica fume and 20% fly ash for cement were made. In the SEM examinations, the formation of ASR gel ,and cracks were not confirmed within reactive grains in mortars in NaOH solution with and without $Ca(OH)_2$ (Figs.6 and 7). EDS spot analyses were made on flakes which mounted on silica surfaces, and at some spots on the polished silica substratum. The flakes were so small that all the results of EDS spot analyses were considered not to represent their compositions. However, judging from the results of spot analyses for relatively large flakes on the polished surfaces of mortars with a replacement of 10% silica fume for cement (Fig.8, Table II.), the flakes were supposed to be precipitates of NaOH.

Table II. Results of EDS analyses for relatively large flakes

Flake No.	Na_2O	Al_2O_3	SiO_2	SO_3	K_2O	CaO	Fe_2O_3
1	72.1	1.2	25.1	1.0	0.1	0.4	0.1
2	91.8	1.5	4.7	1.4	0.1	0.4	0.1
3	80.6	1.3	16.6	0.8	0.2	0.4	0.1
4	85.6	1.1	12.1	0.7	0.1	0.3	0.0

The concentration of sodium in reactive grains was considerably high. These results of EDS analyses and great differences in the SEM image for reactive aggregate grains between mortars with and without a mineral admixture (Figs. 1, 6 and 7) show that considerable amounts of pore solution permeated into reactive grains in admixture-containing mortars, resulting in the precipitation of NaOH during drying processes of samples preparation. The intrusion of considerable amounts of the pore solution into reactive grains in the mortars with an admixture indicates that the diffusion of silica out of reactive aggregate grains under the unlimited supply of NaOH from surrounding solution made the structure of reactive aggregate grains more porous. Another indication that such dissolution of silica also occurred even in standard sand grains used as a non-reactive aggregate was also obtained from the SEM image on smaller sand grains, as shown in Figs. 6 and 7.

The $Ca(OH)_2$ in mortar is considered to increase expansions of mortars by providing ASR gel with calcium, as proposed by Thomas (11). It will also be concluded in another paper by the present authors (15) that reductions in calcium content in ASR gel produced in mortars with fly ash was responsible for reduced expansions of mortars in a closed system without the ingress of water from surroundings. The other function of $Ca(OH)_2$ in duplex films around reactive aggregate grains is to produce expansive ASR gels within reactive aggregate grains by hindering silica from diffusing out of them (8,9). Such a concept on the role of $Ca(OH)_2$ in expansion of concrete due to the ASR reaction has also been proposed by Diamond (9) and Struble (10) on the basis of simple dissolution of the opal in a model mortar system without cement hydration products. No detection of expansive ASR gels in mortars with 10% silica fume and 20% fly ash in the SEM-EDS examinations in this study supports the validity of the function of $Ca(OH)_2$ as a blocker of leakage of silica from reactive aggregate in the ASR expansion of concrete. These results appear to evidence the significance of the local deposits of calcium hydroxide in the interfacial zone surrounding aggregates in the formation of ASR gel in actual concretes, which has been suggested by Chatterji (8) and Diamond (9).

COMPOSITIONS OF ASR GELS AND EXPANSION OF MORTARS AT 20° C

Fig.9 shows expansion curves for mortars immersed in 0.6N NaOH solution with and without $Ca(OH)_2$ after curing in water at 20°C for 56 days. The mortar specimens in the NaOH solutions started expanding around 80 days after immersion. However, mortars in 0.6N NaOH solution with $Ca(OH)2$ expanded at a greater rate after the immersion time of about 135 days. As shown in Fig.9, mortars with fly ash and silica fume showed expansions of about 0.1% even at the immersion time of about one year.

In Fig.10 ,the $(Na_2O + K_2O)/SiO_2$ mole ratios are plotted against CaO/SiO_2 ratios using the data obtained by EDS spot analyses for the gel in cracks within reactive grains in mortars in NaOH solution. In these analyses, five different reactive grains having wide cracks with enough sized gel areas for EDS analyses were selected. The compositions of the gel varied widely even in the massive gel areas with 10 to 20 μm wide and about 150 μm long in cracks in reactive grains. Plots of the $(Na_2O+K_2O)/SiO_2$ and CaO/SiO_2 mole ratios in the gel areas in mortars in NaOH solution with and without $Ca(OH)_2$ are found to widely scatter in their own characteristic patterns. As shown in Fig.10, in all the grains selected, the ranges of $(Na_2O + K_2O)/SiO_2$ ratio in the gels were considerably wide compared to the ranges of CaO/SiO_2 ratio. These results suggest that Ca^{2+} ions as well as Na^+ and K^+ ions diffused into the ASR gels through complicated routes, resulting in great variations in their compositions between the massive gel areas to be selected for EDS analyses.

In Fig.11, the plots of $(Na_2O+K_2O)/SiO_2$ and CaO/SiO_2 ratio for mortars in NaOH solution with $Ca(OH)_2$ are provided. A conspicuous contrast in composition

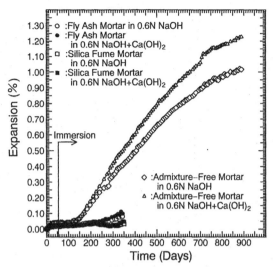

Fig. 9 Expansion curves for mortars in 0.6N NaOH solution with and without Ca(OH)$_2$ at 20° C.

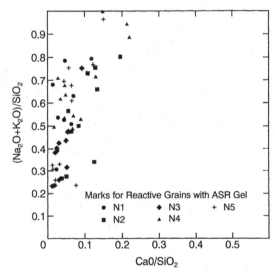

Fig. 10 EDS spot analyses for ASR gel.

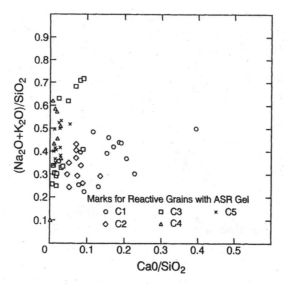

Fig. 11 EDS spot analysis for ASR gels.

between the gels in mortars in NaOH solution with and without $Ca(OH)_2$ is that the range of variations in the $(Na_2O + K_2O)/SiO_2$ ratio for mortars in the solution with $Ca(OH)_2$ was considerably smaller than for mortars in the solution without $Ca(OH)_2$. The formation of such relatively homogeneous gels in mortars in NaOH solution with $Ca(OH)_2$ may be related to the unlimited supply of Ca^{2+} ions from the surrounding solution. The $(Na_2O+K_2O)/SiO_2$ ratios of most plots for the gel in mortars in NaOH solution with $Ca(OH)2$ were below 0.6, but about half of them above 0.6 in mortars in NaOH solution without $Ca(OH)_2$.

In order to more explicitly show the differences in composition between the ASR gels formed in mortars in the two different solutions, the results are plotted in the ternary phase diagram, as shown in Fig.12. The plots in the diagram stand for the average of mol percentages of CaO, (Na_2O+K_2O) and SiO_2 for gels in an individual grain. As indicated in the diagram, the concentrations of CaO in the gels in mortars in NaOH solution with $Ca(OH)_2$ were slightly greater than in mortars in NaOH solution without $Ca(OH)_2$ on the average. On the other hand, the concentrations of alkalis in the gels in mortars in NaOH solution are certainly higher than in the gels in mortars in NaOH solution with $Ca(OH)_2$. As shown in Fig.9, mortars in NaOH solution with $Ca(OH)_2$ steadily expanded at a greater rate than they did in NaOH solution without $Ca(OH)_2$. Since compositions of the gels in several grains overlap between mortars in both of the NaOH solutions, it is ambiguous whether only a little higher concentration of CaO in these gels in mortars in NaOH solution with $Ca(OH)_2$ increased its expansive capacity. However, the increase in alkalis content in the gels in mortars in NaOH solution without $Ca(OH)_2$

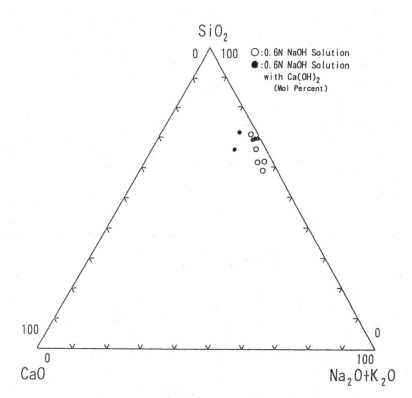

SiO₂

0 ⟋⟍ 100   O : 0. 6N NaOH Solution
            ● : 0. 6N NaOH Solution
                with Ca(OH)₂
                (Mol Percent)

100   0

0   100
CaO   Na₂O+K₂O

Fig. 12 Ternary phase diagram for ASR gels in several reactive grains in mortars immersed in the NaOH solutions for 182 days at 20°C.

must have decreased its expansivity. It has been considered that ASR gels rich in soda become too fluid after a little water uptake to be able to continue exerting much swelling pressure (16). These conclusions obtained in this study are also consistent with a notion that the viscosity of ASR gel relating to expansive stress relaxation decreases with increasing alkalis content (17).

COMPOSITIONS OF ASR GEL AND EXPANSION OF MORTARS AT 38°C

Fig.13 shows expansion curves for mortars without mineral admixture in NaOH solution with and without Ca(OH)₂ at 38°C. Little expansions occurred in all the mortars before immersion in the NaOH solutions. Immediately after immersion in the NaOH solutions, the measured expansions for mortars rapidly increased with time. Around 60 days after immersion, the expansion rate for mortars in NaOH solution started to decrease. Eventually, however, the presence of solid Ca(OH)₂ in NaOH solution had little effects on expansion for mortars at 38°C.

As previously described, reacted portions in grains in mortars immersed in the NaOH solutions for 133 days at 38° C were evacuated. However, in mortar

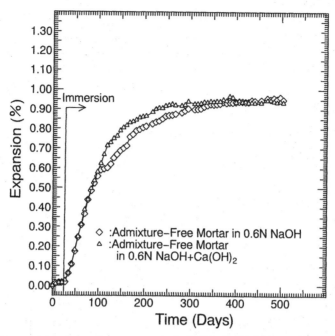

Fig. 13 Expansion curves for mortars in 0.6N NaOH solution with and without Ca(OH)$_2$ at 38°C.

samples for BSE-EDS analyses which were made from mortars immersed in the NaOH solutions for 21 and 63 days, sufficient amounts of the massive ASR gel to be analyzed were found within relatively wide cracks in reactive aggregate grains.

The average compositions for the gel in each reactive aggregate grain are plotted in the ternary phase diagram, as shown in Figs. 14 and 15. Two significant results that there were little differences in the composition of the gel formed between mortars immersed in NaOH solution with and without Ca(OH)$_2$ and that the composition of the gel did not change with time during the period of 21 to 63 days (Fig. 14), were derived from the plot in the ternary phase diagram. Such results indicate that the gel formed at early stages of the chemical reaction expanded absorbing only water without taking in Na$^+$ and K$^+$ ions during the most active progress of expansion, and little Ca^{2+} ions diffused into the gel during the period of 21 to 63 days (Fig.14).

EFFECTS OF TEMPERATURE ON ASR REACTION

In order to compare the composition of gels formed in mortars in the NaOH solutions at 20°C and 38°C at an expansion level of about 0.2%, the average composition of the gel in mortars immersed in the solutions at 20°C and 38°C for

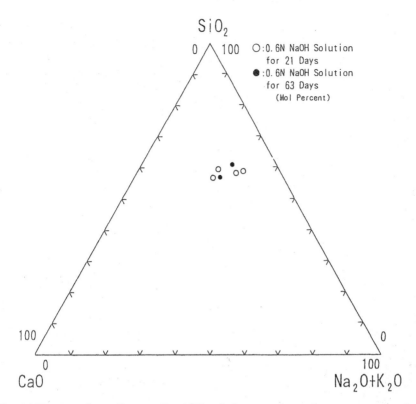

Fig. 14 Ternary phase diagram for ASR gels in several reactive aggregate grains in mortars immersed in 0.6N NaOH solution for 21 and 63 days at 38° C.

183 and 21 days, respectively, are tabulated in the ternary phase diagram, as shown in Fig.16. A conspicuous difference in the composition of gel between mortars stored at the two different temperatures is that the calcium contents of the gel formed at 38°C were considerably greater than at 20°C on an average. Comparatively great amounts of gel must have been produced during short periods at 38°C. Thus, the above result also indicates that Ca^{2+} ions far more rapidly diffused into the gel formed at 38°C. In the rapid formation of the gel, only limited times were given for Ca^{2+} ions to diffuse into the gel. As the result, there might be little differences in the composition between the gels in mortars in NaOH solution with and without $Ca(OH)_2$ at 38°C (Fig.16). The elevation of temperature also accelerated the absorption of water by gel, leading to the rapid expansion of mortars at 38°C (Fig.13). The rapid formation of high alkaline gel followed by the rapid intrusion of water brought about the transformation of gel to sol during relatively short periods, resulting in the termination of expansion of the mortars at early ages as compared to mortars at 20°C.

The ASR gel slowly formed at 20°C, and the calcium contents of the gel were considerably smaller than those of the gel formed at 38°C as a whole (Fig.16).

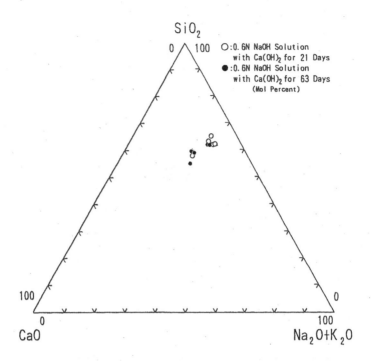

Fig. 15 Ternary phase diagram for ASR gels in several reactive aggregate grains in mortars immersed in 0.6N NaOH solution with Ca(OH)$_2$ for 21 and 63 days at 38°C.

However, taking into consideration the fact that the mortars continued to steadily expand for long times at 20°C, it may be concluded that the calcium content of ASR gel did not so explicitly influence the expansivity of the gel as the alkalis content of it at least within the range of CaO content in the gels formed under a condition of the unlimited supply of a strong NaOH solution. A similar notion on the effect of calcium in synthetic gels on their expansive characteristics was obtained by Struble and Diamond (18).

CONCLUSIONS
There are many fissures in the cristobalite aggregate grains used in this study. The ASR reaction appears to have occurred at silica surfaces along fissures, resulting in the production of the massive ASR gel within wide cracks in the reactive grains. On the basis of the supposition that widening of fissures filled with the gel within the grains lead to the overall expansion of mortars, the following conclusions on the effects of the composition of the ASR gel on the expansion of mortars were obtained.
(1) The intrusion of NaOH solution into reactive aggregate grains without production of the expansive ASR gel in mortars with 10% silica fume and 20% fly ash supports the validity of the supposition that the presence of Ca(OH)$_2$ is the prerequisite for the formation of expansive ASR gel within reactive aggregate in mortars.

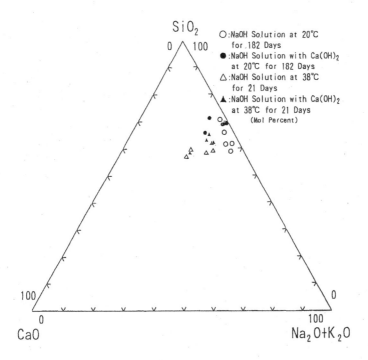

Fig. 16 Ternary phase diagram for ASR gels in several reactive aggregate grains in 0.6N NaOH solution with and without Ca(OH)₂ for 182 days at 20°C and for 21 days at 38°C.

(2) The increase in the alkalis content in ASR gel in mortars decreased its expansivity.
(3) The gel formed at early stages of the chemical reaction expanded absorbing only water without taking in sodium and potassium ions during the most active progress of expansion at 38°C.
(4) The calcium content of ASR gel did not so explicitly influence the expansivity of the gel as the alkali content of it at least within the range of calcium contents in the gel formed in mortars under the unlimited supply of a strong NaOH solution.

REFERENCES

[1] Idorn, G.M.,"Studies of Disintegrating Concrete-Part 1",p. 77 in Progress Report N2, Danish Nat. Inst. of Build. Res. and Acad. of Tech. Sci. Committee on Alkali Reactions in Concrete, Copenhagen, 1961.

[2] Knudsen, T. and Thaulow, N.,"Quantitative Microanalyses of Alkali-Silica Gel in Concrete", Journal of Cement and Concrete Research, 5[5] 443 - 54(1975).

[3] Regourd, M., Hornain, H. and Poitevin, P.,"The Alkali Aggregate Reaction-Concrete Microstructural Evolution" S252/35 in Proc. the 5th Int. Conf. on Alkali-Aggregate Reactions in Concrete, Cape Town,1981.

[4] Diamond, S.,"Alkali Reactions in Concrete-Pore Solution Effects", pp.155-166 in Proc. 6th. Int. Conf. Alkalis in Concrete, Copenhagen,1983.

[5] Powers, T.C. and Steinouer, H.H.,"An Investigation of Some Published

Researches on the Alkali Aggregate Reaction Part 1: The Chemical Reaction and Mechanism of Expansion", Journal of the American Concrete Institute, **26** [6] 497 - 516(1955).

[6] Powers, T.C. and Steinouer, H.H.,"An Interpretation of Some Published Researches on the Alkali Aggregate Reaction Part 2 : A Hypothesis Concerning Safe and Unsafe Reactions with Reactive Silica in Concrete", Journal of American Concrete Institute,**28**[8] 785 - 811(1955).

[7] Kawamura, M., Takemoto, K. and Hasaba, S.,"Application of Quantitative EDXA Analyses and Microhardness Measurements to the Study of Alkali-Silica Reaction Mechanisms", pp.167-174 in Proc. the 6th Int. Conf. Alkalis in Concrete, Copenhagen, 1983.

[8] Chatterji, S., Thawlow, N., Jensen, A.D., and Christensen, P.,"Mechanisms of Accelerating Effects of NaCl and Ca(OH)2 on Alkali-Silica Reaction", pp. 115-119 in Proc. of the 7th Int. Conf. on Alkali-Aggregate Reaction in Concrete, Ottawa, 1986.

[9] Diamond, S., "ASR-Another Look at Mechanisms", pp. 83-94 in Proc. of the 8th Int. Conf. on Alkali-Aggregate Reaction in Concrete, Kyoto, 1989.

[10] Struble, L.J., "The Influence of Cement Pore Solution on Alkali Silica Reaction ", Ph. D. Thesis, Purdue University, 1987.

[11] Thomas, M.D.A., Nixon, P.J., and Pettifer, K.,"The Effect of Pulverized Fuel Ash with a High Total Alkali Content on Alkali Silica Reaction in Concrete Containing Natural U.K. Aggregate", pp. 919-940 in Proc. of the 2nd CANMET/ACI Int. Conf. on Durability of Concrete, Vol.2, A.C.I., Detroit, 1991.

[12] Scrivener, K.L., and Monteiro, P.J.,"The Alkali-Silica Reaction in a Monolithic Opal", Journal of the American Ceramic Society, **77**[11] 2849-56, (1994).

[13] Laing, S.V., Scrivener, K.L. and Pratt, P.L.,"An Investigation of Alkali-Silica Reaction in Seven-Year Old and Model Concretes Using S.E.M. and E.D.S." pp.579-586 in Proc. the 9th Int. Conf. on Alkali-Aggregate in Concrete, London,, 1992.

[14] Lumley, J.S.,"Synthetic Cristobalite as a Reference Reactive Aggregate", pp.561-566 in Proc. of the 8th Intl. Conf. on Alkali-Aggregate Reaction , Kyoto,, 1989.

[15] Kawamura, M., Arano, N. and Terashima, T.,"Mechanisms of the Suppression of ASR Expansion by Fly Ash from a View Point of Gel Compositions" in Proc. of the Sydney Diamond Symposium, Honolulu, 1998.

[16] Diamond, S., Barneyback, R.S. and Struble L.J., "On the Physics and Chemistry of Alkali-Silica Reactions", S252/22 in Proc. of the 5th Intl. Conf. on Alkali-Aggregate Reaction in Concrete, Cape Town,1981.

[17] Helmuth, R. and Stark, D.,"Alkali-Silica Reactivity Mechanisms", pp.131-208 in Material Science of Concrete III, Edited by J. Skalny,The American Ceramic Society, 1992.

[18] Struble, L.J. and Diamond, S., "Swelling Properties of Synthetic Alkali Silica Gels", Journal the American Ceramic Society, **64**[11] 652-55(1981).

MECHANISMS OF THE SUPPRESSION OF ASR EXPANSION BY FLY ASH FROM THE VIEW POINT OF GEL COMPOSITION

Mitsunori Kawamura
Noriyuki Arano
Tsutomu Terashima
Department of Civil Engineering, Kanazawa University, 2-40-20 Kodatsuno, Kanazawa, Japan

ABSTRACT

This study aims at discussing the mechanisms of ASR expansion by fly ash. Fly ash-free mortars indicated a large expansion of 0.18%, but, little expansion occurred in mortars with 20% fly ash at 38°C. BSE images showed that the gels were produced within cracks in the reactive aggregate grains even in mortars with 20% fly ash. The results of EDS measurements showed that the gels in mortars with fly ash had lower contents of CaO than in fly ash-free mortars. It is considered that the gels having low CaO contents in mortars with fly ash resulted from the reduction of Ca^{2+} ions supplied to alkali-silica gels. This reduction in calcium content in gels appears to lead to the decrease of their viscosity. It is concluded that the suppression of ASR expansion by fly ash is caused by the formation of gels having low viscosity. The addition of fly ash to mortars slightly reduced OH^- ion concentration in the pore solution. However, the OH^- ion concentrations were not so low as for the alkali-silica reaction to be completely suppressed.

INTRODUCTION

For the suppression of ASR expansion in hardened concrete, the addition of some mineral admixtures has been accepted as a conventional method. It is reported that the addition of mineral admixture inhibits as the progress of alkali-silica reaction due to the reduction of OH^- ion concentration in the pore solution with pozzolanic reaction (1). However, the relationship between the ASR process and expansion of concrete still remains ambiguous in relation to the mechanisms of the suppression of ASR expansion by the addition of mineral admixtures.

The purpose of this study is to discuss the mechanisms of the suppression of ASR expansion by fly ash from the view point of the composition of alkali-silica gels on the basis of measurements of the expansion of mortars containing fly ash, and the analysis of the pore solution and the alkali-silica gels formed.

EXPERIMENTAL

Material

The reactive aggregate used was calcined flint (C.F.), which was supplied by Blue Circle Industries PLC. This reactive aggregate, with a size fraction of 2.35 mm to 1.00 mm, is mostly pure cristobalite. The reactive aggregate consists of about 96% tetragonal cristobalite, and about 2% quartz and tridymite. A mean crystallite size is approximately 60nm. The density of the C.F. grains was 2.12 (2). The potential alkali reactivity of C.F. determined according to the ASTM Chemical Test, C289 is presented in Table I. The Japanese standard sand, which mainly consists of quartz, was used as a non-reactive aggregate. A high alkali portland cement with the equivalent Na_2O percentage of 1.1 was used. The chemical composition of the high alkali cement and fly ash are given in Table II.

Table I. Alkali Reactivity of the Reactive Aggregate (ASTM Chemical Test)

Rc mmol/l	Sc mmol/l
70	1063

Table II. Chemical Composition of Cement and Fly Ash

	Ig.loss	SiO_2	Al_2O_3	Fe_2O_3	CaO	MgO	Na_2O	K_2O	Na_2Oeq
Cement	1.2	20.1	4.7	3.2	62.5	2.5	0.41	1.1	1.1
Fly ash	1.5	63.1	28.8	2.6	1.1	0.63	0.39	1.0	1.1

Mix Proportion of Mortars

The mix proportion of mortars were aggregate / (cement + F.A.) = 0.50, water / (cement + F.A.) = 0.55, C.F. / (C.F. + non-reactive aggregate) = 0.30, F.A. / (cement + F.A.)= 0, 0.05, 0.10, 0.20, 0.30.

Expansion Test

Mortar bars 40 by 40 by 160 mm were cured in a container maintained at 38°C throughout the expansion tests. Measurements of their length changes with time were started immediately after demoulding.

Pore Solution Analysis

The pore solution was expressed by the high pressure apparatus from mortar cylinders of $\phi 50$ mm by 100 mm. The mortars were prepared at a F.A./(cement + F.A.)ratio of 0 and 0.20. The extracted pore solutions were analyzed for OH^- ion concentration by titration against hydrochloric acid to the phenolphthalein end point. Other ions such as Cl^-, SO_4^{2-}, Na^+, K^+, Ca^{2+} were also analyzed by standard methods. The cation concentrations were equally balanced with the anion concentration in the pore solution.

BSE-EDS Analysis

At the age of 120 days, the polished surfaces of mortars which had been dried in a vacuum drying oven at a room temperature, were used for BSE-EDS analysis. The samples were polished using kerosene as a lubricant. BSE-EDS anlysis was made by means of scanning electron microscope with backscattered electrons (BSE) and energy-dispersive X-ray (EDS) systems. The compositions of ASR gels were accumulated from the spectrum for the K_α peaks of Na, K, Ca, Si, Mg, Al and Fe.

RESULTS AND DISCUSSION

Expansion Behavior of Mortars

Figure 1 shows expansion curves for mortars containing fly ash and the reference at 38 ℃ . Admixture-free mortars showed large expansions immediately after demoulding. Expansion of admixture-free mortars increased with time and reached a value of 0.18% around 60 days. The mortars with fly ash showed decreased expansions. Furthermore, little expansions occurred in the mortars containing 20% fly ash.

Pore solution Composition

Changes in the OH^- ion concentration with time in the pore solutions expressed from fly ash-free and 20% fly ash-bearing mortars are given in Fig. 2. The OH^- ion concentrations in the pore solution with fly ash-free mortars were proportionally decreased with time after the age of 3 days. It is found from Fig.2 that the addition of fly ash to mortars reduced OH^- ion concentrations in the pore solution. However, the OH^- ion concentrations are considered to be high enough to cause the ASR in the mortars with and without fly ash. The ASR must have occurred even in mortars with fly ash showing little expansion.

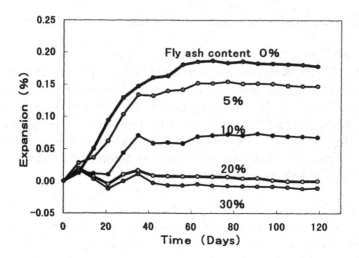

Fig.1. Expansion curves for mortals containing fly ash.

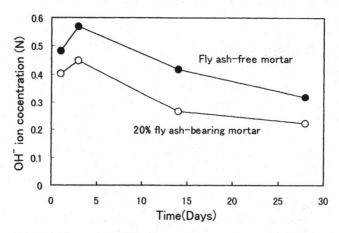

Fig.2. Changes in OH⁻ ion concentration with time in the pore solution in mortars

Alkali-Silica Gel Composition

Figure 3 shows the SEM micrographs for polished surfaces of mortars with and without fly ash at the age of 120 days. BSE micrographs for the section of a reactive aggregate grain and the interface of cement paste-aggregate grain are given in Fig.4. From the SEM and BSE observations, it is found that the alkali-silica gels were produced within the reactive aggregate grains even in mortars with 20% fly ash. EDS mapping images on the same field of BSE micrographs with and without fly ash showed the components of the alkalies, Si and Ca in the gels. It is confirmed that the gels produced were typical alkali-silica gels containing a Ca component.

Chemical composition of alkali-silica gels in fly ash-free and 20% fly ash-bearing mortars from the EDS spot analyze are presented in Table III. EDS spot analyses were carried out on 10~20 spots in each grain in mortars; 4 to 5 reactive aggregate grains were selected for the EDS spot analysis. Table 3 shows the average composition and the ranges of variations of compositions for all analyzed spots. The gels in mortars with 20% fly ash have the chemical compositions of SiO_2 79mol%, (Na_2O+K_2O) 15mol% and CaO 4.1mol%; the gels in fly ash-free mortars : SiO_2, 78mol%, (Na_2O+K_2O) 12mol%, CaO 8.9mol%. The alkali-silica gels in fly ash-bearing mortars are found to have lower CaO contents than in fly ash-free mortars.

Figure 4 shows the relationship between the CaO / SiO_2 and the (Na_2O+K_2O) / SiO_2 mole ratio. The (Na_2O+K_2O) / SiO_2 mole ratios of the gels in fly ash- bearing mortars were similar to those of fly ash-free mortars. The (Na_2O+K_2O) / SiO_2 mole ratios in both mortars were not different from each other. On the other hand, the gels in fly ash-free mortars had larger CaO contents than those in 20% fly ash-bearing mortars.

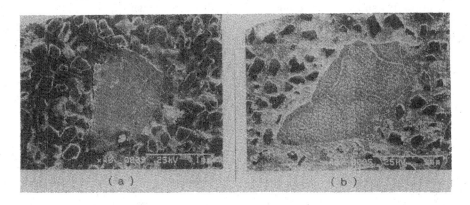

Fig.3. SEM micrographs for polished surface of mortars;
(a) fly ash-free mortar, (b) mortar with 20% fly ash.

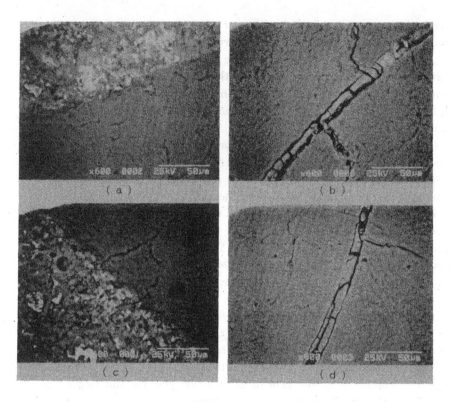

Fig.4. BSE micrographs for polished surface of mortars;
(a)(b) fly ash-free mortar, (c)(d) mortar with 20% fly ash.

Table III. Chemical Composition(mol%) of Alkali-silica Gels by EDS Analysis

| | Gela in fly ash -free mortal | | Gels in 20% fly ash -bearing mortar | |
	Average %	Range %	Average %	Range %
Na_2O	2.9	1.8-3.6	6.4	5.2-7.1
MgO	0.9	0.3-1.7	0.7	0.5-1.0
Al_2O_3	0.5	0.1-1.1	0.8	0.2-2.3
SiO_2	78	67-85	79	74-82
SO_3	0.3	0.1-0.5	0.2	0.1-0.3
K_2O	8.6	7.0-11	8.2	7.3-9.5
CaO	8.6	1.4-18	4.1	1.6-8.0
Fe_2O_3	0.3	0.1-0.5	0.4	0.1-0.7

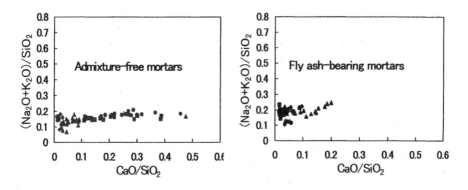

Fig.5.　Chemical composition of alkali-silica gels.

Mechanisms of the Suppression of ASR Expansion by Fly Ash

As described above, the addition of 10% and 20% fly ash greatly decreased the ASR expansions for mortars. It is found from the analysis of the pore solutions (Fig.2) that the alkali-silica reaction must be depressed by a considerable reduction in the OH^- ion concentration. However, the presence of ASR gels was confirmed even in 20% fly ash-bearing mortars showing little expansion. Furthermore, the $(Na_2O+K_2O) / SiO_2$ mole ratios in the gels in mortars with 20% fly ash were a little greater than in admixture-free mortars on an average. These facts indicate that the suppression of ASR expansion due to the addition of fly ash can not be explained only by the reduction in OH^- ion concentration.

It has been demonstrated in another series of SEM-EDS examinations (3) that, in mortars containing the calcined flint grains with many fissures, the pore solution intruded into the reactive grains through fissures and the alkali-silica reaction took place mainly at silica surfaces along the fissures.

As shown in Fig.5, on an average, the CaO contents in gels in mortars with fly ash were considerably lower than in admixture-free mortars. The continuous supply of Ca^{2+} ions may lead to the production of the ASR gels with high CaO contents (4). However, considerable amounts of $Ca(OH)_2$ in the interfacial transition zone around reactive grains must have been consumed by the pozzolanic reaction in mortars with fly ash. The reduction in amount of the $Ca(OH)_2$ as a source of Ca^{2+} ions must have resulted in the production of the gels with lower CaO / SiO_2 mole ratios on an average. The gels with lower CaO contents have lower viscosity, leading to greater expansive stress relaxation. Thomas et al.(5) have reported that the gel found in fly ash-containing concrete generally had a high-alkali and low-calcium content.

We should pay attention to the fact that the addition of fly ash greatly increased only the Na_2O content in the gels formed within reactive grains (Table 3). The reason for the increases in Na_2O content in the gels and its effects on expansivity of the gels are not clear at present. However, considering that the Na_2O content of the fly ash used was slightly less than that of the cement (Table 2) and the K_2O content of the cement was high, this result may be worthy of notice.

The conception concerning the mechanisms of the suppression of ASR expansion by the addition of fly ash seems to contradict that proposed by the senior author in this paper (6). However, it should be noted that the ASR reaction and gel formation processes in mortars containing the calcined flint aggregate with many fissures were different from those in mortars made by the use of an opaline rock used in our previous study (6).

CONCLUSIONS

1) The addition of fly ash to mortars reduced the OH^- ion concentration in the pore solution to some extent, but, little expansion occurred in 20% fly ash-bearing mortars.

2) Alkali-silica gels were produced within cracks in the reactive aggregate grain even in mortars with 20% fly ash.

3) Alkali-silica gels in mortars with 20% fly ash have lower content of CaO than those in fly ash-free mortars, showing the reduction in Ca^{2+} ions supplied to the gels.

4) The suppression of ASR expansion by the addition of fly ash may be caused by the formation of alkali-silica gels having low viscosity due to the reduction in calcium content in gels.

RERERENCES

[1] R.F. Bakker, *5th Int. Conf. On Alkali-Aggregate Reaction*, S252/29 (1981).

[2] J.S. Lumley, *8th Int. Conf. On Alkali-Aggregate Reaction*, 561-566 (1989).

[3] M.Kawamura, N.Arano and T.Terashima, *Proc. of the Sydney Diamond*, Hawaii (1998).

[4] K.L. Scrivenar and P.J. Monteiro, *Journal of the American Ceramic Society*, 11, 2849-2856 (1994).

[5] M.D.A.Thomas, P.J.Nixson and K.Pettifer, *Proc. the 2nd CANMET/ACI Int. Conf. On Durability of Concrete*, Vol.2, A.C.I.,919-940 (1991).

[6] M.Kawamura, K.Takemoto and S.Hasaba, *Proc. of the 6th Int. Conf. Alkali in Concrete*, Copenhagen, 167-174 (1983).

CHLORIDE ION ACTIVITIES IN SYNTHETIC CONCRETE PORE SOLUTIONS: THEORETICAL MODELS AND POTENTIOMETRIC MEASUREMENTS

M.A. Climent, G. de Vera, A. Hidalgo[1], C. Andrade[1] and C. Alonso[1]
Departament D'Enginyeria de la Construcció, Obres Públiques i Infraestructura Urbana. Universitat d'Alacant, E-03080 Alacant, Spain.
[1] Instituto Eduardo Torroja (CSIC), C/Serrano Galvache s/n, 28033 Madrid, Spain.

ABSTRACT
    Concrete pore solution is a concentrated electrolyte, in which the activity of the chemical species has to be properly calculated in order to determine the equilibrium state of such solutions. In this work, chloride activity coefficients in synthetic concrete pore solutions have been obtained potentiometrically using different calibrated Cl⁻ ion selective electrodes; also a discussion about the performance of the different Cl⁻ ISE in alkaline media is made. Experimental results are compared with the calculated ones using the Pitzer specific ionic-interaction model. A good agreement between them is found. Solutions with alkalinities about KOH 0.5m have a practically constant chloride activity coefficient of 0.60 for the Cl⁻ concentration range between 0.001 to 4m. These low values are due to the introduction of KOH, that results in a decreasing of the solvent local viscosity around Cl⁻ ions and an increasing of ion-ion interactions.

INTRODUCTION
    The knowledge of the composition of concrete internal pore solution and of the chemical or electrochemical equilibria taking place between this electrolyte and solid phases of concrete or embedded steel is of great interest in the study of reinforced concrete durability. In order to proceed to its study, it results of crucial importance the correct establishment of the activities of the different ions, due to concentrations can not be used in a so high ionic strength typical of the high pore solution alkalinity.
    The law of mass action defines equilibrium constants in terms of activities of the species participating in the reaction. Activity of solution component i (a_i) is related to its concentration by the activity coefficient (f_i) as,

$$a_i = f_i \cdot c_i \qquad [1]$$

In a real electrolyte solution the separation of f_i from the unit value is due to ion-ion and ion-solvent interactions, that changes the effective concentration with which component i participates in chemical or electrochemical equilibria. For any particular solution, activity coefficients approach unity as the solution becomes more diluted. Since concentration of a solution component is usually expressed on the molal, molar or mole fraction scale, three different activity scales may be defined. In this work we have chosen the molal (m) scale due to the availability of experimental and calculated literature data. The activity coefficient of a particular component in the molal scale is generally represented by γ_i.

The activities of species present in solution can be experimentally determined by several methods. The most important among them are measurements of colligative properties. Nevertheless, in dealing with electrolyte solutions, some difficulties appear for measuring the activities of the different ions brought in solution by electrolyte dissociation, since it is impossible to dissolve either only positive or only negative ions. For a salt $M_{v+}A_{v-}$ which dissociates in solution to yield v_+ cations (M^{z+}) and v_- anions (A^{z-}) a mean activity coefficient ($\gamma_\pm$) may be experimentally determined. This mean coefficient is related to the particular activity coefficients of the cation (γ_+) and the anion (γ_-) by the equation:

$$\gamma_\pm^v = \gamma_+^{v_+} \cdot \gamma_-^{v_-} \qquad [2]$$

where $v = v_+ + v_-$ is the total number of ions produced by salt dissociation.

Electrochemical cells allow to determine particular ionic activity coefficients in salt solutions since it is possible to obtain half cells that give potentiometric response to variations in the activity of an ion, such as the ionic selective electrodes (ISE). Nevertheless these measurements need a previous calibration of the cells with standard solutions of known ionic activity. This fact is due to the difficulty in obtaining an standard ideal ISE and to uncertainties in the liquid junction and reference electrode potentials (1).

The aim of this work is to calculate and to determine potentiometrically the activity coefficient of chloride ion (Cl^-) in alkaline solutions simulating concrete inner pore solution. These data may be useful for those studying the interactions of Cl^- with reinforced concrete, for instance chloride binding or adsorption, determination of free Cl^- content thresholds to induce steel corrosion, etc. Moreover these measurements allow further assessments about the performance of chloride ISE in alkaline medium, for instance in analyzing solutions obtained by pore pressing (2-8) or using these electrodes as embeddable sensors in concrete (9-11).

EXPERIMENTAL

Test solutions were made with PA grade reagents and Millipore Milli-Q$_{PLUS}$ 185 ultrapure water. Three chloride solutions series were tested: NaCl, NaCl + KOH 0.5m and NaCl + KOH 0.5m + Ca(OH)$_2$ saturated, with 0.001, 0.01, 0.1, 0.2, 0.5, 1, 2, 4 and 5 NaCl molalities.

Two Cl$^-$ selective electrodes were used in this work: an Ingold type 15 213 3000 (referred as ISE A) and a Ag/AgCl electrode (referred as ISE B) prepared in the following way (12): AgCl was deposited on a 0.5 mm diameter silver wire from a 0.1M HCl solution by anodizing at a current density of 0.4 mA/cm^2, with another Ag wire acting as a cathode. The electric charge passed through the cell was 720 mC/cm^2. Ag wires were previously cleaned in ammonia for three hours and in distilled water for a night. The reference electrode was a calomel with saturated KCl, i. e. 4.2M (30), as internal solution (SCE). So, the electrochemical measuring cell may be expressed as

$$\text{SCE // NaCl (m) test solution / Cl}^- \text{ ISE} \qquad [3]$$

Solutions were saturated with solid AgCl in order to attain stable potential readings (13, 14), stirred and maintained at 25 ± 0.1°C in a thermostated jacket vessel. The SCE contacted the solution through a Luggin capillary filled with the same test solution, in order to avoid Cl$^-$ contamination. Once thermal equilibrium was reached the stirring was stopped and the potential measured with a ± 0.1 mV precision Crison 2002 mV-meter. Stable potential readings were achieved within ten minutes.

THEORETICAL METHODS OF CALCULATION OF SINGLE ION ACTIVITY COEFFICIENTS

Single ion activity coefficients for several alkaline and alkaline earth chloride aqueous solutions with concentrations up to 6m were calculated by Bates et al. (15). These authors followed some conventions (16, 17), and applied the hydration theory developed by Stokes and Robinson (18). This theory, takes into account the effect of ion-solvent interaction in terms of "hydration" in addition to the coulombic interactions considered by the Debye-Hückel theory (19). Of course, there are spectroscopic evidences (20) that the kinetic unit of the solute in many electrolyte solutions is an ion with several relatively firmly attached water molecules. The Pitzer approach, or the specific ionic-interaction theory (21,22,23,24,25,26), initially developed by Guggenheim (27), takes into account long-range (attractive and repulsive) and short-range electrostatic forces, the net effect of solvation, London's dispersion forces, and any other effects of similar range; for instance, permanent dipoles or multipoles. General equations are derived from statistical mechanics for electrolyte theory, that yields thermodynamic functions from knowledge of the interionic potentials of mean force and the radial distribution functions; so activity coefficients are expressed in virial expansion. This model is the first one to take into account the association of chloride ions with alkali and alkali earth metal ions, and it has proven to be useful in calculating activity coefficients in electrolyte mixtures.

At this point it should be necessary to define what ion pair means; ion pairs are flickering entities in that two given ions may be free at one moment, may be associated for a brief period, and then become free again. Still, at any instant a given fraction of the ions in a solution are paired and the time-averaged concentration of pairs will remains constant as long as the temperature, pressure, and stoichiometric composition of the solution are invariant. One may, therefore, use the methods of equilibrium thermodynamics to study ion association; then to calculate association constants in a multicomponent system two types of information are required; the free activity coefficient and the free concentration of each ion, as well as the stoichiometric concentration of all ions, or ionic force.

In this work, activity coefficients of free chloride ions have been calculated using the program PHRQPITZ (28). This program is a FORTRAN 77 computer program that makes geochemical calculations in brines and other electrolyte solutions to high concentrations. The Pitzer treatment of the aqueous model is based largely on the equations as presented by Harvie and Weare (29) and Harvie and others (30). An interactive input code called PITZINPT was used to construct input data sets to PHRQPITZ.

RESULTS AND DISCUSSION
Calibration of cells for determining chloride ion activities.

The experimental potential difference measured in cell [3] may be divided into the sum of four terms:

$$E_{obs} = \left(E_{ISE}^0 - E_{ref}\right) - \frac{R \cdot T}{F \cdot \log e} \cdot \log a_{Cl^-} - E_j + N \qquad [4]$$

where E_{ISE}^0 is the ISE standard potential at unit chloride activity, taken here as the standard potential of the redox couple AgCl/Ag, i. e. 0.22233 V vs. SHE (31); E_{ref} is the reference electrode potential, i. e. 0.2412 V vs. SHE (30); E_j is the liquid junction potential , the potential difference created by the liquid contact between the reference electrode inner solution (β phase) and NaCl test solution (α phase). This last term may be calculated by the Henderson equation (13, 32):

$$E_j = \phi^\beta - \phi^\alpha = \frac{\sum_i \dfrac{|z_i|}{z_i} \cdot u_i \cdot \left(c_i^\beta - c_i^\alpha\right)}{\sum_i |z_i| \cdot u_i \cdot \left(c_i^\beta - c_i^\alpha\right)} \cdot \frac{R \cdot T}{F} \cdot \ln \frac{\sum_i |z_i| \cdot u_i \cdot c_i^\alpha}{\sum_i |z_i| \cdot u_i \cdot c_i^\beta} \qquad [5]$$

where z_i, u_i and c_i are the charge, the ionic mobilities and the molar concentrations respectively of the ionic species in solution. In applying equation [5] we have assumed the common practice of substituting ionic mobilities by equivalent ionic conductivities extrapolated to infinite dilution, λ^0_i (32). This procedure has been shown to give adequate results for halide solutions with equal concentrations to those studied in this work (14).

The previous history of an electrode may affect considerably its potentiometric response thus preventing direct activity measurements. For this reason a constant N has been introduced in equation [4] to account for this separation from the ideal ISE behaviour.

Cell [3] may be calibrated for determining chloride single activities by measuring its potential (E_{obs}) for a series of NaCl solutions whose a_{Cl^-} are known. A linear regression of the plot of (E_{obs} + E_j) vs. log a_{Cl^-} yield the electrode slope (S), which must be close to its nernstian value (59.15 mV), and constant N from the origin intercept. This procedure is similar to those adopted by Shatkay and Lerman (13) and by Bagg and Rechnitz (14) for normalising the potentials measured with different halide ISE.

Table I shows the relevant data for calibration of both Cl⁻ ISE used in this work with NaCl aqueous solutions. An statistical estimation of the experimental error has been made with a 95% confidence interval. Second column contains the theoretical Cl⁻ activities calculated with Pitzer's model, which are in good agreement with the values of γ_{Cl^-} obtained by Bates, Staples and Robinson (15) applying the Hydration Theory. The linear regressions described in the precedent paragraph applied to the observed potential of both electrodes yield slopes very close to the nernstian value, small N constants and good correlation coefficients (r), indicating an almost ideal behaviour of both ISE in aqueous NaCl solutions. It is worth noting that ISE B prepared by electrolysis of a silver wire (see experimental section) is not stable in NaCl solutions of m > 2. In these media ISE B shows a fast deterioration process evidenced by a progressive falling of its potential to more negative values (it reaches -134 mV vs. SCE in 10 min.), and an appearance change from its characteristic dark colour to white. In this sense surface damage of solid membrane chloride ISE has been reported (14) for prolonged immersion in concentrated NaCl solutions, above 1m. It was explained as due to AgCl leaching by formation of soluble silver chloride complexes at high Cl^- concentrations (14).

TABLE I. Data relative to calibration of Cl^- ISE with NaCl solutions of known Cl^- activities (see text for details)

m_{NaCl}	- log a_{Cl^-} (theor.)	E_j (mV)	E_{obs} (mV vs. SCE) ISE A	E_{obs} (mV vs. SCE) ISE B	- log a_{Cl^-} (experimental) ISE A	- log a_{Cl^-} (experimental) ISE B
0.001	3.016	4.1	158.3	155.0	2.99±0.02	2.99±0.03
0.01	2.045	2.9	103.5	101.5	2.07±0.02	2.07±0.02
0.1	1.115	1.5	47.8	47.3	1.13±0.01	1.13±0.02
0.2	0.844	0.9	30.6	31.2	0.84±0.01	0.85±0.02
0.5	0.489	-0.1	10.5	11.5	0.49±0.01	0.50±0.02
1	0.220	-1.1	-4.7	-4.6	0.22±0.01	0.21±0.02
2	-0.059	-2.4	-19.6	-21.2	-0.05±0.01	-0.09±0.02
4	-0.363	-4.0	-37.5	-	-0.37±0.01	-
5	-0.470	-4.6	-43.7	-	-0.48±0.01	-
		S	60.6±0.7	59±1		
		N	0.2±0.9	0±2		
		R	0.99992	0.99982		

Experimental and calculated Cl^- activity coefficients in synthetic concrete inner pore solutions.

Once calibrated a particular cell (ISE + reference electrode), it may be used for determining Cl^- activities through equation [4], after calculating the corresponding E_j through equation [5]. On the other hand Pitzer's model allows to calculate Cl^- activity coefficients in mixture electrolytes. This provides a possibility of mutual validation of both methods (experimental and theoretical) of deriving γ_{Cl^-}.

Expression of concrete pore fluid (2, 33) is up to now the most reliable technique to study this otherwise inaccessible internal solution. Several authors (2-6) have investigated the composition of expressed solutions from pure cement pastes showing that they are mainly composed of alkaline hydroxides (especially KOH and NaOH in a lower proportion), with other species in much lower concentrations: SO_4^{2-}, Ca^{2+}, AlO_2^-, SiO_2 ... (2).

The hydroxide ion concentration found in Portland cement paste pore expressed solutions is high, ranging from 0.1M to 0.7M (2-6), while blended cement pastes usually produce lower OH^- contents (2-4). Mortars and concretes also yield less concentrated hydroxide solutions, ranging from 0.04M to 0.3M (6-8), but chloride contamination induces an increase in OH^- content in the expressed solutions (6-8), reaching values of about 0.4M to 0.55M for certain Cl^- containing ordinary Portland cement mortars (8). Taking all these facts into account, besides to the objective of assessing the performance of Cl^- ISE in alkaline media, we have chosen as test solutions KOH 0.5m and KOH 0.5m saturated with $Ca(OH)_2$.

Table II shows the Cl^- activity coefficients theoretically calculated through Pitzer's model and derived from the experimental potential measurements taken with the electrochemical cells corresponding to both ISE. A statistical stimation of the experimental error has been made with a 95% confidence interval.

TABLE II. Chloride activity coefficients derived from experimental data and theoretically calculated for synthetic concrete inner pore solutions.

m_{NaCl}	E_j	E_{obs} (mV vs. SCE)		γ_{Cl^-} (experimental)		γ_{Cl^-}
	(mV)	ISE A	ISE B	ISE A	ISE B	(theor.)
			NaCl + KOH 0.5m			
0.001	-4.6	146.0	139.7	2.25±0.09	2.6±0.2	0.6389
0.01	-4.6	117.7	112.2	0.66±0.02	0.75±0.04	0.6377
0.1	-4.6	61.6	50.2	0.56±0.02	0.84±0.04	0.6272
0.2	-4.7	42.1	31.8	0.58±0.01	0.86±0.03	0.6176
0.5	-4.8	18.6	6.8	0.57±0.01	0.91±0.03	0.5974
1	-5.1	4.0	-2.1	0.50±0.01	0.65±0.02	0.5791
2	-5.6	-14.2	-22.1	0.51±0.01	0.72±0.03	0.5679
4	-6.4	-34.8	-	0.58±0.01	-	0.5854
5	-6.7	-41.3	-	0.60±0.01	-	0.6036

m_{NaCl}	E_j	E_{obs} (mV vs. SCE)		γ_{Cl^-} (experimental)		γ_{Cl^-}
	(mV)	ISE A	ISE B	ISE A	ISE B	(theor.)
			NaCl + KOH 0.5m + $Ca(OH)_2$ sat.			
0.001	-4.6	138.1	123.4	3.0±0.1	4.9±0.3	0.6389
0.01	-4.6	114.5	110.9	0.74±0.03	0.79±0.05	0.6377
0.1	-4.6	59.0	49.0	0.61±0.02	0.88±0.04	0.6272
0.2	-4.7	42.0	27.3	0.59±0.01	1.02±0.04	0.6176
0.5	-4.8	18.7	7.8	0.57±0.01	0.87±0.03	0.5974
1	-5.1	1.7	-8.7	0.55±0.01	0.84±0.03	0.5791
2	-5.6	-15.5	-23.1	0.54±0.01	0.74±0.03	0.5679
4	-6.4	-33.9	-	0.56±0.01	-	0.5854
5	-6.7	-42.1	-	0.62±0.01	-	0.6036

A good agreement between calculated and experimental values of γ_{Cl^-} may be appreciated for ISE A in both electrolytic media, except for the less concentrated Cl⁻ solutions, equal or below 0.01m. This fact may be explained by the sensitivity of chloride ISE to OH⁻ ions. Commercial chloride ISE based on solid membranes are about 100 times more sensitive to Cl⁻ than to OH⁻ ions (1), but the potentiometric interference of the latter may be appreciable for the less concentrated Cl⁻ solutions in Table II. This fact must be considered when using Cl⁻ ISE to determine potentiometrically low chloride concentrations in concrete pore pressed solutions. Another point of interest in Table II is that ISE B, prepared by anodising a silver wire, behaves worse in alkaline media than the commercial solid membrane ISE A. The former reads systematically lower measured potentials than ISE B in both solutions series, thus yielding Cl⁻ activity coefficients higher than those theoretically calculated. This fact may be ascribed to a slow deterioration process of ISE B in alkaline media similar to that reported for Ag/AgCl electrodes embedded in Portland cement mortar specimens with different NaCl contents (11). The possible deteriorations suffered by ISE A and B, after performing the potential measurements corresponding to both solutions series in Table II, were investigated by checking their potentiometric responses in a pure NaCl solution of 1m: ISE A and B showed potentials that differed only in 2 mV and 5 mV respectively from those read before these measurements series (Table 1). This difference points out again to a slow deterioration process of ISE B in alkaline media, while the commercial solid membrane ISE A is practically not affected by interaction with these basic solutions, at least with the short contact time exposures used in this work.

CONCLUSIONS

Concrete pore solutions with alkalinities about KOH 0.5m have a low and practically constant value of chloride activity coefficient for a wide range of Cl⁻ concentrations; being this value due to the introduction of K^+. This structure breaker ion, with a hydration picture of little structural order, produces a decreasing on the local viscosity of the solvent, and then an increasing of ionic interactions with chlorides reducing their mobility.

Chloride activity coefficients in concrete inner pore solution may be calculated theoretically through Pitzer's model, provided that the solution composition is fully known, or may be potentiometrically determined using Cl⁻ ionic selective electrodes. The practical electrochemical cells (ISE + reference electrode) must be calibrated with a series of NaCl solutions of known Cl⁻ activities.

Some limitations to the experimental potentiometric determinations are the following:

-Chloride activities derived from potentials measured in dilute Cl^- alkaline solutions ($m_{NaCl} < 0.01$) are excessively high due to interference of OH^- ions on the ISE potentiometric response.

-Ag/AgCl electrodes prepared by anodising a silver wire, in the way described in this work, behave worse in alkaline media than commercial solid membrane ISE. The former yield higher γ_{Cl^-} than those corresponding to solid membrane ISE, which are in agreement with those calculated through Pitzer's model. Moreover Ag/AgCl wire electrodes are not stable in highly concentrated Cl^- solutions ($m_{NaCl} > 2$).

ACKNOWLEDGMENT

This work was partially financed by the Conselleria d'Educació i Ciència de la Generalitat Valenciana through the project GV-3224/95 and by the CICYT Project MAT96-0539.

REFERENCES

[1] J. Koryta, *Anal. Chim. Acta*, **61**, 329(1972).

[2] P. Longuet, L. Burglen and A. Zelwer, *Rev. Matér. Constr. Trav. Publics*, **676**, 3541(1973).

[3] S. Diamond, *Cem. Conc. Res.*, **11**, 383(1981).

[4] K. Byfors, C. M. Hansson and J. Tritthart, *Cem. Conc. Res.*, **16**, 760(1986).

[5] J. Tritthart, *Cem. Conc. Res.*, **19** 586 and 683(1989); *ibid.*, **22**, 129(1992).

[6] C. L. Page, P. Lambert and P. R. W. Vassie, *Mater. Struct.*, **24**, 243(1991).

[7] M. N. Haque and O. A. Kayyali, *Cem. Conc.Res.*, **25**, 531(1995).

[8] M. Maslehuddin, C. L. Page, Rasheeduzzafar and A. I. Al-Mana, *Corrosion of Reinforcement in Concrete Construction*, Edited by C. L. Page, P. B. Bamforth and J. W. Figg, The Royal Society of Chemistry, Cambridge, U. K., 1996.

[9] B. Elsener, L. Zimmermann, D. Flückige, D. Bürchler and H. Böhni, *Chloride Penetration into Concrete*, RILEM Workshop, Paris, Oct. 1995.

[10] C. P. Atkins, J. D. Scantlebury, P. J. Nedwell and S. P. Blatch, *Cem. Conc. Res.*, **26**, 319(1996).

[11] M. A. Climent-Llorca, E. Viqueira-Pérez and M. M. López-Atalaya, *Cem. Conc. Res.*, **26**, 1157(1996).

[12] D. J. G. Ives and G. J. Janz, *Reference electrodes*, ch. 4, Academic Press, New York, 1961.

[13] A. Shatkay and A. Lerman, *Anal. Chem.*, **41**, 514(1969).

[14] J. Bagg and G. A. Rechnitz, *Anal. Chem.*, **45**, 271(1973).

[15] R. G. Bates, B. R. Staples and R.A. Robinson, *Anal. Chem.*, **42**, 867(1970).

[16] R. G. Bates and E. A. Guggenheim, *Pure Appl. Chem.*, **1**, 163(1960).

[17]D. A. MacInness, *J. Amer. Chem. Soc.*, **41**, 1086(1919).

[18]R. H. Stokes and R. A. Robinson, *J. Amer. Chem. Soc.*, **70**, 1870(1948).

[19]P. Debye and E. Hückel, *Phys. Z. (Leipzig)*, **24**, p.185 and 305(1923).

[20]F.P. Daly, C.W. Brown, *J. Phys. Chem.*, **76**, 3664(1972).

[21]Pitzer, K. S., *J. Phys. Chem.*, **77**(2), 268(1973).

[22]Pitzer, K. S., "Theory: ion interaction approach"; p.157 in *Activity coefficients in electrolyte solutions*, Vol.1, Edited by Pytkowicz, CRC Press, Florida, 1979.

[23]Pitzer, K. S., Mayorga, G., *J. Phys. Chem.*, **77**(19), 268(1973).

[24]Pitzer, K. S., Mayorga, G., *J. Sol. Chem.* **3**(7), 539(1974).

[25]Pitzer, K. S., Peterson, J. R., Silvester, L. F., *J. Sol. Chem.*, **7**(1), 45(1977).

[26]Pitzer, K. S., Silvester, L. F., *J. Phys. Chem.*, **82**, 1239(1978).

[27]Guggenheim, E. A., *Phil. Mag.*, **19**, 588(1935).

[28]L.N. Plummer, D.L. Parkhurst, G.W. Fleming, and S.A. Dunkle, "A computer program incorporating Pitzer's equations for calculation of geochemical reactions in brines", *Water resources investigations report 88-4153*, U.S. Geological Survey, Reston, 1988.

[29]C.E. Harvie, J.H. Weare, *Geochimica et Cosmochimica Acta*, **44**, 981(1980).

[30]C.E. Harvie, N. Moller, J. H. Weare, *Geochimica et Cosmochimica Acta*, **48**, 723(1984).

[31]*Handbook of Chemistry and Physics*, 66th Edition, Edited by R. C. Weast, CRC Press, Boca Raton, Florida 1985-86.

[32]A. J. Bard and L. R. Faulkner, *Electrochemical Methods*, Edited by John Wiley and Sons, New York, 1980.

[33]R. S. Barneyback and S. Diamond, *Cem. Conc. Res.*, **11**, 279(1981).

APPENDIX

Concentrations in this work are given in the molal scale but they can be interconverted with the molar scale by the following equations:

$$m_i = \frac{1000 \cdot c_i}{1000 \cdot \rho - \sum_j c_j \cdot M_j} \quad [6] \qquad c_i = \frac{1000 \cdot \rho \cdot m_i}{1000 + \sum_j m_j \cdot M_j} \quad [7]$$

being the sumatory applies for all species in solution except solvent and where,

m_i is the molal concentration of species i in mol/kg

c_i is the molar concentration of species i in mol/dm^3

ρ is the solution density in kg/dm^3

M_j is the molar weight of species j in g/mol

DELAYED ETTRINGITE FORMATION; EFFECT OF CLINKER PARTICLE SIZE AND COMPOSITION ON EXPANSION OF MORTAR BARS

P.E. Grattan-Bellew,* J.J. Beaudoin,* & V-G Vallée**
*Institute for Research in Construction,
National Research Council Canada,
Ottawa, ON, Canada K1A 0R6.
** Department of Civil Engineering, University of Ottawa,
Ottawa, ON, Canada K1N 6N5

ABSTRACT

The effect of Blaine fineness, C_3A and alkali contents of cements on expansion of mortar bars was investigated. The mortar bars were cured at 95°C and subsequently exposed to cycles of wetting and drying before being stored at 23°C, in lime water, for 59 days. The SO_3 content of the cements was kept constant at 4.5% to minimize the number of variables. The rate of expansion of the mortar bars was found to be inversely proportional to the fineness of the cements. There is a pessimum C_3A content of about 10.3% , and a pessimum SO_3/Al_2O_3 ratio of about 0.85 leading to maximum expansion of the mortar bars. There was a linear correlation between mass and expansion over the period from 17 to 59 days.

INTRODUCTION

Premature deterioration of steam cured, portland cement concrete has been attributed to the delayed formation of ettringite after the concrete has hardened and been exposed for several years to moist conditions (1-8). When portland cement concrete is cured under ambient conditions, the first reactions to occur are the hydration of tricalcium aluminate (C_3A) and its reaction with sulfates in the pore solution to form ettringite ($3CaO. Al_2O_3. 3 CaSO_4. 32H_2O$). In heat cured concrete, instead of reacting with the aluminate phase, the sulfate is thought to enter the lattice of the hydrating tricalcium silicate (C_3S) to form the so called "Phase X" of Kalousek & Adams(9). In heat cured concretes/mortars the hydration of C_3S is accelerated (10).

There is considerable confusion in the literature concerning the use of the terms delayed and secondary ettringite and in some instances the two are used interchangeably(11). In this document, the term delayed ettringite is confined to ettringite formed in heat cured concrete, after it has hardened and cooled and subsequently been exposed to moist conditions. Delayed ettringite is thought to form in micro-cracks resulting from the heat treatment of the fresh concrete (12). Such ettringite is frequently observed at the paste-aggregate interface, (3,5,6,8,13). Lewis Scrivener & Kelham (5) showed that no ettringite was present initially in mortar, after heat curing, but that it formed subsequently when the mortar was exposed to moisture. Further discussion of the mechanism of delayed ettringite formation lies outside the scope of this paper.

Research by Grabowski et al.(14) and Fu (15) indicated that expansion of mortar bars due to ettringite formation was influenced by the composition of the cement and also possibly by the type of aggregate used. Grabowski et al. showed that portland cements have a pessimum SO_3/Al_2O_3 molar ratio at which maximum expansion is observed. A similar pessimum was observed by Heinz & Ludwig (1). Delayed ettringite in field concretes appears to mainly affect heat cured concretes made with ASTM Type 30 cements, or their equivalents, i.e. high early strength cements which are commonly used in pre-cast concrete to accelerate strength development and optimize the turn around time in the casting yard. Fu (16) only observed expansion of mortar bars made using Type 30 cements. Type 10 and Type 50 cements produced no expansion under the same experimental conditions that produced excessive expansion with the Type 30 cements.

There is currently no universal or standardized method of testing cement-aggregate combinations for premature deterioration due to delayed ettringite formation. Heinz & Ludwig (1) developed a test using mortar bars heat-cured to temperatures varying from 75°C to 100°C. No expansion was observed at 75°C; maximum expansion was observed at 100°C. The disadvantage of this test is that some samples did not start expanding until after one year of storage in water at 20°C. Kelham (17) also used mortar bars in evaluating the expansion potential of German cements. He only observed significant expansion at a temperature of about 90°C. The time required for the experiments was about 150 days.

Grabowski et al. (14) used the "Duggan Test" to evaluate the effect of aggregate and cement composition on deterioration of concrete due to ettringite formation resulting from heat treatment. In this test concrete is cast and hydrated under ambient conditions. Cores are taken from the hardened concrete which are then subjected to three cycles of dry heat at 82°C followed by cooling in distilled water at 21°C, the "Duggan Cycle". Subsequently the cores are stored in distilled water and length change is monitored for 90 days. This test method does not evaluate deterioration due to DEF because ettringite would have formed in the normal way when the concrete was cured under ambient conditions. Following the Duggan Cycle the early formed ettringite or monosulfoaluminate goes into solution and is re-precipitated, as secondary ettringite, in cracks formed by thermal cycling.

Fu (15) developed a test method for the Stability of Portland Cement Mortars Moist-Cured at High Temperatures . The mortar bars are prepared using silica sand and portland cement with a cement/sand ratio of 0.36 and a w/c of 0.48. After molding, the specimens are pre-cured for 1 hour in a fog room at 23°C before being heated to 95°C for 12 hours. They are then cooled to ambient temperature during 4 hours. The mortar bars are then demolded and stored in water for 6 hours at 23°C prior to taking the zero length reading. This is followed by heating to 85°C for 24 hours and subsequent immersion in limewater. The samples are stored in lime water and the length is monitored for 56 days. This protocol appears to be the most practical of the methods proposed. However, it has not yet been demonstrated that cements causing expansion in the accelerated test would in fact cause deleterious expansion in steam cured field concrete exposed to excessively high temperatures (>~70°C).

The accelerated test developed by Fu was selected for this investigation. The reasons for this selection are: It appears be the most rapid of the proposed methods; and to extend the investigations commenced by Fu. The experiments described here were designed to investigate the effects of cement composition and particle size on expansion due to DEF in the test method proposed by Fu (15) The correlation between the results of this test and the performance of steam cured field concretes is not addressed, as this would require a long term field study.

EXPERIMENTAL PROCEDURE

Materials

Cements : In the experiments designed to investigate the effect of cement composition and Blaine fineness on expansion due to DEF, a low alkali cement clinker ground to three Blaine finenesses was used, Table 1. This clinker was selected because of its low alkali and C_3A contents which would permit these components to be increased by addition of C_3A and or alkali. It was decided to limit the variables by keeping the SO_3 constant, at 4.5%, which is the average of the North American Type III cements (19). Earlier work by Fu (15) showed that, using his proposed test method, significant expansion only occurred with Type III cements. The SO_3 in the clinkers was increased to 4.5% by addition of gypsum. Kelham (17) observed maximum expansion with the same range of SO_3 contents (4.1 to 4.5%). Glasser et al. (18) calculated a threshold SO_3 of 4 to 5% which would lead to significant expansion. These results suggest that the critical SO_3 value for delayed ettringite formation is about 4.5%

The C_3A content of the cement is also thought to affect the formation of delayed ettringite in concrete (10). The average C_3A content of North American Type III cements is 10.3% (19). It was decided to increase the C_3A content of the clinkers, from their average 7.4%, to 10.3% and 12%, the maximum in North American cements (19). The C_3A content of the clinkers was increased by the addition of pure C_3A (made by CTL Laboratories). Kelham (17) showed that the alkali content of the cement also affects expansion

Table 1. Mineralogical composition of clinker

Clinker #	Fineness cm^2/g	Mineralogical composition %			
		C_3S	C_2S	C_3A	C_4AF
#1	305	66.94	14.08	7.51	7.43
#2	425	64.17	16.54	7.46	7.46
#3	452	67.38	13.23	7.29	7.40

Oxide composition (mass %)

Clinker	SiO_2	CaO	Al_2O_3	Fe_2O_3	MgO	Na_2O	K_2O	SO_3	LOI	Na_2O Eqv.
#1	22.52	66.60	4.39	2.44	2.10	0.18	0.46	0.18	0.52	0.48
#2	22.65	66.15	4.38	2.45	2.08	0.21	0.49	0.18	0.77	0.53
#3	22.34	66.22	4.30	2.43	2.06	0.18	0.46	0.17	1.06	0.48

due to DEF. Accordingly, it was decided to make one set of mortar bars in which the alkali content of the cement was increased from the mean of the clinkers used, 0.5%, to 0.9%. The alkali was increased by the addition of mixed potassium and sodium sulfates, in the same ratio as is present in the clinker. Five cements were made from the three clinkers by adding gypsum and by varying the C_3A and alkali contents, Table 2.

Table 2. Compositions of cements made from the three clinkers

Variable	Blaine of clinker	Cement #1	Cement #2	Cement #3	Cement #1A	Cement #2A
(SO_3%)		4.5	4.5	4.5	4.5	4.5
C_3A%	305	7.51	10.3	12	7.51	10.3
C_3A%	425	7.46	10.3	12	7.46	10.3
C_3A%	452	7.29	10.3	12	7.29	10.3
Na_2O Eqv.		0.5	0.5	0.5	0.9	0.9

Fine aggregates: In all the mixtures, the aggregates were graded as shown in Table 3. A pure quartz sand, blended with coarse grained crushed crystal quartz, was used in all experiments.

Mixing of mortar

The mortars were mixed following the procedure specified in ASTM C 305 except that in the case of the mixtures containing only the coarse fractions of quartz, some modification was required. A minimum of 4 mortar bars 25.4 x 25.4 x 152 mm were made from each mixture.

Table 3. Aggregate grading

Sieve sizes	Percent retained %	Mass (g)
-5.00 + 2.50 mm	10	82.5
-2.5 + 1.25 mm	10	82.5
-1.25 mm + 630 μm	30	247.5
-630 + 315 μm	40	330.0
-315 μm	10	82.5
Total	100	825.0

Curing
The curing cycle followed is that proposed by Fu (15). After filling the molds, they were placed in a fog room at 23°C for one hour. The molds were then transferred to sealed containers and placed in an oven. The heaters were turned on so that the temperature rose to 95°C in one hour. The samples were maintained at 95°C for 12 hours when the heaters were turned off and the oven allowed to cool for 4 hours. The mortar bars were then demolded and placed in lime water at 23°C for 6 hours prior to measuring the initial lengths.

Thermal cycling
Following the initial length measurement, the samples were placed in an oven at room temperature, set to reach 85°C in one hour. The 85°C temperature was maintained for 24 hours. After cooling the samples were again immersed in lime water for 24 hours. This constituted one thermal cycle. The samples were subjected to 3 thermal cycles to accelerate expansion during subsequent storage in lime water. After three thermal cycles, the mortar bars were then returned to storage at 23°C, in lime water, and length change was monitored for at least 56 days.

RESULTS

Effect of Blaine fineness on expansion
The effect of changes in the fineness of the clinker on the rate of expansion of mortar bars made with all the cements is shown in Figure 1. The rate of expansion is used because, once the main expansive phase of the reaction has commenced, the rate is independent of time. The rate is obtained from the slope of the regression line fitted to a plot of percentage expansion versus the square root of time in days. For all the mixtures tested, expansion was found to be inversely proportional to the fineness of the clinker. In all cases, the maximum rate of expansion was obtained with the 305 Blaine clinker.

Effect of C_3A and alkali content of clinker on expansion
There appears to be a pessimum C_3A content of about 10% which leads to

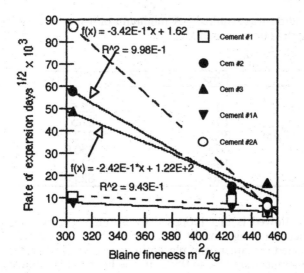

Figure 1. Effect of Blaine fineness on the rate of expansion of mortar bars exposed to the Fu thermal cycle. Cements #1 & #1A contain 7. 3% C₃ A; Cement 3 contains 12% C₃A. Cements #1A & #2A contain 0.9% alkali, the remainder contain 0.5%.

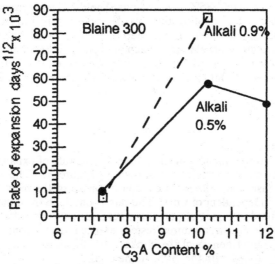

Figure 2. Change in rate of expansion with C_3A and alkali contents of cements.

maximum expansion, Figure 2. In the mortars made with the clinker containing 10.3% C3A, the rate of expansion was 50% higher when the alkali content of the cement was 0.9% compared to 0.5%.

Effect of SO3/Al2O3 ratio on the rate of expansion
There appears to be a pessimum SO3/Al2O3 ratio of about 0.84 (molar ratio ~1.03) at which the rate of expansion is a maximum, Figure 3.

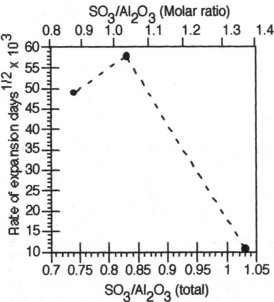

Figure 3. Effect of SO3/Al2O3 on the rate of expansion of mortar bars made with cement containing 10.3% C3A.

Correlation between gain in mass and expansion
There is a linear correlation between the percentage change in mass and the percentage expansion of the mortar bars from the age of 17 days, when expansion commences, onwards, Figure 4.

DISCUSSION

Blaine fineness
The mean Blaine fineness of all North American Type III cements is 548 kg/m^2, that of Type I cements 374 kg/m^2 (19). Fu (15) only observed expansion in laboratory mortar tests with Type III cements. Type III cements are also normally used in steam cured precast concrete. These observations appear to indicate that the fineness of the cement could be a factor in the

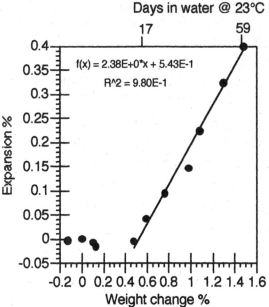

Figure 4. Correlation between increase in mass and expansion of mortar
subjected to the Fu cycle and subsequently stored in water.

formation of DEF. The results obtained in this investigation, that expansion is
inversely proportional to the fineness of the cement, Figure 1, is therefore
unexpected. Kelham (17) showed that using cements made with five UK
clinkers, there was a large increase in expansion when the fineness of the
cement was increased from 250 to 350 m²/kg. However, there was only a small
increase in expansion when the fineness was further increased to 450 m²/kg.
The change in the amount of expansion, when the fineness of the cement is
increased from around 350 to 450 m²/kg, may depend on the particular brand of
cement under investigation. It is also possible that due to the fineness of the
cements with Blaine values of 425 and 452, significant hydration of C3A and
gypsum to form ettringite may occur before the temperature becomes high
enough to increase the solubility of ettringite, i.e. within the first one and one
half hours after casting. This could effect the amount of DEF formed. An x-ray
diffraction pattern taken of fresh paste, made in a specially designed cell, one
hour after casting, showed a weak broad peak at about 0.96 nm which is about
the same d-spacing as was observed for ettringite by Lewis et al (5). This
evidence appears to confirm the formation of a small amount of ettringite in the
paste after one hour of hydration at 23°C.

C3A Content
It is evident from Figure 2 that there is a pessimum C3A content of about

10% in these cements which gives rise to the maximum expansion. Cements with a C3A content of about 7.5% had very low rates of expansion. This critical C3A content for expansion due to DEF is important for the precast concrete industry because the mean C3A content of high early strength cements (ASTM Type III) is 10.39±1.17% (standard deviation) which appears to be at the pessimum. By contrast, little expansion would be observed with ASTM Type V, North American sulfate resisting cements which have a mean C3A content of 3.53±1.21%(19). Unfortunately, Type V cements would not produce the high early strength needed by the precast industry.

SO3/Al2O3 ratio

The pessimum SO3/Al2O3 ratio of ~0.85, Figure 3a, of the cement causing maximum expansion confirms the earlier value of 0.79 - 0.88 observed by Heinz & Ludwig (1). When the SO3/Al2O3 mass percentage ratio is converted to a molar ratio, the pessimum value is in reasonable agreement with that of Heinz & Ludwig (1), (0.79-0.88)· These results are also in reasonable agreement with those of Grabowski et al.(14), although, as discussed previously, their experiments probably generate secondary rather than delayed ettringite. It is concluded from these experiments that there is a pessimum SO3/Al2O3 ratio of about 0.85 which results in maximum expansion, at least under laboratory conditions. However, it is not known if this pessimum ratio would also hold for steam cured field concrete.

Effect of alkali content of cements

The increased rate of expansion observed when the alkali content of the cement was increased to 0.9% from 0.5% may be related to the greater availability of the SO3 in the high alkali cement due to the extra alkali being added as highly soluble sodium and potassium sulfates, rather than as gypsum, which supplies most of the SO3 in the low alkali cement, Figure 2. Alkali aggregate reaction, as a contributory factor in the expansion, can probably be ruled out because the aggregate used, well crystalline quartz, is not reactive. Kelham (17) also observed increased expansion with higher alkali cements. Glasser (18) showed, by calculation, that as the alkali contents of cement pore solutions increase, ettringite solubility increases. This would lead to a greater potential for DEF. Glasser (18) proposed limiting the alkali content of the cement as one of his recommendations to minimize concrete deterioration due to DEF.

When feasible, careful selection of a low alkali Type III cement with a C3A content below the pessimum value and an SO3/Al2O3 ratio less than 0.85 may help reduce the incidence of DEF when concrete is subjected to excessive heat during accelerated curing. An example of such a cement is #1288 In Gebhardt's survey (19), which contains 0.2% Na2O equivalent alkali, 8.9% C3 A and has an SO3/Al2O3 ratio of 0.66.

The results of all the experiments are summarized in Figure 5 which shows the effect of C3A content, Blaine fineness and alkali content of cement on the rate of expansion of mortar bars subjected to the Fu thermal cycles, due to DEF formation. The effects of individual components on expansion are described below.

Correlation between Expansion and Gain in Mass of Mortar Bars

If it is assumed that most of the expansion of the mortar bars is due to the formation and growth of ettringite. The linear correlation between expansion and mass gain of the mortar bars, Figure 4, showing maximum expansion between 17 and 59 days is an indicator that much of the mass gain may be due to the conversion of C3A to ettringite.

$$C_3A + 3\ CaSO_4.2H_2O + 26\ H_2O \longrightarrow C_3A.\ 3\ CaSO_4.\ 32\ H_2O$$

Molecular weights: $270 + 516.5 + 468.4 \longrightarrow 1255$
or: 1 g C3A + 1.9 g gypsum + 1.8 g water --> 4.7 g ettringite.

Each of the most expansive mortar bars is made with 51 g cement containing 10.3% C3A, i.e. 5.25 g C3A. Assuming that 85% of the C3A is converted into ettringite, the amount converted is 4.46 g which is rounded to 4.5 g for ease of calculation. One gram of C3A combines with 1.8 g water to produce ettringite, hence 4.5 g C3A will combine with 8.1 g of water. On the assumption that only 85% of the C3A converts to ettringite, this is the maximum possible mass gain. The mortar made with quartz aggregate with a mean size of 630 μm, showing an expansion of 0.837%; its mass gain was 6.7 g which is of the same order of magnitude as the calculated value (20). These results indicate that the observed expansion of the mortar bars in these experiments is most likely due to the formation and growth of delayed ettringite. This conclusion is supported by the good correlation which was observed between the rate of expansion of the mortar bars and the intensity of the ettringite peak at 0.560 nm in the x-ray diffractogram of the mortar (20). Fu et al (21) showed that the crystallization of ettringite at crack tips generated sufficient pressure to cause expansion of mortar.

CONCLUSIONS

Blaine fineness

In the range of cement finenesses evaluated, 300 - 450 m^2 /kg, the rate of expansion was found to be inversely proportional to the fineness of the cement. The reduced expansion observed with the finer cements may be due to the formation of primary ettringite during the approximately 1.5 hours before the temperature would have risen sufficiently to increase the solubility of ettringite and hence prevent its crystallization.

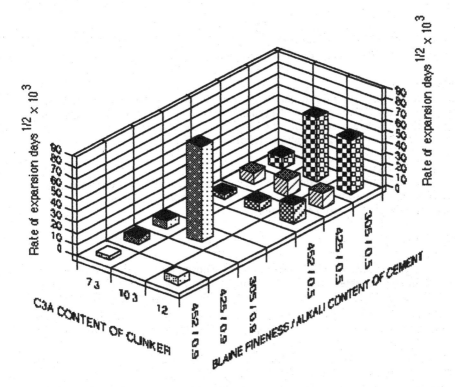

Figure 5. Summary figure showing effect of Blaine fineness, C3A and alkali contents on the rate of expansion of mortar bars subjected to the Fu thermal cycle and subsequent storage in water.

C3A

in the cements investigated, it was found that there is a pessimum C3A content of about 10% which leads to maximum expansion of mortar bars, subjected to the Fu thermal cycle, due to DEF. This pessimum value is of importance because it corresponds to the mean C3A content of North American high early strength Type III cements which are typically used in the precast industry.

SO3/Al2O3 ratio

The pessimum SO3/Al2O3 ratio of ~0.85, and the molar SO3/Al2O3 ratio of ~1.1, leading to maximum expansion of mortar bars, confirms those of previous investigators.

Alkali
The rate of expansion of mortar bars was found to increase with the alkali content of the cement as predicted by Glasser (18)·

Mass gain
The mass gain of the mortar bars was found to be proportional to expansion during the main expansive phase of the DEF reaction. Calculations showed that much of the mass gain could be due to the uptake of 26 moles of water by reaction of C3A, gypsum and water to form ettringite.

Recommendations
Field experiments are needed to verify that the results of the accelerated laboratory experiments are applicable to precast concrete subject to excessive temperatures >~70°C, during steam curing.

REFERENCES

[1]Heinz, D., & Ludwig, U., "Mechanism of Subsequent Ettringite Formation in Mortars and Concretes after Heat Treatment," in *Proc.. 8th Intern. Congress on the Chem. of Cem., Rio de Janeiro, Brazil,* Vol. 5, 189-194 (1986).

[2]Heinz, D., Ludwig, U., & Rudiger, I., "Delayed Ettringite Formation in Heat Treated Mortars and Concrete," in *Concrete Precast Plant & Technology,* 11, 56-61 (1989).

[3]Lawrence, C.D., Daizel, J.A., & Hobbs, D.W., British Cement Association Interim Technical Note 12, pp. 43 (1990).

[4]Yang, R., Lawrence, C.D., & Sharp, J.H., "Sulfate attack arising from delayed ettringite formation," *Cem. & Concr. Res.* 11, 1649-1659 (1996).

[5]Lewis, M.C., Scrivener, K.L., & Kelham, S., "Heat Curing and Delayed Ettringite Formation," *Mat. Res. Soc. Symposia Proc.* Vol. 370, 67-76 (1995).

[6]Siedel, H., Hempel, S., & Hempel, R., "Secondary Ettringite Formation in Heat Treated Portland Cement Concrete: Influence of Different w/c Ratios and Heat Treatment Temperatures," *Cem. & Concr. Res.* 23, 453-461 (1993).

[7]Wieker, W., & Herr, R., "Some Problems Concerning the Chemistry of Portland Cement," *Zeit. f. Chem.* 29, 305-352 (1989).

[8]Diamond, S., Ong, S., & Bonen, D., "Characteristics of Secondary Ettringite Deposited in Steam Cured Concretes Undergoing ASR," *Proc. 16th Intl. Conf. on Cement Microscopy, Richmond Virginia, USA,* p. 294-305, (1994).

[9]Kalousek, G.L., Curtis, W., Davis, W., & Schmertz, W.E., "An Investigation of Hydrating Cements and Related Hydrous Solids by Differential Thermal Analysis," "*Proc. Am. Concr. Inst.* 45, 693-712 (1949).

[10]Fu, Y., Ping, G., Ping, X., & Beaudoin, J.J., "A Kinetic Study of Delayed Ettringite Formation in Hydrated Portland Cement Paste," *Cem. & Concr. Res.* 25, 63-70 (1995)

[11]Odler, I., "Letter to the Editor Ettringite Nomenclature," *Cem. & Concr. Res.* 27, 473 (1997).

[12]Sylla, H-M., "Reactions in Cement Stone Due to heat treatment," *Beton,* 38, 449-454 (1988).

[13]Shayan, A., Quick, G.W., "Microscopic features of Cracked and Uncracked Concrete Railway Sleepers," *ACI Materials J.* 89, 348-361 (1992).

[14]Grabowski, E., Czarnecki, B., Gillott, J.E., Duggan, C.R., & Scott, J.F., "Rapid Test of Concrete Expansivity Due to Internal Sulfate Attack," *ACI Materials J.* 89, 469-480 (1992).

[15]Fu, Yan, "Delayed Ettringite Formation in Portland Cement Products," Ph.D. Thesis, p.199., Dept. Civil Engineering, University of Ottawa, (1996).

[16]Fu, Y., & Beaudoin, J.J., "Mechanisms of Delayed Ettringite Formation in Portland Cement Systems," *ACI Materials J.,* July-August 327-333 (1996).

[17]Kelham, S., "The Effect of Cement Composition and Fineness on Expansion Associated with Delayed Ettringite Formation," *Cement Concrete & Composites* 18, 171-179 (1996).

[18]Glasser, F.P.l, "The Role of Sulfate Mineralogy and Cure temperature in Delayed Ettringite Formation," *Cement Concrete & Composites,* 18, 187-193 (1996).

[19]Gebhardt, R.F., "Survey of North American Cements: 1994," *Cement Concrete & Aggregates,* 17, 145-189 (1995).

[20]Grattan-Bellew, P.E., Beaudoin, J.J. & Vallée, V-G., *"Effect of Aggregate Particle Size and Composition on expansion of Mortar Bars Due to Delayed Ettringite Formation,"* Accepted for publication in *Cem. & Concr. Res.* (1998).

[21]Fu, Y., Ping, X., Ping, G., & Beaudoin, J.J. "Significance of Pre-Existing Cracks on Nucleation of Secondary Ettringite in Steam Cured Cement Paste," *Cem. & Concr. Res.* 24, 1015-124 (1994).

SOLIDIFICATION OF SELECTED HEAVY METALS INTO CEMENT MORTARS

Adolf Bajza, Ildikó Rouseková, Stanislav Unčík and Vladimír Pavlík
Slovak University of Technology
Faculty of Civil Engineering
Radlinského 11
813 68 Bratislava
Slovak Republic

ABSTRACT

Results achieved from the solidification of heavy metals such as Cr(total), Cr(+6), Ni, Cu, Zn, Fe into cement mortars are presented. Two types of waste sludges from galvanic processes and three courses of their pretreatment were used. The dosages of sludges were 2.5, 5, 10, and 15% (dry matter) of the cement weight. Technical properties such as the compressive strength, dynamic modulus of elasticity, and shrinkage as well as the leachability of heavy metals into buffer solutions with pH 4, 7, and 13 were determined. Microstructure of cement composites was evaluated using MIP and SEM.

INTRODUCTION

Industry produces various types of wastes, including wastes containing hazardous metals. Metals are the only hazardous pollutants which cannot be destroyed or altered by chemical or thermal methods; this is in contrast, for example, to cyanides or some organic pollutants. To prevent their reentry into the environment, they must be converted into their most insoluble form[1-3]. Many of them (e.g., As, Sb, Be, Cd, Cr, Co, Cu, Hg, Mo, Ni, Pb, Ag, Zn) are hazardous to humans and other forms of life. However, some of them in very low quantities are vital to life and are required for the normal growth, development, and maintenance. Therefore, the slow leaching of metals at a controlled rate need not be detrimental to human health and the environment[1].

The migration of metals into the environment must be prevented by suitable treatment (stabilization and solidification). The main processes of metals stabilization (fixation) are pH control, redox potential control, and precipitation (hydroxide, sulfide, silicate). Sometimes, solidification follows the stabilization. However, very often the solidification technology is the same as the technology of stabilization. Stabilization processes can be basically organic and inorganic. Organic processes have been applied in the area of radioactive wastes solidification - the physical confinement, for example by urea - formaldehyde, bitumen, or polymerisation systems. Inorganic processes enable production of nontoxic, environmentally safe materials. The most important processes are those using systems with Portland cement, lime - fly ash, lime - Portland cement, Portland cement - fly ash, Portland cement - lime, Portland cement - sodium silicate.

This contribution presents a part of the results achieved during research into the solidification of sludges from electroplating processes into building materials.

METHODS
Materials used
Two types of *waste sludges* from galvanic processes were used for the experiments (Table I).

Portland cement CEM I 42.5 was used (Table II), along with a *siliceous sand* with a continuous grading curve.

Pretreatment and testing of waste sludges
Heavy metals in the sludges from the electroplating processes were fixed by a combination of the following procedures:
1. adjustment of pH of the sludge to the value of 5.9,
2. stabilization of the sludge moisture content to the laboratory conditions, followed by homogenization and grinding,
3. treatment of the sludge by 20 % NaOH solution to a pH value in the range from 6.9 to 7.1,
4. chemical fixation of heavy metals by exposure to a 15 % Na_2S solution for 8 hours; the chemical agent was applied as a 1.2 multiple of the heavy metals equivalent,
5. chemical fixation of heavy metals by water containing 40 g/l of $Ca(OH)_2$ in suspension for 8 hours; the chemical agent was applied as a 1.2 times the amount of heavy metals equivalent,
6. modification of the water suspension of sludges stabilized with $Ca(OH)_2$ by addition of bentonite and admixture W 32 (supplied by FZB Umweltlabor GmbH, Berlin); the dosage of these components was 1 % of dry matter.

Note to points 4 and 5: The excess of Na_2O and $Ca(OH)_2$ was chosen in order to reduce the solubility of the heavy metal precipitates to the lowest level possible.

Table I. Composition of waste sludges

Component		Type of sludge	
		B	S
Dry matter content	[%]	87.8	44.3
Conductivity	[mS/m]	-	371.6
Cyanides total	[mg/kg of dry matter]	0.02	0.02
Cyanides easily releasable	[mg/kg of dry matter]	0.02	0.02
Antimony	[mg/kg of dry matter]	-	5.53
Arsenic	[mg/kg of dry matter]	-	11.5
Chromium total	[mg/kg of dry matter]	1 820	1 550
Chromium (+6)	[mg/kg of dry matter]	0.5	0.02
Cadmium	[mg/kg of dry matter]	-	3.32
Cobalt	[mg/kg of dry matter]	78.6	7.86
Copper	[mg/kg of dry matter]	942	24 800
Nickel	[mg/kg of dry matter]	2 750	4 910
Lead	[mg/kg of dry matter]	-	12.2
Mercury	[mg/kg of dry matter]	-	1.25
Selenium	[mg/kg of dry matter]	-	5.73
Thallium	[mg/kg of dry matter]	-	5.73
Zinc	[mg/kg of dry matter]	487 000	77 000
Iron	[mg/kg of dry matter]	-	19 500

Table II. Composition and properties of Portland cement

Chemical composition		Physical properties		
Loss on ignition	1.94 %	Specific surface area [m^2/kg]		380
SiO_2	19.51 %	Normal consistency [%]		28
Al_2O_3	5.24 %	Setting time Initial		2 h 05 min.
Fe_2O_3	3.02 %	Final		3 h 10 min.
CaO	62.44 %	Soundness: L.CH. test [mm]		1
MgO	2.62 %	Compressive strength [MPa]		
SO_3	2.22 %	after 2 days		29
Na_2O	0.25 %	after 28 days		48
K_2O	0.61 %			
Cr_2O_3	0.06 %			
Free lime	0.34 %			

Mineralogical composition					
$CaSO_4$	3.77 %	C_2S	10.99 %	C_4AF	9.18 %
C_3S	60.00 %	C_3A	8.78 %		

The summary of the methods used for stabilization and subsequently solidification of the sludges is given in Table III. The stabilized sludges were vacuum filtrated, and were then used as the admixtures into cement mortars. Samples obtained by filtration were tested for composition and leachability. The filter cakes were analyzed according to the criteria of the Order in Council of the Slovak Republic No. 606/92[4]. The leaching of the filter cakes was carried out for 24 hours by using buffer solutions with pH values of 4, 7 and 13.

The leachates were analyzed using STN (Slovak Technical Standard) and EPA (Environmental Protection Agency) standards.

Table III. Methods of stabilization and solidification

Type of sludge	Designation of samples	Course of stabilization	Solidification
B	BI	1., 2., 3., 4.	
	BII	1., 2., 3., 5., 6.	
	SI	1., 2., 3., 4.	Cement composite
S	SII	1., 2., 3., 5., 6.	
	SO	1., 2.	

Preparation, curing, and testing of cement composites

The investigation was carried out in two stages. The aim of the *first stage* was the determination of the optimum dosages of sludge. For this purpose samples from cement mortars with sludge dosages of 2.5, 5, 10 and 15 % dry matter by weight of cement were prepared. Control samples from standard cement mortars without sludge additions were also prepared. All cement mortars used had a constant flow diameter of 140 mm. The size of the cylindrical samples was 30 x 30 mm. The decisive criterion for the determination of the optimum dosage of sludges was the 28 day compressive strength of the composites.

In the *second stage* only the cement composites with a sludge dosage of 5 % by weight of cement were used. A series of prismatic specimens, 40 x 40 x 160 mm, were prepared. After two days of moist curing the specimens were demolded and subsequently stored in a moist atmosphere. Bulk density, flexural strength and compressive strength were determined after 3, 28, and 308 days of curing. MIP, SEM, and leaching tests were carried out on selected samples. The dynamic modulus of elasticity and the length changes were continuously determined throughout the duration of the experiments. Workability of mortars was determined according to standard STN EN 452-2.

Leaching tests were done on dried samples which were ground to a particle size below 0.8 mm. Buffer solutions with pH values of 4, 7 and 13 as leaching media were used.

RESULTS AND DISCUSSION
Waste sludges
The results of analyses of filter cakes from stabilized sludges are given in Table IV.

Table IV. Concentration of heavy metals in filter cakes

Element	Type of sludge B		S	
	Designation of samples			
	BI	BII	SI	SII
	Concentration [mg/kg of dry matter]			
Chromium total	1 760	1 850	1 495	1 535
Chromium (+6)	0.5	0.5	0.02	0.02
Cadmium	-	-	-	3.23
Cobalt	55.7	68.3	7.85	7.69
Copper	953	858	25 000	25 000
Nickel	2 259	2 690	4 915	5 000
Zinc	485 000	485 000	77 000	77 000
Iron	28 000	27 300	20 000	19 500
Dry matter content [%]	41.6	42.2	53.7	48.4
pH	8.35	8.55	8.91	9.41

Characteristic data of the leachates from the stabilized galvanic sludges at pH values of 4, 7, and 13 are given in Table V.

The data in Table V show that these methods of heavy metals fixation were effective, and only small quantities of metals were leached from filter cakes. While the amount of leached $Cr(+6)$ varied only insignificantly with the pH values of buffers, the concentration of other heavy metals was noticeably influenced by the pH.

The given data confirm the relationship between the solubility of metal hydroxides and sulphides and pH values of the solutions, and the dominant influence of the amphoteric character of metals on the leaching at extreme pH values. This behaviour is characteristic especially for zinc, which represents the dominant metal in both waste sludges. The different heavy metals fixation methods in galvanic sludges do not show any significant differences in the leachability tests.

Table V. Characteristic data of the leachates from filter cakes

Designation of samples	pH of buffer	Heavy metal content	[mg/l]			
		Cr_{total}	Cr(+6)	Co	Cu	Zn
BI	4.0	0.304	0.19	0.305	1.385	38.58
	7.0	0.251	0.16	0.062	<0.1	0.318
	13.0	0.901	0.30	0.431	2.023	1.007
BII	4.0	0.554	0.32	0.242	1.092	26.03
	7.0	0.342	0.23	0.013	0.164	<0.2
	13.0	0.930	<0.02	0.681	2.690	2.150
SI	4.0	<0.1	<0.02	0.004	0.629	0.221
	7.0	<0.1	<0.02	0.005	0.232	<0.2
	13.0	0.141	<0.02	n	0.848	8.70
SII	4.0	0.100	<0.02	<0.004	0.885	11.34
	7.0	<0.1	<0.02	0.009	0.129	<0.2
	13.0	0.411	<0.02	n	0.855	9.376

n - content of Co in leachates with pH 13.0 undetectable

Cement composites

Compressive strength and flexural strength. The effect of sludge dosage on 28 day compressive strength of cement composites is shown in Fig. 1. These data were used to choose the optimum dosages of waste sludges for the cement mortars.

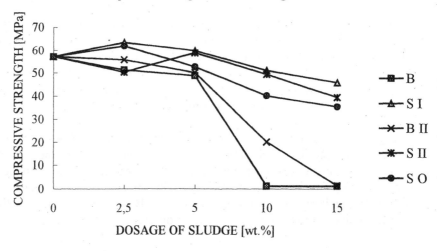

Fig. 1 Relationship between 28 days compressive strength of cement mortars and dosage of waste sludges

The decisive criterion was the sludge dosage at which the decrease in the compressive strength of the specimens compared to those of specimens without sludge (control specimens) was less than 15 %. From this, a sludge dosage of 5 wt.% dry matter by weight of the cement was taken as optimum.

The relationship between the compressive strength of cement mortars containing 5 wt.% of sludge and curing time is shown in Fig. 2. The best results were achieved when sludge S was used. The smaller values of the early compressive strength of composites containing sludge B indicate, that the rate of cement hydration was retarded as a consequence of the very high content of Zn in the sludge[5].

The flexural strength development of the cement composites was the same as that of the compressive strength development.

<u>Dynamic modulus of elasticity</u>. Fig. 3 shows that while the type of waste sludge markedly affects the values of dynamic modulus of elasticity, the effect of the stabilization method is not so great.

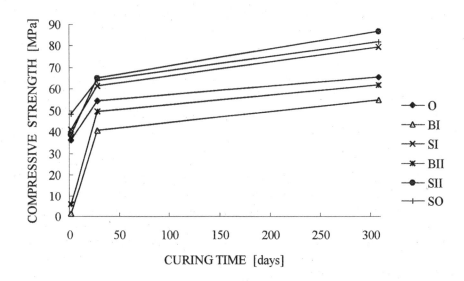

Fig. 2 Relationship between compressive strength and curing time of cement mortars (dosage of sludge 5 wt.% dry matter by weight of cement)

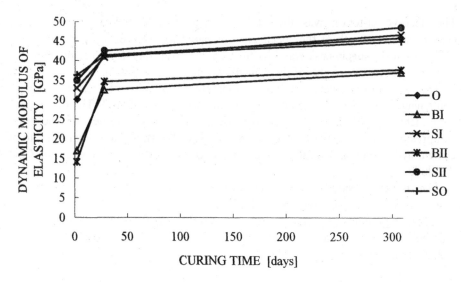

Fig. 3 Relationship between dynamic modulus of elasticity and curing time of
cement mortars (dosage of sludge 5 wt.% dry matter by weight of cement)

The early values of the dynamic modulus of elasticity of composites containing waste sludge B (similar to the compressive strength) are considerably lower than those containing sludge S or the control mortars.

Length changes. The length changes (shrinkage) of composites with various waste sludges at given stages of curing are markedly lower than those of control composites (without sludge admixture). The type of sludge or the stabilization method had practically no effect on the shrinkage of cement mortars (Fig. 4).

Pore structure. The results of the mercury intrusion porosimetry tests on cement composites with various dosages of waste sludges are given in Table VI. The data illustrate that the density (compactness) of cement composites decreases with increasing dosage of sludge. The comparison of the compressive strength of mortars with various dosages of sludges (Fig. 1) and their pore structure (Table VI) shows a good correlation - higher density of the structure corresponds to higher compressive strength.

It was found that dosages of sludge above 5 % generally led to a degradation of all technical properties of the cement mortars.

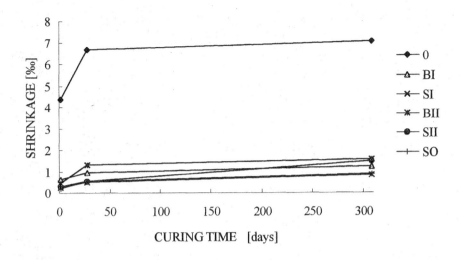

Fig. 4 Shrinkage of cement mortars without and with admixture of different type of sludges (dosage of sludge 5 wt.% dry matter by weight of cement)

Table VI. Pore structure of cement mortars with various dosages of waste sludges

Design. of samples	Dosage of sludges [wt.%]	Pore volume <7500 nm [mm³/g]	Median of micropores [nm]	Bulk density of dry samples [kg/m³]	Total porosity [%]	Specific surface area [m²/g]
	2.5	42.29	53	2166	9.16	2.212
B II	5	53.27	49	2057	10.95	3.081
	15	102.59	252	2001	20.52	1.873
	2.5	46.50	55	2128	9.89	2.351
S II	5	44.13	53	2164	9.55	2.174
	15	62.29	60	2097	13.07	2.131
	2.5	36.72	49	2196	8.06	1.831
SO	5	42.93	56	2099	9.01	1.948
	15	59.00	58	2092	12.34	1.880

Table VII. Pore structure of cement mortars without and with 5 wt.% of various sludges

Design. of samples	Dosage of sludges [wt.%]	Age of samples [days]	Pore volume <7500 nm [mm^3/g]	Median of micropores [nm]	Bulk density [kg/m^3]	Total porosity [%]	Specific surface area [m^2/g]
		2	56.16	69	2082	11.69	2.268
O	0	28	45.95	52	2189	10.06	2.009
		308	31.44	18	2202	6.92	4.209
B I		28	66.88	248	2055	13.74	2.049
		308	42.26	191	2128	8.25	3.494
		2	59.17	131	2068	12.24	2.321
S I		28	51.76	64	2179	11.28	2.321
	5	308	32.15	41	2250	7.23	3.611
B II		28	67.72	64	2024	13.71	3.433
		308	42.12	59	2102	8.85	3.996
		2	55.25	70	2042	12.74	2.024
S II		28	45.27	50	2187	9.90	2.359
		308	21.64	26	2231	4.83	3.595
SO		28	53.74	48	2064	11.09	2.892
		308	29.84	21	2135	7.31	4.230

The data in Table VII indicate densification of composite structure with curing time. The effect of structure changes on the technical properties of the composites was most significant in the case of sludge S. The highest density was reached with composite SII.

Scanning electron microscopy.   SEM evaluation (e.g., Figs. 5 and 6) demonstrated that   addition of 5 % pretreated sludge by the cement weight, with the heavy metals stabilized in an insoluble form, did not have a   visible effect on the microstructure of cement mortars. The main hydration product observed in mortars without and with addition of sludges was CSH gel type III[6]. Massive $Ca(OH)_2$ deposits growing through the CSH gel were noted.

It should be mentioned that Chen et al.[7], emphasized the importance of solubility of compounds containing heavy metal. He reported that Cr in insoluble form did not   affect hydration of β-C_2S; soluble Cr(3+), on the other hand, inhibited hydration of β-C_2S significantly. Ivey et al.[8], using SEM, had found that Cr in soluble form hindered formation of ettringite. Kindness et al.[3] found normal morphology of hydrated OPC spiken with 5000 ppm of chromium with no Cr found to be concentrated in any hydrated phase. At higher concentrations chromium was widely dispersed throughout all the hydrates and particularly those containing Ca, Al and SO_4. While Kindness et al.[3] reported that C-S-H was not found to absorb Cr, Tashiro et al.[9] reported that Cr(3+), when added in high concentration, can enter into solid solution with C-S-H. There are indications in the literature that some heavy metals become incorporated into the C-S-H. Further, Nocun-Wczelik et al.[10] have reported that, in the presence of heavy metal compounds, the highly disordered C-S-H formation is promoted.

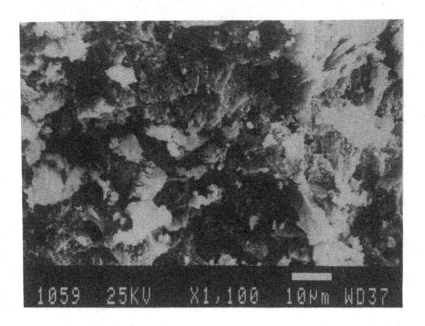

Fig. 5   Scanning electron micrograph of cement mortar without addition of sludge (control sample, 28 days)

Leachates of heavy metals from cement composites. The efficiency of the fixation of the heavy metals from the waste sludges into the cement matrix was evaluated from the leaching tests results. The concentrations of heavy metals in the leachates are given in Table VIII.

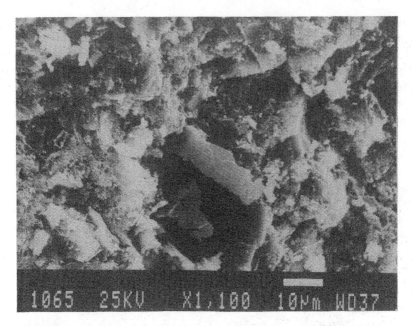

Fig. 6 Scanning electron micrograph of cement mortar with 5 % dosage of SII
sludge - 28 days

At present, there are no valid regulations in the Slovak Republic that would
limit the heavy metals content in building materials (in new products). The content
of heavy metals is, therefore evaluated according to regulations for handling
wastes. These regulations classify wastes according to the limit concentrations of
contaminants in the water leachates[6]. Common housing wastes are in Class
I.Wastes belonging to the leachability Classes II and III must be placed in the
special landfills.

The concentrations of Cu, Ni, Zn and Fe determined in the leachates (Table
VIII) would permit the inclusion of all tested composites into leachability Class I.
The contents of Cr_{total} and $Cr(+6)$ are questionable. The regulations require for
Class I the concentration of Cr_{total} to be less than 0.05 mg/l and $Cr(+6)$ less than
0.02 mg/l.

From the overall evaluation of the leaching tests of the cement composites
containing sludges, it should be emphasized that the content of Cr_{total} in the water
leachated from control samples "O*" (without sludge) was the same as from the
samples BI, SI, BII, SII, and the content of $Cr(+6)$ in the water leachates from
samples "O*" was even higher than from samples BI - 308d, SII - 28d and SII -
308d.

Table VIII. Leachates from the cement mortars without and with 5wt.% of various sludges

Samples design.	Samples age [days]	pH of buffer	Cr_{total}	$Cr(6+)$	Cu	Ni	Zn	Fe
					Heavy metal contents [mg/l]			
0*	28	4.0	0.85	0.83	<0.05	<0.01	<0.2	<0.5
		7.0	<0.1	0.1	<0.05	<0.01	<0.2	<0.5
		13.0	1.35	1.32	<0.05	<0.01	<0.2	<0.5
	308	4.0	0.42	0.22	0.01	<0.01	0.1	<0.1
		7.0	0.08	0.08	<0.01	<0.01	<0.1	<0.1
		13.0	0.39	0.30	<0.01	<0.01	<0.1	<0.1
BI	28	4.0	0.384	<0.02	<0.05	0.07	21.6	1.44
		7.0	<0.1	0.07	<0.05	<0.01	<0.2	<0.5
		13.0	0.75	0.73	<0.05	<0.01	<0.094	<0.5
	308	4.0	0.31	<0.02	0.04	0.042	20	0.87
		7.0	0.09	0.07	<0.01	<0.01	<0.08	<0.1
		13.0	0.65	0.65	<0.01	<0.01	0.9	<0.1
SI	28	4.0	0.44	<0.02	<0.05	0.05	0.81	0.73
		7.0	<0.1	0.05	<0.05	<0.01	<0.2	<0.5
		13.0	1.26	0.91	<0.05	<0.01	<0.2	<0.5
	308	4.0	0.4	<0.02	<0.01	<0.01	<0.1	<0.1
		7.0	<0.1	0.05	<0.01	<0.01	<0.1	<0.1
		13.0	0.37	0.13	<0.01	<0.01	<0.1	<0.1
BII	28	4.0	<0.01	<0.02	<0.05	0.02	1.92	<0.5
		7.0	<0.1	0.11	<0.05	0.1	<0.2	<0.5
		13.0	0.26	0.31	<0.05	<0.01	0.42	<0.5
	308	4.0	<0.1	<0.02	0.05	0.015	1.85	<0.5
		7.0	<0.1	<0.02	<0.01	<0.01	<0.1	<0.1
		13.0	0.25	0.25	<0.01	<0.01	0.4	<0.1
SII	28	4.0	<0.1	<0.02	<0.05	<0.01	0.24	<0.5
		7.0	<0.1	<0.02	<0.05	<0.01	<0.2	<0.5
		13.0	0.63	0.62	<0.05	<0.01	<0.2	<0.5
	308	4.0	<0.1	<0.02	0.04	<0.01	0.2	0.15
		7.0	<0.1	<0.02	<0.01	<0.01	<0.1	<0.1
		13.0	0.6	0.38	<0.01	<0.01	<0.1	<0.1
SO	28	4.0	0.73	0.27	0.36	0.08	1.5	1.08
		7.0	0.15	0.14	<0.05	<0.01	<0.2	<0.5
		13.0	1.10	1.01	<0.05	<0.01	<0.2	<0.5
	308	4.0	0.49	0.19	0.35	0.07	1.5	1.05
		7.0	0.15	0.15	<0.01	<0.01	<0.1	<0.1
		13.0	0.61	0.51	<0.01	<0.01	<0.1	<0.1

*Cement mortar without sludge

This leads to the conclusion that the chemically pretreated sludges also partly fixed the chromium from the binder (cement).

As expected the leachability of heavy metals from the composites generally decreased with the curing time of the specimens.

Both methods used for heavy metal stabilization were found to be suitable. The combination of bentonite + W32 admixture was more effective, however, than the fixation by sulphides.

CONCLUSION

The data showed the possibility of the fixation of the waste sludges from galvanic processes into cement composites.

The suitable pretreatment of waste sludge and using at its optimum dosage led to comparable technical properties of the composites without and with sludge additions. Additional data are being sought but, based on the above and available technical literature, the application of this knowledge in practice is not only a technical, but also a legislative, problem.

ACKNOWLEDGMENTS

The authors are grateful to the European Commission - Joint Research Center for the financial support of the Copernicus project CIPA-CT94-0181. The authors would like to thank FZB Unweltlabor GmbH, Berlin and Techmix Ltd, Brno for close cooperation. The presented data represent only part of the data obtained.

REFERENCES

[1]J.R. Conner, *Chemical Fixation and Solidification of Hazardous Wastes*. Van Nostrand Reinhold, New York, 1990.

[2]*Chemistry and Microstructure of Solidified Waste Forms*. Edited by Roger D. Spence. Oak Ridge National Laboratory, Lewis Publishers, Tennessee, 1993.

[3]A. Kindness, A. Macias, and F.P. Glasser, „Immobilization of Chromium in Cement Matrices", *Waste Management*, **40** [1] 3-11 (1994).

[4]*Order in Council of the Slovak Republic No. 606/92. Law Gazette.Wastes.*

[5]I. Jawed, J. Skalny and J.F. Young, „Hydration of Portland Cement"; pp. 237-317 in *Structure and Performance of Cements*. Edited by P. Barnes. Applied Science Publs. Ltd., Barking, 1983.

[6]S. Diamond, „Cement Paste Microstructure. An Overview at Several Levels"; pp. 2-30 in *Hydraulic Cement Pastes: Their Structure and Properties*. Proc. Conf. University of Sheffield, Cement and Concrete Association, London, 1976.

[7]J.H. Chen, Y.Y.Wang and C.C. Wan, „MAS/NMR Studies of the Hydration Process of β-C_2S in the Presence of Chromium", *Cem. Concr. Res.*, **24** [2] 319-324 (1994).

[8]D.G. Ivey, R.B. Heimann, M. Neuwirth, S. Shumborski, D. Conrad, R.J. Mikula and W.W. Lam, „Electron Microscopy of Heavy Metal Waste in Cement Matrices", *J. Mater. Sci.*, **25** 5055-5062 (1990).

[9]C. Tashiro, H. Takahashi, M. Kanaya, I. Hirakida and R. Yoshida, „Hardening Property of Cement Mortar Adding Heavy Metal Compounds and Solubility of Heavy Metal from Hardened Mortar", *Cem. Concr. Res.*, **7** [2] 283-290 (1977).

[10]W. Nocun-Wczelik and J. Malolepszy, „Studies on Immobilization of Heavy Metals in Cement Paste - C-S-H Leaching Behaviour", 4iv043, 8 pp in *Performance and Durability of Cementitious Materials*. Proc. 10th Intl. Congr. Chem. Cem., Vol. 4. Edited by H. Justnes. SINTEF Civ. Eng. and Environmental Eng., Cem. and Concr., Trondheim, 1997.

THE ROLE OF CALCIUM IN ALKALI-SILICA REACTION

M.D.A. Thomas
Department of Civil Engineering,
University of Toronto,
35 St. George St., Toronto, Ont. M5S 1A4

ABSTRACT
    This paper presents results from experimental studies aimed at elucidating the role of calcium in the alkali-silica reaction. Mortar bars containing reactive aggregate were manufactured with various cementitious systems such that the form and availability of calcium covered a wide range. These systems included (i) portland cement with high levels of pozzolan, (ii) various combinations of calcium hydroxide and pozzolan, (iii) portland cement with mortar bars carbonated prior to exposure, and (iv) high-alumina cement. Following curing (and carbonation in some cases) mortar bars were immersed in solutions of alkali hydroxide at varying temperatures. In addition to length measurements, mortars were periodically subjected to optical and scanning electron microscopy (with energy dispersive x-ray analysis), thermal gravimetric analysis, x-ray diffraction analysis, and pore solution expression and analysis. The results show that significant expansion only occurs when sufficient calcium is available (e.g. as calcium hydroxide). Systems without $Ca(OH)_2$ showed little expansion despite evidence that alkali-silica reaction had occurred. Pore solution analysis revealed that in the absence of $Ca(OH)_2$ the reacted silica remained in solution, with up to 0.5 M SiO_2 being found in solution in some cases. These data confirm that calcium is required to promote expansion due to alkali-silica reaction. The precise role of calcium in the reaction is discussed.

INTRODUCTION
    Classical theories describing alkali-silica reaction (ASR) in concrete generally do not consider calcium in the primary role of reaction and expansion (1-3). However, a number of workers have suggested that the presence of $Ca(OH)_2$ is required for damaging reaction to occur (4-10). The precise part played by calcium is unclear, although the following mechanisms have been proposed:

- Calcium may replace alkalis in the reaction product thereby regenerating alkalis for further reaction (1)

- $Ca(OH)_2$ may act as a buffer maintaining a high level of OH^- in solution (8)

- High calcium concentrations in the pore solution prevent the diffusion of silica away from reacting aggregate particles (5,6)

- If calcium is not available reactive silica may merely dissolve in alkali hydroxide solution without causing damage (4)

- The formation of calcium-rich gels is necessary to cause expansion either directly or through the formation of a semi-permeable membrane around reactive aggregate particles (9,10).

This paper presents data from various experimental studies that have been carried out at either the Building Research Establishment (U.K.) or the University of Toronto (Canada), and which further demonstrate the importance of calcium availability in determining the degree of deleterious reaction.

STUDIES WITH FLY ASH CONCRETE

Previously published studies by the author have dealt with the role of $Ca(OH)_2$ in the expansion of concrete containing fly ash (9-11). In a study of 7-year-old concretes containing reactive flint sand, evidence of alkali-silica gel was found in fly ash concretes with no associated expansion or cracking (9). Energy-dispersive x-ray analysis showed this gel to be relatively low in calcium (Ca/Si atomic ratio ≈ 0.30) compared to the gel found in the paste and filling cracks and voids in the badly deteriorated OPC concrete (Ca/Si atomic ratio ≈ 0.90). It was postulated that the low-calcium gel is relatively fluid and could be accommodated within the cement matrix, whereas the higher- calcium gel found in the OPC concrete is more viscous and is less able to dissipate when it swells on contact with water. The lower availability of calcium in the fly ash concrete, due to consumption of $Ca(OH)_2$ by the pozzolanic reaction, is thought to prevent the formation of the high- calcium reaction product.

Microstructural studies of fly ash concretes containing reactive flint sand and immersed in alkaline solutions at elevated temperature for extended periods of time (e.g. 1M NaOH at 80°C for up to 44 months) confirmed the importance of calcium in developing expansive reactions (10,11). In these tests, fly ash was effective in suppressing the expansion of concrete despite the abundance of reactive silica and alkalis. Considerable evidence of reaction was found in the fly ash concrete in the form of decomposed flint particles and alkali silica reaction product. However, the gel found was characteristically low in calcium and was found to have penetrated into the cement matrix surrounding the original site of reaction without leading to disruption. In OPC concrete, reaction rims high in calcium were found to form around many of the reacting flint grains and these were invariably sites of expansion. It was hypothesized that the high-calcium rim was either the source of expansive forces itself or formed a semi-permeable membrane, which prevented the diffusion of alkali-silica solution away from the reacting grain thereby leading to osmotic pressure build-up.

STUDIES ON CARBONATED MORTARS

Mortar bars (25x25x250mm) were cast with low-alkali Portland cement (< 0.60 Na_2O_e) and Spratt aggregate (a siliceous limestone) in accordance with ASTM C 227. Mortar bars were then cured in limewater at 38°C for 28 days to ensure a high degree of hydration. After curing some of the mortar bars were stored in a CO_2-rich atmosphere at 20°C and 65%RH until complete carbonation of the specimen, as determined by periodically spraying a freshly fractured surface with phenolphthalein solution, had occurred. Other mortar bars were stored in a CO_2-free atmosphere for the same duration. Mortar bars were then immersed in 1M NaOH at 80°C for up to 1 year. Length change, pore solution composition and microstructural features were observed intermittently throughout exposure. Pore solution was extracted using the method described by Barneyback and Diamond (12) and microstructural examinations were carried out using an Hitachi 'environmental' scanning electron microscope (ESEM) and energy-dispersive x-ray analysis (EDS).

Figure 1 shows the expansion of normal and carbonated mortar bars with reactive Spratt aggregate and non-carbonated bars with a non-deleteriously-reactive carbonate aggregate (Nelson). The carbonated mortar shows much reduced expansion compared to the non-carbonated mortar containing the same aggregate. Indeed, the process of carbonation reduces the expansion to a level below or similar to that shown by mortar containing a non-deleteriously-reactive aggregate.

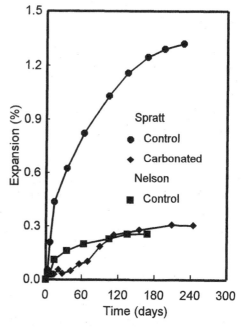

Figure 1 Effect of Carbonation on the Expansion of Mortar Bars

Figure 2 shows the change in pore solution composition with time for carbonated mortar bars containing reactive Spratt aggregate. The composition shown at a time of zero days refers to the solution expressed from bars after 28 days curing in limewater. The sodium ion content of the pore solution is approximately in equilibrium with the external host solution after 90 days. However, the hydroxyl ion concentration in the pore solution was consistently below the 1000 mM/L in the host solution at all ages. Significant concentrations of silica (SiO_2 = 410 mM/L) were found in the pore solution of the carbonated mortar bar after 90 days immersion. This compares with concentrations of SiO_2 < 35 mM/L determined for the pore solution extracted from the non-carbonated mortar containing Spratt aggregate. The host solution containing the carbonated mortar was found to contain significant quantities of silica (SiO_2 = 28 mM/L) after 270 days, whereas no silica was detected in the solution containing the non-carbonated mortar.

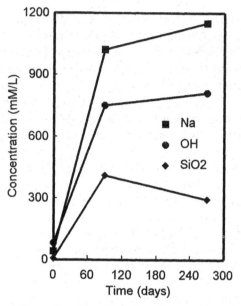

Figure 2   Pore Solution Composition in Carbonated Mortar

After 270 days the carbonated mortar was found to contain a number of fine cracks throughout the cement paste although the aggregate was essentially intact. Alkali-silica reaction product was found filling voids and small fissures in the cement matrix and within intact limestone particles. EDS analysis showed the product to be generally high in alkali ([Na+K]/[Si] = 0.2 – 0.6) and low in calcium ([Ca]/[Si] < 0.3). The non-carbonated mortar containing Spratt aggregate was extensively cracked after 270 days with gel-filled cracks running through aggregate particles and the cement paste. The composition of the gel found in the mortar sample paste varied widely and in many locations was determined to have relatively high calcium contents ([Ca]/[Si] > 1.0).

STUDIES ON HIGH-ALUMINA CEMENT MORTARS

Mortar bars (25x25x250mm) were cast with high-alumina cement (HAC) and either reactive Spratt aggregate or non-reactive Nelson aggregate. Mixing and proportioning procedures followed ASTM C 227. Mortar bars were cured in limewater at 38°C for 28 days prior to immersion in 1M NaOH at 80°C.

Expansion results are shown in Figure 3. HAC mortars showed reduced expansion compared with OPC mortar bars with reactive Spratt aggregate, but greater expansion than OPC mortars with non-reactive Nelson aggregate. Surprisingly the presence of reactive aggregate reduced the expansion of the HAC mortar.

Analysis showed that the pore solution of both HAC mortars had reached equilibrium with the external host solution after 90 days with both Na^+ and OH^- concentrations being close to 1000 mM/L. Silica contents in the pore solution were found to be $SiO_2 = 32$ mM/L and $SiO_2 < 1$ mM/L for HAC mortars with Spratt and Nelson aggregate, respectively.

SEM analysis showed only trace amounts of reaction product within intact aggregate particles in mortars containing Spratt aggregate. This product had a very low calcium content ([Ca]/[Si] < 0.1). X-ray diffraction analysis revealed the presence of small quantities of strätlingite (C_2ASH_8) in the HAC mortars containing reactive aggregate.

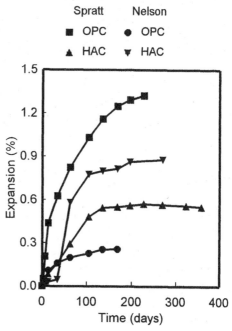

Figure 3 Expansion of Mortar Bars with High-Alumina Cement

High-alumina cement was selected for study as it does not produce $Ca(OH)_2$ during hydration and it was thought that CaO would be less available than in ordinary Portland cement. In hindsight, HAC was not a good choice for these tests as the hydrated aluminates may be attacked by alkali hydroxides (alkaline hydrolysis) (13) masking any effects due to ASR.

STUDIES WITH LIME-FLY ASH BINDERS

Recently a series of mortar specimens have been cast using alkali-activated lime-fly ash as the cementitious component (i.e. no Portland cement). In these tests Spratt aggregate was combined with $Ca(OH)_2$, low-lime fly ash from bituminous coal (lime-fly ash ratio from 0.7 to 9.3), and NaOH (1.3% by mass of lime + fly ash). The aggregate-cementitious material ratio was 2.25 and the water content was adjusted to maintain a constant flow. Mortar bars were cured at 80°C over water for 7 days prior to immersion in 1M NaOH at 80°C.

Expansion results following 28 days immersion in NaOH solution are shown in Figure 4. The expansion decreases as the lime-fly ash ratio decreases indicating that the magnitude of expansion is dependent on the availability of $Ca(OH)_2$. Pore solution expressed from mortar bars at 28 days were analyzed as follows:

Ash-Lime	Na^+	K^+	OH^-
0.7	1.18	0.00	1.09
9.3	1.02	0.01	0.19

It is suspected that the mass balance in the solution from the sample with the highest ash-lime ratio is maintained by silica. Unfortunately, insufficient solution was retrieved to permit analysis.

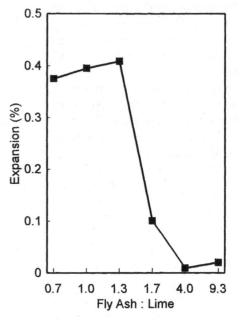

Figure 4   Expansion of Lime-Fly Ash Mortar Bars

EFFECT OF LIME ON THE PESSIMUM LEVEL OF FLINT SAND

Mortar bars containing various levels of flint sand (from the Thames Valley in the U.K.) were tested in accordance with ASTM C 1260 (except that the duration of test was extended beyond 14 days). This source of reactive sand is known to display pessimum effects in concrete with the maximum expansion being observed when the sand makes up between 20 to 30% of the total aggregate (14). Significant expansion is generally not observed in concrete when the flint sand makes up more than 60% of the aggregate. Expansion results for various levels of flint sand (blended with non-reactive carboniferous limestone) are shown in Figure 5.

It is somewhat surprising that the accelerated mortar bar test (ASTM C 1260) is sensitive to pessimum effects. The pessimum effect supposedly reflects the sensitivity of expansion to the alkali-reactive silica ratio in the system (14). If the reactive silica content is very high, rapid consumption of alkali occurs whilst the concrete is still plastic and insufficient alkali remains to promote deleterious reaction and expansion in the hardened concrete. In the accelerated test, an inexhaustible external supply of alkalis is provided to fuel further reaction and expansion; thus one would expect pessimum effects to be overcome. Figure 5 shows that this is clearly not the case as a clear pessimum level exists around 30% flint sand. The addition of $Ca(OH)_2$ to the mortar at the mixing stage has a marked effect on the expansion, especially when the level of flint sand is above the normal pessimum proportion. Deleterious expansion and cracking is observed for all the mortars containing flint and $Ca(OH)_2$ even when the flint sand represents 100% of the aggregate. Furthermore, the pessimum content appears to shift to higher values as the level of $Ca(OH)_2$ increases.

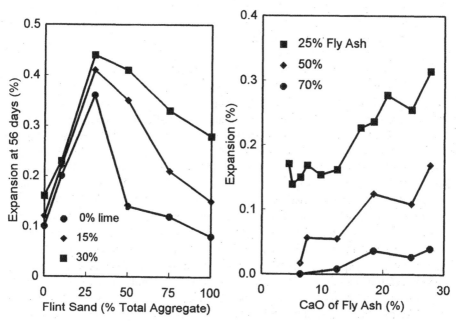

Figure 5   Effect of Lime on the Pessimum Flint Content

Figure 6   Effect of Fly Ash Composition on the Expansion of Mortars

THE EFFECT OF FLY ASH COMPOSITION ON EXPANSION

As part of a wider program looking at the effects of fly ash composition on ASR, mortar bar expansion tests were carried out using a wide range of fly ashes. Fly ashes with calcium contents in the range 4 to 28% CaO were used together with low alkali cement (Na_2O_e < 0.60%) and different reactive aggregates to produce mortar bars. These specimens were cured for 28 days in limewater at 38°C, prior to immersion in 1M NaOH at the same temperature. The aim of these tests was to evaluate the role of fly ash on ASR beyond pore solution effects; i.e. the storage conditions were intended to compensate for any effect the fly ash might have on the pore solution alkalinity. Mortar bars were produced with a high water-cementitious material (W/CM = 0.70) to facilitate rapid equilibrium between the pore solution in the mortar and the external host solution.

Figure 6 shows the expansion of mortar bars after approximately 12 months storage in NaOH solution. The expansion of the mortar bars is clearly dependent on both the level of fly ash and the calcium content of the ash.

DISCUSSION

The various studies reported above demonstrate that the presence of reactive silica and alkali hydroxides is a necessary but not sufficient condition for damaging ASR to occur. It is apparent that calcium plays an important role in producing expansion and that in the absence of available calcium, reaction may proceed without causing expansion.

The dependence on $Ca(OH)_2$ for the promotion of damaging AAR is not a recent phenomenon. In 1952, Conrow (15) suggested that the expansion of concretes containing a siliceous sand-gravel may be related to the quantity of $Ca(OH)_2$ produced by the cement and that the beneficial effect of pozzolan is related to its ability to react with $Ca(OH)_2$. In a discussion of this paper, Mather stated an observation that concrete that has undergone ASR is characterized by " ... *materially reduced quantities of crystalline calcium hydroxide.*" He suggests that $Ca(OH)_2$ may be consumed by the reaction and that "... *the mere consumption of calcium hydroxide by reaction with a pozzolan is sufficient to explain the beneficial effects of pozzolans in preventing abnormal expansion.*"

Chatterji (5) showed that the expansion of mortar bars containing opaline silica, immersed in saturated NaCl at 50°C could be eliminated if $Ca(OH)_2$ was removed prior to exposure. Lime removal was achieved by first leaching the bars in concentrated $CaCl_2$ or by the addition of a highly siliceous pozzolan (diatomite). In addition, mortar bars with very low cement contents (aggregate/cement = 19) did not expand and this is attributed to insufficient $Ca(OH)_2$ in these specimens. Significant quantities of silica gel were observed to precipitate on the surfaces of mortar bars with low cement content. Chatterji (5,16) has proposed an ASR expansion mechanism with the role of $Ca(OH)_2$ clearly defined. He suggested that expansion occurs in concrete when the net amount of material entering a reactive silica grain (K^+, Na^+, Ca^{2+}, OH^-) exceeds the amount of material leaving the grain (Si^{4+}). The calcium concentration in the pore solution surrounding the grain controls the rate of diffusion of silica away from the reactive site and at high levels of calcium the migration of silica is prevented, leading to expansion of the reactive grain.

Tang et al. (17) suggest that the preventative effect of mineral admixtures is related to their acidic oxide content (SiO_2 + Al_2O_3 + Fe_2O_3) and their ability to remove $Ca(OH)_2$ and lower the "basicity" of the system. They found that adding

10% CaO to mortar mixes containing opal and various admixtures increased the expansion and explained this on the basis of the increased "basicity" [CaO/(SiO_2 + Al_2O_3 + Fe_2O_3)] of the cement. Similar behaviour was reported by Thomas (10,11); the addition of $Ca(OH)_2$ to fly ash concretes containing reactive flint was found to substantially increase the expansion of specimens stored in alkaline solution. The absence of expansion in fly ash concretes despite evidence of reaction has been attributed to the formation of a fluid or soluble reaction product in the absence of $Ca(OH)_2$ (9,10).

Other workers have suggested that the incorporation of calcium into the reaction product is actually beneficial in terms of reducing the tendency for expansion to occur. Powers and Steinour (18,19) postulated that the ratio of lime to alkali determined the type of reaction product formed. At low concentrations of $Ca(OH)_2$ an expansive or "unsafe" alkali-silica product is formed but if sufficient $Ca(OH)_2$ was present a non-expansive or "safe" calcium-alkali-silica complex is formed. Similar findings were recently reported by Wilson et al (20), who suggested that the expansive forces were reduced if high Ca^{2+} concentrations could be maintained in the pore solution. They suggested that the high concentrations of Ca^{2+} could be achieved by the use of pozzolans. Support for this can be found in the work of Kawamura et al. (21) who found that the presence of fly ash increased the penetration of alkali and, especially, calcium ions into opal particles. They concluded that fly ash does not actually inhibit alkali-silica reaction but facilitates the mobilization of calcium into the reacting particles leading to the formation of reaction products with higher calcium contents. It is suggested that the formation of higher calcium gels is responsible for the absence of expansion.

Struble and Diamond (22,23) suggested that the explanation of Powers and Steinour (18,19) regarding gel composition was an oversimplification. They studied the swelling properties of a series of synthetic alkali-silicate (Na_2O-SiO_2-H_2O) and lime-alkali-silicate (CaO-Na_2O-SiO_2-H_2O) gels of varying composition. No consistent relationship between the composition of the gel and either the free expansion or swelling pressure was found. Furthermore, the presence of calcium had no significant effect on the free expansion or the swelling pressure generated by the gel. Calcium gels behaved in a similar manner to gels of the same Na_2O/SiO_2 ratio without calcium. However, calcium gels were observed to remain solid after testing whereas the sodium-silicate gels generally became fluid. The molecular structure of some of these synthetic gels was found to be unstable, and the gel became fluid to an extent that it was able to pass through the semipermeable membrane used to constrain gel movement.

In a further study of opal in "model" pore solutions, Struble (24) demonstrated that significant quantities of silica ($\geq 2M$) can remain in solution with alkali hydroxides in the absence of cement hydration products (calcium) without the formation of alkali-silicate gel. However, in mortars containing cement together with the opal, the silica content of the pore solution was more than 2000 times lower (<0.001M), and gel was observed to form. This would imply that silica simply dissolves in alkali hydroxide solution and does not form alkali-silicate gel in the absence of calcium (4).

Further support of the solubility of silica in the absence of calcium is found in the work of Kilgour (25). She found low-calcium ashes to be partially soluble (losing 20% soluble mass in 6 months) when exposed to alkali hydroxyl solution (1g ash in 100 ml of 0.7M NaOH-KOH) in the absence of calcium. The principal soluble phase was analyzed as silica which remained in solution (no evidence of gel formation). Under the same conditions except with addition of calcium hydroxide, an increase in mass was observed and attributed to the formation of a reaction product analogous to ASR gel.

The results of this author's studies do not support the theory that the incorporation of calcium in the reaction product suppresses expansion. Nor do they support the finding that fly ash increases the rate of penetration of calcium into reacting aggregate particles resulting in the formation of a 'safe' product. On the contrary, damaging expansion has only been found in association with high-calcium gels and the use of sufficient levels of fly ash appears to prevent both the production of high-calcium gel and the development of expansion (10,11). The consumption of readily available $Ca(OH)_2$, either by pozzolanic reaction to form CSH or carbonation to form $CaCO_3$, prevents significant expansion from occurring even in the presence of reactive silica and alkali hydroxides. The beneficial effects of pozzolans can be negated by simply providing additional $Ca(OH)_2$ at the mixing stage.

Previous studies with fly ash indicated that, in the absence of $Ca(OH)_2$, the reaction product is an extremely fluid alkali-silica gel that is readily accommodated within the cement matrix. The more recent tests on carbonated mortars suggest that, without calcium, significant quantities of reactive silica can simply dissolve in the highly alkaline pore solution. The silica concentration in the pore solution of the carbonated mortar was found to be over 400 mM/L SiO_2. This may be lower than the 2 molar silica concentrations observed by Struble (24), but the conditions of test were very different in the two studies.

The fact that pessimum behaviour is observed in expansion tests where the alkalis are supplied from an inexhaustible external reservoir suggests that the effect is not just a function of the ratio of alkalis to reactive silica in the system. The ability of admixed $Ca(OH)_2$ to induce expansion in mortar bars with reactive silica contents above the normal pessimum, and the tendency for increased $Ca(OH)_2$ to raise the pessimum content, indicates that pessimum behaviour is a function of the available $CaO\text{-}Na_2O_e\text{-}SiO_2$ in the system. In tests where mortar bars are immersed in alkali hydroxide solutions, the pessimum level will likely be strongly dependent on the availability of the $Ca(OH)_2$.

It is not clear precisely what role the calcium in fly ash plays in determining the expansion due to ASR. Under test conditions representative of field conditions (i.e. no external alkalis), the composition of the fly ash likely effects the pore solution alkalinity and hence the potential for reaction. However, in the test conditions used in the study reported here, the alkalinity of the pore solution is controlled by the host solution. Thus, the fly ash clearly has a role beyond mere pore solution effects and the results suggest that the efficacy of the fly ash in controlling expansion in these tests is linked to the ability of the ash to reduce the availability of calcium.

$Ca(OH)_2$ is rarely completely consumed by pozzolanic reaction with fly ash unless relatively large ash levels are used. However, the use of fly ash can produce localized $Ca(OH)_2$ depletion and increased density of hydration products in the interfacial zone. These changes decrease the buffer of OH^- ions and reduce ionic mobility in this region and it has been suggested (26) that these effects are of paramount importance for reducing the alkali-silica reaction in concrete. Furthermore, the complete removal of $Ca(OH)_2$ may not be necessary to prevent damaging expansion. It is possible that there is a limiting $Ca(OH)_2$ content below which further reaction or expansion does not occur. It is likely that, at a given level of replacement, low-calcium fly ash is more effective in tying up calcium than high-calcium fly ash. This may explain the reduced efficiency of high-calcium fly ash in controlling expansion of mortar bars immersed in alkaline solutions. However, further studies are required to determine the role of fly ash composition.

REFERENCES

1. Hansen, W.C. *Journal of the American Concrete Institute,* Vol. 15 (3), 1944, pp. 213-27.

2. McGowan, J.K.,Vivian, H.E. *Australian Journal of Applied Science,* Vol. 3, 1952, pp. 228-232.

3. Dent Glasser, L.S. *Cement and Concrete Research,* Vol. 9, 1979, pp. 515-517.

4. Diamond, S. *Proceedings of the 8th International Conference on Alkali-Aggregate Reaction,* (Ed. K.Okada et al), Kyoto, 1989, pp. 83-94.

5. Chatterji, S. *Cement and Concrete Research,* Vol. 9 (2), 1979, pp. 185-188.

6. Chatterji, S., Clausson-Kass, N. F. *Cement and Concrete Research,* Vol. 14 (6), 1984, pp. 816-818.

7. Tang, M., Han, S. *Proceedings of the 7th International Conference on the Chemistry of Cement*, Vol. II, 1980, p. 94.

8. Wang, H., Gillott, J.E. *Cement and Concrete Research*, Vol. 21 (4), 1991, pp. 647-654.

9. Thomas, M.D.A., Nixon, P.J., Pettifer, K. *Proceedings of the 2nd CANMET/ACI International Conference on Durability of Concrete*, (Ed. V.M. Malhotra), Vol. 2, American Concrete Institute, Detroit, 1991, pp. 919-940.

10. Bleszynski, R.F., Thomas, M.D.A. *Advanced Cement Based Materials*, Vol. 7, 1998, pp. 66-78.

11. Thomas, M.D.A. *Proceedings of the 17th International Conference on Cement Microscopy*, Calgary, Canada, 1995.

12. Barneyback Jr., R.S. and Diamond, S. *Cement and Concrete Research*, Vol. 11, 1991, pp. 279-285.

13. Taylor, H.F.W. *Cement Chemistry*, Academic Press, London, 1990, 475 p.

14. Nixon, P.J., Page, C.L., Hardcastle, J., Canham, I., Pettifer, K. *Proceedings of the Eighth International Conference on Alkali-Aggregate Reaction*, (Ed. K. Okada et al.), Kyoto, 1989, pp 129-134.

15. Conrow, A.D. *Proceedings of the American Society for Testing and Materials*, Vol. 52, 1952, pp. 1205-1227.

16. Chatterji, S. *Proceedings of the 8th International Conference on Alkali-Aggregate Reaction*, (Ed. K.Okada et al), Kyoto, 1989, pp. 101-105.

17. Tang, M., Ye, Y.F., Yuan, M.Q., Zheng, S.H. *Cement and Concrete Research*, Vol. 13, 1983, pp. 171-176.

18. Powers, T.C., Steinour, H.H. *Journal of the American Concrete Institute*, Vol. 26 (6), 1955, pp. 497-516.

19. Powers, T.C., Steinour, H.H. *Journal of the American Concrete Institute*, 26 (8), 1955, pp 785-811.

20. Wilson, M., Cabrera, J.G., Zou, Y. *Advances in Cement Research*, Vol. 6 (23), 1994, pp. 117-125.

21. Kawamura, M., Takemoto, K., Hasaba, S. *Transactions of the Japan Concrete Institute*, Vol. 5, 1983, pp. 91-96.

22. Struble, L.J., Diamond, S. *Journal of the American Ceramic Society*, 64 (11), 1981, pp. 652-656.

23. Struble, L.J., Diamond, S. *Cement and Concrete Research,* 11 (4), 1981, pp 611-617.

24. Struble, L.J. *The Influence of Cement Pore Solution on Alkali Silica Reaction.* PhD Thesis, Purdue University, 1987.

25. Kilgour, C.L. *Composition and Properties of Indiana Fly Ashes.* PhD Thesis, Purdue University, Indiana, 1988.

26. Larbi, J.A., Bijen, J.M.J.M. *Proceedings of the 4th International Conference on Fly Ash, Silica Fume, Slag and Natural Pozzolans in Concrete,* (Ed. V.M. Malhotra), ACI SP132, Vol. 1, American Concrete Institute, Detroit, 1992, pp. 655-669.

MICROSTRUCTURE, ELASTIC PROPERTIES, AND FREEZE-THAW DURABILITY OF HIGH-PERFORMANCE CONCRETE

Walid Baalbaki and Pierre-Claude Aïtcin
Department of Civil Engineering
Université de Sherbrooke
Sherbrooke, QC (Canada) J1K 2R1

Guanshu Li and Micheline Moranville
Laboratoire d'Études et de Recherches sur les Matériaux
72-74 avenue Gambetta
93170 Bagnolet (France)

ABSTRACT

SEM observation of fractured surface of concrete can explain why the $E'_c = \phi$ (f'_c) relationships commonly found in codes, and which perform well for so-called normal-strength concrete, fail to accurately predict the elastic modulus of high-performance concrete. In high-performance concrete, the bond between hydrated cement paste and aggregate is so complete that the elastic properties of the aggregate significantly influence those of the high-performance concrete. In normal-strength concrete, however, in which the bond between aggregates and the hydrated cement paste is very loose, it is the elastic properties of the very porous hydrated cement paste that mostly determines the elastic properties of concrete.

When concrete subjected to freezing-thawing cycles under saturated conditions is observed under a SEM, significant amount of nonexpansive ettringite can be observed in laboratory specimens and field samples, especially in the carbonated area. This ettringite is essentially found in air bubbles and microcracks. When the freeze-thaw cycles occur in a semi-dry condition, ettringite is less prone to develop within the concrete. Consequently freezing and thawing in concrete is somewhat more than a simple increase of a volume increase of water as it freezes that can be overcome by introducing an air-bubble system with an appropriate spacing factor. Freeze-thaw cycles are also accompanied by a series of ionic solubilization and precipitation, which precipitate secondary ettringite into air bubbles and preexisting microcracks, especially in carbonated areas.

INTRODUCTION

In recent years, concrete technology has evolved remarkably, transforming concrete from essentially a commodity product to a more sophisticated construction material, and even, in some cases, into a high-tech material that has more in common with ceramics than what we usually think of as concrete (1). Most of the improved characteristics of this new concrete are not chance occurrences, but rather result from the use of very efficient admixtures, from a better understanding of the interaction between admixtures and portland cement, silica fume, and the different supplementary cementitious materials currently used when making concrete (2).

These advances would not have been so rapid and extensive without the use of sophisticated investigating techniques that complemented observation with the naked eye and optical microscope, which was as far as we could go until the number of years ago. Today the materials used to make portland-cement-based materials are tested and characterized with a variety of techniques that includes SEM, TEM, NMR, IR spectrometry, GLC, and ESCA (3). We now have a fair understanding of these materials at the microscopic level and how they react. In a few cases, we are even beginning to improve our knowledge of concrete on the nanometer scale with the Atomic Force Microscope, AFM (4, 5).

Clearly, our understanding of the microstructure of calcium silicate hydrate is not as good as that of two other closely related silicate hydrates: kaolinite, which is the aluminum counterpart of calcium silicate hydrate, and chrysotile, its magnesium counterpart. It is amazing to see how a small difference in the diameter of Ca^{2+}, Mg^{2+} and Al^{3+} results in drastically different structures and properties in these hydrated silicate minerals.

Sidney Diamond was one of the first of us to foresee the potential of using electron microscopy to better understand hydrated cement paste and, thereby have a clearer understanding of concrete (6). His pioneering work started in the seventies when he described the different types of C-S-H and resulted in making SEM observation a must when a particular behavior of concrete has to be understood and explained, when a diagnostic has to be given on some concrete distresses, or when some new developments in the area of admixtures or supplementary cementitious material is taking place (7).

We would not be going too far in saying that the microstructural knowledge derived through SEM observation underlies the recent successes in relating macroscopic properties and behavior to concrete microstructure.

As the mechanical properties and the durability of any material are the macrostructural reflections of its microstructure, Sidney's contribution in this domain must be emphasized. This paper aims at showing how SEM observations have been very useful in the development and practical applications of high-performance concrete over the last 20 years and in understanding of freeze-thaw mechanism in concrete.

NORMAL-STRENGTH AND HIGH-PERFORMANCE CONCRETE

The concrete community continues to argue about whether "high strength" and "high-performance" are appropriate expressions in referring to concrete. The authors believe that such discussions are frivolous for scientists. In reality, this kind of concrete should be known as low water/cement ratio or even better low water/binder ratio concrete because portland cement is no longer the only cementitious material that goes into concrete. All the characteristics and properties common to this concrete derive from a water/binder ratio that is significantly lower than in so-called normal-strength concrete. This difference in the water/binder ratio radically impacts concrete microstructure; these microstructure differences can account for the differences in the macro-properties of low and high water/binder ratio concrete (8, 9).

ESSENTIAL DIFFERENCES BETWEEN THE MICROSTRUCTURE OF NORMAL-STRENGTH AND HIGH-PERFORMANCE CONCRETE

It is surprising that so many people still speak of normal-strength concrete when referring to concrete in the 15-to-30-MPa (2000 to 4500 psi) range. Obviously, they have never observed such concrete under a scanning electron microscope, because they would have been horrified to see how poor its microstructure is (see Figure 1). The amazing thing is that this type of concrete has any compressive strength at all when tested under uniaxial compressive load!

Concrete in this range has a very open microstructure, characterized by very porous contact between the coarse aggregate and hydrated cement paste. This interface often shows gaps filled with large portlandite crystals, long ettringite needles, and large monosulfoaluminate crystals. These crystals have no trouble growing around the aggregates in this very weak zone, which is currently referred to as the transition zone. A close observation of this transition zone also reveals that portlandite crystals can be oriented at the surface of the aggregate (see Figure 1).

It is not also surprising that this weak concrete (from a compressive strength point of view) is not durable and is so vulnerable to attack by external aggressive agents. There are so many wide-open routes for these agents to penetrate concrete and develop their detrimental action.

Figure 1 also illustrates why coarse aggregate has practically no influence on the elastic modulus of low water/cement ratio concrete: the contact between the hydrated cement paste and coarse aggregate is so poor that there is practically no stress transfer at this level.

Figure 2, in contrast, clearly shows that HPC durability derives from the compactness of the hydrated cement paste and that the tightness of the bond between the coarse aggregate and hydrated cement paste allows the coarse aggregate

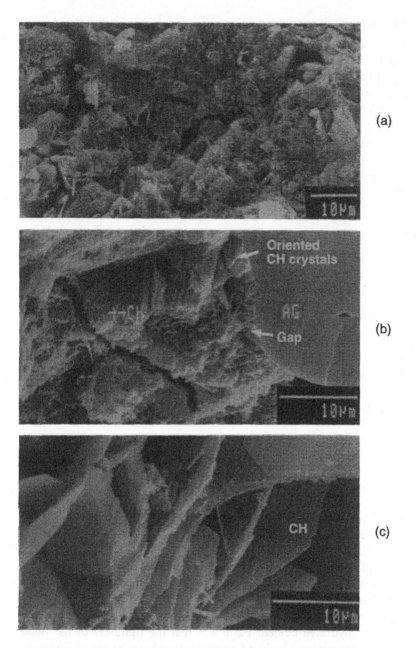

Figure 1. Microstructure of high water/cement ratio concrete: (a) high porosity and heterogeneity of the matrix, (b) orientated crystal of Ca(OH)$_2$ on aggregate, (c) CH crystals

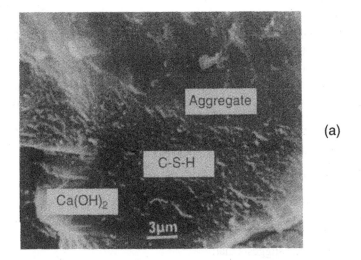

(a)

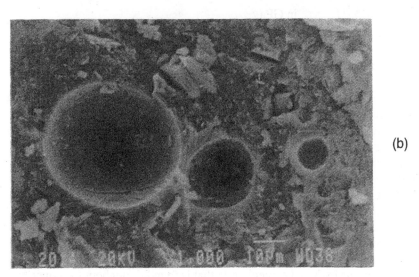

(b)

Figure 2. Microstructure of a high performance concrete: low porosity and homogeneity of the matrix

to affect the concrete's elastic properties. The two pictures in Figure 2 show the great homogeneity and low porosity of the matrix. There are no apparent crystals, and the bonding between the matrix and the aggregate is excellent. One picture shows some fine cracks, of unknown origin, through the matrix and air bubbles.

HOW THESE OBSERVED MICROSTRUCTURAL DIFFERENCES CAN EXPLAIN THE DIFFERENCES IN THE ELASTIC PROPERTIES OF HIGH AND LOW WATER/BINDER RATIOS CONCRETE

In the past, in order to simplify the design of concrete structures, structural engineers were successful in linking most of the mechanical properties of high water/binder ratio concrete to concrete compressive strength. In particular, the codes link concrete elastic modulus E'_c to concrete compressive strength $'_c$, through different empirical formulas (10).

The European code CEB 1990 suggests that:

$$E'_c = 10 \left(f'_c + 8 \right)^{1/3} \quad \text{in GPa}$$

The ACI code suggests that:

$$E'_c = 3.32 \sqrt{f'_c} + 6.9 \text{ in GPa}$$

The Canadian code suggests that:

$$E'_c = 5 \sqrt{f'_c} \quad \text{in GPa}$$

The Norwegian code suggests that:

$$E'_c = 9.5 \left(f'_c \right)^{1/3} \frac{\rho}{2400} \quad \text{in GPa with } \rho \text{ in kg/m}^3.$$

Essentially, these relationships show that concrete elastic modulus (difficult to measure) can be deduced from concrete compressive strength (easy to measure). The determination of the elastic modulus involves the simultaneous recording of a load and the resulting strain.

Experience shows that as long as water/binder ratio of concrete is greater than 0.50, it is found that the predictive value of these formulas is quite acceptable with an error not greater than 10 to 20 percent which is satisfying for most designers.

Microstructural observation of these concrete (see Figure 1) can explain how concrete elastic modulus can be linked to f'_c in what appears, at first glance, to be a heterogeneous, porous material. In fact, high water/binder ratio concrete is more homogeneous than it appears on the macroscopic level due to the overall weakness of the very porous hydrated cement paste that is found throughout the concrete and more particularly in the transition zone. Moreover, in concrete with a high water/binder ratio, aggregate does not participate to any great degree when an axial compressive load is applied to the concrete, which means that the aggregate's elastic characteristics contribute little to the concrete's overall elastic response. Therefore, concrete compressive strength and elastic modulus are almost exclusively dependent on the quality of the hydrated cement paste, which means that they can be linked together with the kind of relationships found in current codes.

This is not the case with low water/binder ratio concrete, in which, as shown in Figure 2, there is very close bonding between the aggregate and the very dense hydrated cement paste; there is practically no weak transition zone. Therefore, it is not surprising to see that the elastic response of such concrete is largely influenced by the elastic properties of the component aggregate, as shown in Figure 3 (11). Nilsen and Aïtcin (12) have shown that the kind of relationship between $'_c$ and E'_C suggested in current codes is of no value with low/binder concrete because two HPC with the same compressive strength can have quite different E'_c. For example, the modulus of elasticity of a 100-MPa lightweight concrete can be 25 GPa compared to 65 GPa for a 100-MPa heavyweight concrete with ilmenite as its coarse and fine aggregate.

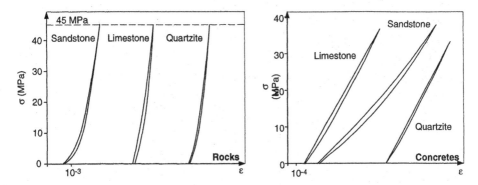

Figure 3. Stress-strain curves of different rocks and of high-performance concretes made with crushed coarse aggregates made out of these rocks

It was also found that Poisson's ratio of low water/cement or low water/binder ratio concrete was greatly influenced by the Poisson's ratio of the aggregate.

Recently, in an effort to develop a more realistic mechanical model for concrete based on what is observed on concrete microstructure, W. Baalbaki proposed two mechanical models, one of which is shown in Figure 4. Based on his first model, Baalbaki proposes the following formula for predicting concrete elastic modulus (13)

$$E'_c = ag_1 E_1 + \frac{(1-ag_1)^2}{\dfrac{(1-a)}{E_1}g_1 + \dfrac{g_2}{E_2}}$$

where E_1 and E_2 are the elastic moduli of the mortar and the aggregate, g_1 and g_2 their respective proportion in the concrete and, a, a parameter depending on the E_1 and E_2 ratios and the type of aggregate used.

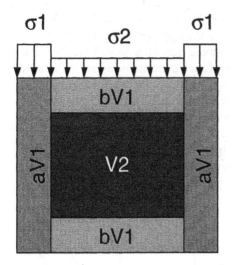

bV1	hydrated cement paste in the transition zone
aV1	hydrated cement paste
V2	volume of the coarse aggregate

Figure 4. W. Baalbaki model for concrete

Baalbaki developed a second formula that yields the E'_c value when the elastic modulus of the paste, E_p, and the elastic modulus of the aggregate, E_g, are known:

$$E'_c = \left(g^* - g\right)E_p + \frac{\left(1 + g - g^*\right)}{\dfrac{g}{E_g} + \dfrac{\left(1 - g^*\right)}{E_p}}$$

where $g^* = 1 - 0.47\left(\dfrac{d}{D}\right)^{0.2}$, d being the minimal diameter of the fine aggregate and D being the maximum size of the coarse aggregate, and g is the proportion of aggregate in the paste. Figure 5 shows that the values published by several researchers fit quite well with the values predicted with model (13).

MICROSTRUCTURE OF CONCRETE EXPOSED TO FREEZING AND THAWING

Freezing and thawing is a major cause of distress in concrete. Despite the facts that the macroscale effects of freezing and thawing can be easily recognized, that the beneficial effects of entraining air in concrete to make concrete "freeze-thaw resistant" is indubitable, and that the role of air bubbles during freezing and thawing is more or less well understood, SEM observations of concrete that have been subjected to freezing and thawing show that there are still some physicochemical phenomena occurring during freezing and thawing that are not accounted for by the current theories on the action of freeze-thaw cycles in concrete. For example, the presence of ettringite in air bubbles or microcracks developed during the freeze-thaw cycles has never been clearly explained (14).

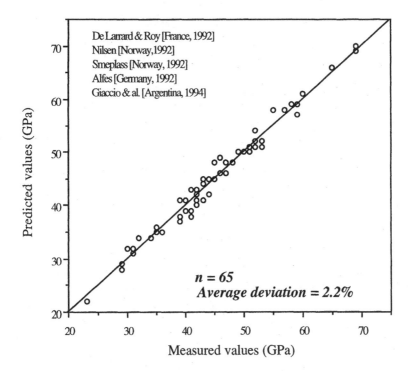

Figure 5. W. Baalbaki model for concrete

In order to illustrate this last point, the concrete microstructure of three concretes will be presented and discussed.

The first is a 60-MPa (8700-psi) air-entrained high-performance silica-fume concrete that was cast in 1980 as part of an experimental sidewalk. It was one of the first uses of silica fume in concrete in North America, and several papers have been published on the results of this experiment (15, 16).

It is interesting to look at the microstructure of this air-entrained concrete characterized by a quite low air content (2.6 percent on the hardened concrete), a quite good spacing factor (215 μm; 8.5×10^{-3} in), and a quite high air-void specific surface 36 mm^{-1} (914 in^{-1}). Over 18 years, it has withstood an average of 50 natural freeze-thaw cycles as well as to an unknown (but large) number of artificial freeze-thaw ones through the repeated use of deicing salts. The chloride-ion permeability of this concrete remains under 800 Coulombs. The concrete in question is located at the entrance of the social center of a plant with 200 workers going in and out.

The second concrete is an air-entrained high-performance concrete that passed successfully 500 freeze-thaw cycles in conformity to Procedure A (freezing and thawing in water) of ASTM C666. This high-performance concrete tested at 57.4 MPa (8300 psi) at 28 days. Its air content was 8.6%, its spacing factor 195 µm (7.8 × 10⁻³ in), and its air void specific surface 16 mm⁻¹ Consequently, this concrete is freeze-thaw resistant according to the most severe acceptance standards. For example, Canadian CSA A23.1 requires that, to be considered freeze-thaw resistant, a concrete must have an average spacing factor under 230 µm with no individual value higher than 260 µm. Therefore, this air-entrained high-performance concrete qualified as a freeze thaw resistant concrete in all respects.

At the end of the 300 cycles, there was some room available in the freeze-thaw testing machine, so the concrete specimens were left in the machine to determine their long-term behavior under repeated freeze-thaw cycles. These "freeze-thaw resistant" specimens failed after about 1900 cycles and were destroyed after 1960 cycles.

The last concrete is a 40-MPa silica-fume concrete used in the footings of a barn built in 1981. The outer faces of the footing have been exposed to freezing and thawing in saturated conditions with no deicing-salt exposure during the last 17 years. These conditions correspond roughly to the conditions required under ASTM C666 Procedure A, although the rate of freezing and thawing was much lower. The inner faces of the footing were subjected to the same number of the freeze-thaw cycles (the barn is not heated) but under semi-dry conditions. The concrete remains in very good shape; its chloride-ion permeability is 650 coulombs. The microstructure of this concrete, which has been subjected to two different types of freeze-thaw cycles will be compared, specifically focusing on the carbonated area as well as deeper into the uncarbonated areas.

SEM observation of the 18-year old air-entrained high-performance silica-fume concrete (Figures 6 and 7)

Figure 6 (a) shows the presence of ettringite fibers in a pore co-existing with portlandite crystals. Most probably, the pore solution was high in calcium. The type of ettringite present in this pore is nonexpansive. Figure 6 (b) shows ettringite crystals in a crack; Figure 7 some ettringite in a pore.

This silica-fume air-entrained high-performance concrete, which continues to perform outstandingly after 18 years of freezing and thawing in the field with significant exposure to deicing salts, evidenced few ettringite crystals in its air voids and in the few microcracks that have been observed. The ettringite crystals are of a nonexpansive type; those found in the cracks were deposited there after formation.

SEM observation of "freeze-thaw resistant" high-performance concrete after 1960 freeze-thaw cycles

Figure 8 shows the accumulation of nonexpansive ettringite crystals in a pore and in a crack around a pore. This observation could support the usefulness of

Procedure A of ASTM C666 in simulating freeze-thaw field conditions, because of the usual development of nonexpansive ettringite crystals in field concrete exposed to freezing and thawing as well as the laboratory concrete tested under accelerated test conditions.

Figure 6. Air-entrained, high-performance silica-fume concrete used to build an experimental sidewalk 18 years ago

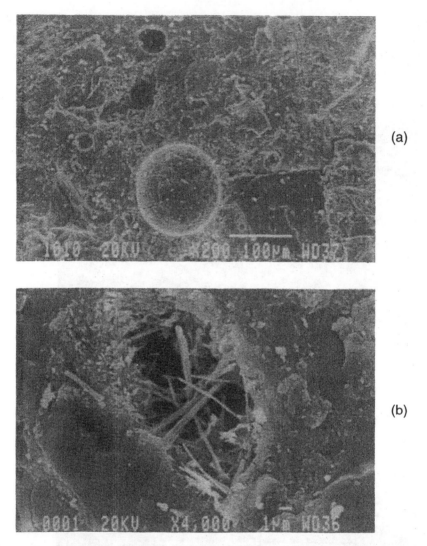

(a)

(b)

Figure 7. General aspect and few ettringite crystals in a pore of the air-entrained, high-performance silica-fume concrete used to build an experimental sidewalk 18 years ago

<u>SEM observation of air-entrained silica-fume concrete submitted to freeze-thaw cycles in saturated and dry conditions</u>

Figure 9 presents two typical air bubbles filled with ettringite crystals in the carbonated area on the top of the outer face of the footing. The dense hydrated cement paste usually found in silica-fume concrete surrounds these bubbles.

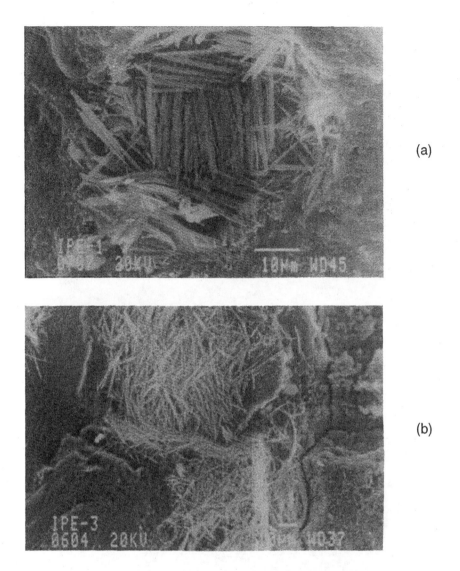

(a)

(b)

Figure 8. Large deposits of ettringite crystals in the freeze-thaw resistant high-performance concrete after its failure after 1960 freezing and thawing cycles

Figure 10 represents a typical fractured surface of this air-entrained concrete from the noncarbonated part at the bottom of outer face of the footing. The system of air bubbles within the dense hydrated cement paste is well distributed, although, in this case, the inside of the air bubbles is usually smooth and free of ettringite deposits. Most probably, the concrete in this part of the footing was not subjected to as many freeze-thaw cycles because the snow cover acted as natural thermal insulation.

Figure 9. Tiny air bubbles filled with ettringite in the carbonated part of the external footing

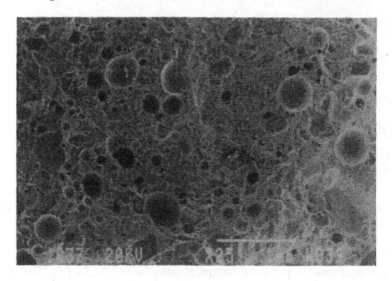

Figure 10. Air bubbles free of ettringite in the non carbonated part of the outside footing

Figure 11 presents a typical air bubble found in the carbonated area of the inside footing. This time, the inside of the bubble with minute calcite crystals, instead of ettringite needles.

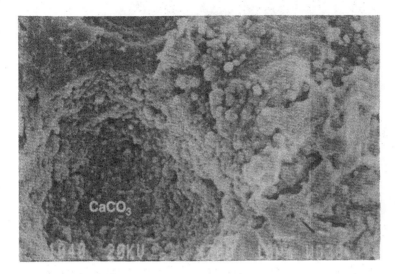

Figure 11. In the carbonated part of the inside footing air bubbles are no more coated with ettringite needles but rather with minute calcite crystals

Finally, Figure 12 presents the typical aspect of the fractured surface of a concrete sample taken from the bottom of the inside part of the footing. This picture clearly shows that the inside of the air bubbles is usually free of any deposited crystals. These observations tend to prove that there is water movement towards the air bubbles during freezing and thawing, as indicated by the different crystalline coatings. The water involved, however, isn't pure but rather rich in ions moving towards the bubbles and the cracks. These ionic species interact to form essentially nonexpansive ettringite crystals or calcite crystals in more or less great numbers, depending on the local conditions around the air bubble or the crack.

Some argue that the ettringite very often found in such concretes could be responsible for the cracking. It has further been argued that air bubbles provide sites where ettringite crystals could develop without any deleterious action on concrete microstructure. In fact, SEM observation of non-air-entrained and air-entrained concrete submitted to freeze-all cycles under saturated conditions reveals the number of large ettringite crystals in air bubbles and cracks. But a closer look at the morphological characteristics of these ettringite crystals shows that they have been precipitated into the cracks and air bubbles during the freezing cycles. In other words, they are not responsible for the cracks. They grew and developed from ions that went into solution during the thawing cycles and precipitated in subsequent freezing cycles.

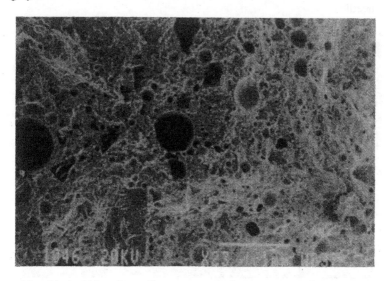

Figure 12. Air bubbles free of any crystals in the non carbonated part of the inside footing

CONCLUSION

Observation of concrete microstructure is invaluable when the properties and behavior of this complex material have to be explained and understood. SEM observation tends to show that oversimplifying concrete behavior by using simple or simplistic formulas may have advantages from a technological standpoint, it is in actuality much more complex. Of course, since concrete obeys the law of physics, chemistry, and thermodynamics, it is up to us to use correctly these laws to interpret images from SEM and, in the very near future, from the Atomic Force microscope. Then we will be able to develop more realistic models that convey mathematically what we observe. This will enable us to quantify the influence of concrete microstructure features at a macroscopic level in terms of mechanical and durability properties, as suggested by a doctoral candidate that you, Sidney, supervised during of your stay at the University of British Columbia (17).

ACKNOWLEDGMENTS

The authors are grateful to Irène Kelsey Lévesque, who took the SEM photos. Part of this research was carried out in various research projects conducted by the Centre of Excellence on High-performance Concrete Network.

REFERENCES

[1]P. Richard and M.H. Cheyrezy, "Reactive Powder Concrete with High Ductility and 200-800 MPa Compressive Strength," pp. 507-518 in *ACI SP144*, 1994.

[2]P.-C. Aïtcin and A.M. Neville, "High-Performance Concrete Demystified," *Concrete International*, **15** [1] 21-26 (1993).

[3]H.F.W. Taylor, "Cement Chemistry," Thomas Telford, London (1997).

[4]J.-P. Goudonnet, "The Direct Measurement of Bonding Forces Between Hydrates" (in French), *Journées de l'ATILH*, Paris, France, 1998.

[5]H. Van Damme,"The Physics of Bonding Forces Between Hydrates and How to Act at the Molecular Level" (in French), *Journées de l'ATILH*, Paris, France, 1998.

[6]S. Diamond, "Cement Paste Microstructure—An Overview at Several Levels," pp. 2-30 in *Hydraulic Cement Pastes: Their Structures and Properties Proceedings*, Tapton Hall, University of Sheffield, Designed and published by the Cement and Concrete Association, 1976.

[7]S. Diamond, "Historical Development of Microstructural Studies in Cement and Concrete Research," *Conference in Tribute to Micheline Moranville Regourd*, Sherbrooke, QC, Canada, Edited by P.-C. Aïtcin (1994).

[8]M. Regourd, "Microstructure of High Strength Cement Paste Systems," pp. 3-17 in *Materials Research Society*, **42**, Edited by J.F. Young, 1985.

[9]P.-C. Aïtcin, M. Regourd, and C. Bédard, "Microstructural Studies of a 135 MPa Ready-Mix Concrete. pp. 164-179 in *Fifth International Conference on Cement Microscopy*, Nashville, TN, USA, 1983.

[10]P.-C. Aïtcin, *High-performance Concrete*, E & FN SPON, London, 1998.

[11]W. Baalbaki, B. Benmokrane, O. Chaallal, and P.-C. Aïtcin, "Influence of Coarse Aggregate on Elastic Properties of High-performance Concrete," *ACI Materials Journal*, **88** [5] 499-503 (1991).

[12]Nilsen, A.U., and P.-C. Aïtcin, "Properties of High-Strength Concrete Containing Light, Normal and Heavyweight Aggregates," *Cement, Concrete and Aggregates*, **14** [1] 8-12 (1992).

[13]W. Baalbaki, "Contribution to the Prediction of Elastic Modulus of Concrete," Ph.D. Thesis (in French), Université de Sherbrooke, Sherbrooke, Québec, Canada, 1997.

[14]R.J. Detwiler and L.J. Powers-Couche, "Effect of Ettringite on Frost Resistance," *Concrete Technology Today*, Portland Cement Association, Skokie, USA, IL 60077, USA., **18** [3] 1-4 (1997).

[15]M. Regourd, B. Mortureaux, P.-C. Aïtcin, and P. Pinsonneault, "Microstructure of Field Concrete Combining Silica Fume," pp. 249-260 in *Proceedings of the 4th International Cement Microscopy Conference ICMMA*, 1982.

[16]M. Lachemi, G. Li, A. Tagnit-Hamou, and P.-C. Aïtcin, "Long-term Performance of Silica-fume Concrete," *Concrete International*, **20** [1] 59-65 (1998).

[17]M. Nehdi, "Microfiller Effect on Rheology, Microstructure and Mechanical Properties of High-Performance Concrete," Ph.D. thesis, University of British Columbia (supervised by S. Mindess), 1998.

AN EVIDENCE OF REDUCTION IN FIXATION OF CHLORIDE IONS IN THE CEMENT PASTE-AGGREGATE INTERFACIAL ZONE

Mitsuru Saito
Department of Civil Engineering, Kanazawa Institute of Technology
Nonoichi, Ishikawa 921, Japan

Mitsunori Kawamura
Department of Civil Engineering, Kanazawa University
Kanazawa, Ishikawa 920, Japan

ABSTRACT

Chloride profiles across the cement paste-aggregate interfacial zone in model concrete specimens exposed to NaCl solution were obtained by X-ray fluorescence analysis for elucidating the effects of the interfacial zone on the chloride permeability of concrete. The model concrete disks were made using a rock prism and the cement paste with a water:cement ratio of 0.35. Three types of rocks, i.e., andesite, granite, and limestone, were used as aggregates. The zones up to about 2000 μm away from the interface were analyzed. It was found from the X-ray fluorescence analysis that the zones up to about 20 μm away from the interface in the specimens after the completion of the accelerated chloride ingress test showed half to two thirds of the chloride content of the bulk cement paste. Calculations carried out on the assumption that the relationship between bound and free chloride was expressed by the Langmuir adsorption isotherm, exhibited that bound chloride contents in the interfacial transition zone must have been extremely lower than those in the bulk cement paste.

INTRODUCTION

The durability of reinforced concrete structures exposed to saline environments depends greatly on the ingress of chloride ions from the surroundings. Since most of the cement paste phase in concrete is occupied by the interfacial transition zone developed around aggregate grains, as found by Diamond et al. (1), the chloride ingress characteristics of the zone exert a direct influence upon the corrosion of reinforcing steel in the concrete structures.

Tognon and Cangiano (2) observed the preferential penetration of a NaCl solution through the cement paste-aggregate interfacial zone which was attributable to the existence of microscopic pores in the zone. Experimental results showing an increase in the diffusivity of chloride ions in the interfacial zone were also obtained by Bretton et al. (3). In contrast, the present authors indirectly inferred from the result of a rapid chloride permeability test that the interfacial zone between the cement pastes with a low water:cement ratio and the rock aggregates favorably contributed to the reduction in the chloride permeability of concrete (4). These differences appear to arise from the complicated nature of the interfacial zone in concrete. At any rate, more informations are needed on the chloride ingress characteristics of the interfacial zone. In this study, chloride profiles across the interfacial zone between the cement paste with a low water:cement ratio and the rock aggregates in model concrete specimens exposed to NaCl solution were investigated for elucidating the role of the interfacial zone in the chloride permeability of concrete.

EXPERIMENTAL

Materials

The cement was an ordinary portland cement. Its specific gravity and Blaine fineness were 3.15 and 3250 cm^2/g, respectively. Three types of rocks, i.e., andesite, granite, and limestone, were used as model aggregates. The chemical composition of the cement and the three types of rocks is given in Table I. Specific gravity and absorption capacity of the rocks are presented in Table II.

Table I. Chemical composition of cement and rocks used (%)

	Cement	Andesite	Granite	Limestone
Ig. loss	1.5	1.7	0.2	43.1
SiO_2	21.6	60.3	76.4	0.2
Al_2O_3	4.4	17.3	12.8	0.1
Fe_2O_3	2.9	5.6	1.3	0.0
CaO	65.2	6.4	0.0	52.1
MgO	1.0	4.3	0.2	0.0
SO_3	2.2	0.0	–	–
Na_2O	0.26	3.82	4.40	0.12
K_2O	0.51	0.89	4.60	0.02
Total	99.57	100.31	99.90	95.64

Table II. Specific gravity and absorption capacity of rocks

	Andesite	Granite	Limestone
Specific gravity	2.60	2.63	2.68
Absorption capacity (%)	2.4	0.1	0.8

Preparation of Specimens and Chloride Ingress Tests

Two series of chloride ingress tests were carried out in this study consisting of (I) a rapid chloride permeability test by the application of an electrical potential gradient and (II) a diffusion cell test by a difference in concentration. A total of twelve model concrete disks were prepared to determine chloride profiles across the interfacial zone in the specimens; six specimens were used for the former test series (Series I), six specimens for the latter test series (Series II).

In Series I, the model concrete disk, ϕ 100 × 30 mm, was made by embedding a 12 × 12 × 30 mm rock prism in the cement paste matrix with a water:cement ratio of 0.35, as shown in Fig. 1(a). The rock prisms were cut from a rock block by the use of a diamond blade saw. The surfaces of rock prisms were polished by hand on smooth iron plates using water and two carborundum powders of # 60 and # 180. In the preparation of the specimen, a rock prism was held at the predetermined position in a steel mould using a special device. An vibrator with a ϕ 10 mm vibrating head was used for the compaction of specimens. The model concrete specimens cast were stored in a room maintained at 20 °C and 85 % relative humidity for 24 hours in the moulds, and then cured under saturated $Ca(OH)_2$ solution at 20 °C for 5 weeks. After curing, half the specimens were

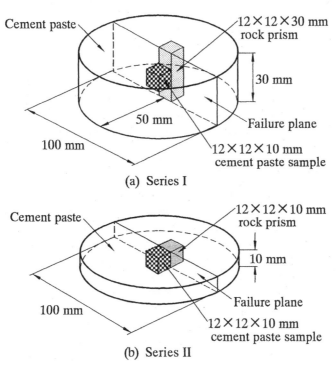

(a) Series I

(b) Series II

Fig. 1. Disk-like model concrete specimens.

used for the rapid chloride permeability test, the rest for the reference.

The rapid chloride permeability test was carried out according to AASHTO T277. The potential and test duration adopted were 30 V and 6 hours, respectively. A vacuum conditioning process for saturating specimens with water was not carried out, because the specimens had not been dried. Experiments were conducted in a room kept at 20 ℃. After completion of the AASHTO T277 test, the model concrete disks were broken by flexural loads to produce cement paste samples including the portions close to the interface for the X-ray analysis. A $12 \times 12 \times 10$ mm cement paste sample was cut from the cement paste region shown in Fig. 1(a) with a dry diamond blade saw.

In Series II, the model concrete specimen was a ϕ 100×10 mm disk which was made using a $12 \times 12 \times 10$ mm rock prism and the cement paste with a water:cement ratio of 0.35, as shown in Fig. 1(b). The preparation and curing methods of the specimens in Series II were the same as those in Series I. After curing, half the model concrete disks were fitted into the diffusion cells similar to those recommended by AASHTO T277. One side of the cell contained 3 % NaCl and saturated $Ca(OH)_2$ solution, while the other side was filled with saturated $Ca(OH)_2$ solution. The duration of the diffusion test was 4 weeks. Other disks were moreover cured under saturated $Ca(OH)_2$ solution for 4 weeks as the reference. A $12 \times 12 \times 10$ mm cement paste sample for the X-ray analysis was prepared from the cement paste region shown in Fig. 1(b) using the same way as that in Series I.

X-ray Analysis Procedures

The X-ray fluorescence and diffraction analysis was applied to determine chlorine concentration and to confirm the existence of Friedel's salt in the portions close to the interface. The X-ray analyses were conducted at the face of about 20 μm, 40 μm, 80 μm, 150 μm, 300 μm, 600 μm, 1000 μm, 1500 μm, and 2000 μm away from the interface in the cement paste samples. Surfaces of the cement paste samples to be analyzed were prepared by careful abrasion with # 240 and # 400 emery papers. The weight of materials removed by abrasion was measured with a balance with 0.05 mg sensitivity. The distances from the interface to the surfaces to be analyzed were calculated on the basis of the weight of materials removed and the density of bulk cement paste.

The X-ray fluorescence analyses were made using the Shimadzu X-ray fluorescence spectrometer. The accelerating voltage and electric current intensity used was 40 kV and 50 mA, respectively. The density of the samples has an effect on the fluorescent X-ray intensity, when the analyzable depth for the samples exceeds their thickness. On the other hand, the effect disappears, when the samples have enough thickness compared to the analyzable depth. The experimental conditions were selected so that the analyzable depth was several μm. A minimum space of the faces of the sample to be analyzed was about 20 μm as described above. Therefore, the fluorescent X-ray intensity obtained may not be affected by variations in density between different portions from the interface.

RESULTS AND DISCUSSION

Chloride Profiles across the Interfacial Zone in the Specimens after the Accelerated Chloride Ingress Test

Fig. 2 shows plots of the fluorescent X-ray intensity for chlorine obtained by X-ray fluorescence analysis across the interfacial zone in the model concrete specimens after the accelerated chloride ingress test. In all the types of rock aggregates, the X-ray intensity for chlorine across the interfacial zone in the specimens indicated the presence of chlorine in all the portions analyzed. In the andesite model concrete specimen (Fig. 2(a)), a relatively low chlorine intensity is found in the portions up to about 20 μm away from the interface. The intensity increases with distance from the interface, and the portions deeper than about 80 μm show a high content of chlorine. Similar chlorine profiles are also obtained in the granite and limestone model concrete specimens, as shown in Fig. 2(b) and (c). When the chlorine content is evaluated on the basis of levels of the chlorine intensity of the reference specimens, the portions close to the interface in the model concrete specimens show about half or two thirds of the chloride content of the portions deeper than about 100 μm. Since the thickness of the interfacial zone estimated from the degree of preferential orientation of $Ca(OH)_2$ crystals is about 50 μm, the plots in Fig. 2 indicate that the interfacial zones are regions with lower chloride contents.

The chlorine detected by the X-ray fluorescence analysis originally existed as bound chloride ions on pore surfaces and free chloride ions in pore solution within cement paste. The chloride was physically and chemically bound on pore surfaces. All the X-ray diffraction patterns for portions up to about 2000 μm away from the interface in this series showed no Friedel's salt. Sergi et al. (5) confirmed that the relationship between bound and free chloride in cement paste could be approximately represented by the Langmuir adsorption isotherm. Calculations were carried out to estimate the bound and free chloride in the interfacial zone and the bulk cement paste on the assumption that the Langmuir adsorption isotherm can be applied to the degree of fixation of chloride ions in the cement paste-aggregate interfacial zone.

The following additional assumptions were adopted for the calculations:

[1] The chloride ion concentration in the pore solution in the interfacial zone and the bulk cement paste was 0.513 mole/l, which was the same as that in the diffusion cells.

[2] The porosity of the interfacial zone adjacent to the interface and the bulk cement paste was 30 % and 10 %, respectively.

[3] The specific gravity of solid substances within the interfacial zone and the bulk cement paste was 2.19.

By using the Assumption [1] and the Langmuir adsorption isotherm, the quantity of the bound chloride ions of 0.277 mmole/g cement was obtained. The Assumption [2] was based on the previous study (6). Scrivener et al. (6) obtained the porosity of portions across the interfacial zone in concrete specimens using SEM image analysis. According to their results, the highest value of porosity was about 30 % at the regions close to the

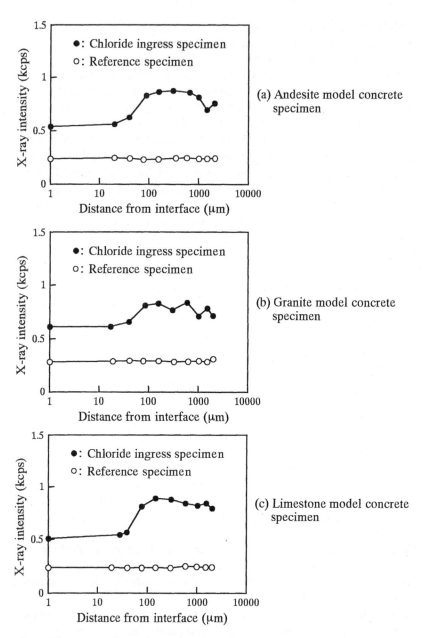

Fig. 2. Fluorescent X-ray intensity for chlorine across the interfacial zone in the specimens after the accelerated chloride ingress test.

Table III. Free, bound, and total chloride per unit volume of the interfacial zone and the bulk cement paste

	Interfacial zone	Bulk cement paste
Porosity (%)	30	10
Unit weight of solid substances (g/cm³)	1.53	1.97
Free chloride content (mg/cm³)	5.5	1.8
Bound chloride content (mg/cm³)	15.0	19.3
Total chloride content (mg/cm³)	20.5	21.1

interface, and the porosity of portions deeper than about 30 µm was about 10 %. The specific gravity shown in the Assumption [3] was estimated from the porosity of the bulk cement paste in the Assumption [2] and an average value of specific gravity of the bulk cement paste.

The free, bound, and total chloride per unit volume of the interfacial zone and the bulk cement paste calculated based on the assumptions described above are given in Table III. From the results given in Table III, it is found that there are considerable differences in the free chloride and the bound chloride between the interfacial zone and the bulk cement paste, and the total chloride content in the interfacial zone is nearly equal to that in the bulk cement paste. This result is not consistent with the finding that the interfacial zone exhibits about half or two thirds of the chloride content of the bulk cement paste (Fig. 2). It is likely that the Assumption [1], [2], and [3] adopted for the calculations are reasonable. Therefore, this discrepancy seems to suggest that the chloride binding capacity of the interfacial zone is less than that of the bulk cement paste, because the chloride ion concentration of the pore solution in the interfacial zone should be identical to that in the bulk cement paste. Additional calculations indicate that a decrease to a third in the chloride binding capacity of the interfacial zone results in a decrease to a half in the total chloride content of the zone. The supposition that the bound chloride content in the interfacial transition zone must be extremely lower than that in the bulk cement paste can reasonably explain the result obtained from the X-ray fluorescence analysis in this study. Larsen (7) compared the chloride binding capacity of three types of mortars with that of the corresponding cement pastes, and revealed that the chloride binding capacity was reduced by the presence of aggregate grains. This finding supports the conclusion that the interfacial transition zone shows extremely less bound chloride content than the bulk cement paste.

Chloride Profiles across the Interfacial Zone in the Specimens
after the Usual Chloride Diffusion Test
    The fluorescent X-ray intensity for chlorine across the interfacial zone in the model concrete specimens subjected to the usual diffusion cell test is plotted against the distance from the interface, as shown in Fig. 3. All the X-ray diffraction patterns across the

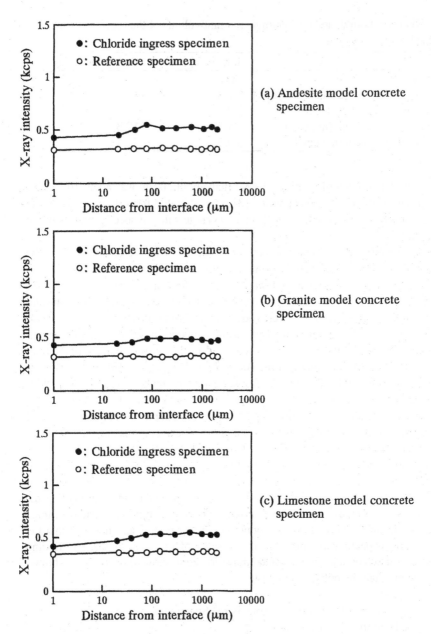

Fig. 3. Fluorescent X-ray intensity for chlorine across the interfacial zone in the specimens after the usual chloride diffusion test.

interface zone in this series showed no Friedel's salt. It is found from Fig. 3 that the fluorescent X-ray intensity in the portions up to about 20 μm away from the interface is lower than that in the portions deeper than about 80 μm in all the types of specimens. Fig. 2 and Fig. 3 show that the chlorine intensity for the interfacial zone in specimens in Series I is higher than that for the interfacial zone in specimens in Series II. It is apparent from the comparison of Fig. 2 with Fig. 3 that the chloride content of the interfacial zone is also lower than that of the bulk cement paste in the specimens in Series II.

CONCLUSIONS

The following conclusions were drawn from this study.

1. The interfacial zones in the model concrete specimens after the completion of the accelerated chloride ingress test showed about half to two thirds of the chloride contents of the bulk cement paste.
2. The chloride contents in the interfacial zone in the model concrete specimens after the usual chloride diffusion test were also lower than those in the bulk cement paste.
3. The calculations carried out based on the assumption that the relationship between bound and free chloride is expressed by the Langmuir adsorption isotherm, exhibit that bound chloride contents in the interfacial zone must be extremely lower than those in the bulk cement paste.

REFERENCES

[1] S. Diamond, S. Mindess and J. Lovell, " On the spacing between aggregate grains in concrete and the dimensions of the aureole de transition," Proc. RILEM Colloq., Toulouse, C.42-46 (1982).

[2] G.P. Tognon and S. Cangiano, " Interface phenomena and durability of concrete," Proc. 7th Int. Congress on the Chemistry of Cement, Paris, 3, VII.133 -38 (1980).

[3] D. Bretton, J.-P. Ollivier and G. Ballivy, " Diffusivite des ions chlore dans la zone de transition entre pate de ciment et roche granitique," Proc. RILEM Int. Conference, Interfaces in Cementitious Composites, Toulouse, 269-78 (1992).

[4] M. Saito and M. Kawamura, " Effects of the cement paste-aggregate interfacial zone on the chloride permeability of concrete with a low water:cement ratio," Proc. Int. Symp., Brittle Matrix Composites 4, Warsaw, 484-91 (1994).

[5] G. Sergi, S.W. Yu and C.L. Page, " Diffusion of chloride and hydroxyl ions in cementitious materials exposed to a saline environment," Magazine of Concr. Res., 44, 63 -69 (1992).

[6] K.L. Scrivener, A.K. Crumble and P.L. Pratt, " A study of the interfacial region between cement paste and aggregate in concrete," Proc. Mat. Res. Soc. Symp., Boston, 114, 87-95 (1987).

[7] C.K. Larsen, " Effect of type of aggregate, temperature and drying/rewetting on chloride binding and pore solution composition," Proc. RILEM Int. Workshop, Chloride Penetration into Concrete, St-Remy-les-Chevreuse, 27-35 (1995).

A CHEMO-PHYSICO-MECHANICAL APPROACH TO ASR

Weihua Jin and Christian Meyer
Columbia University
Department of Civil Engineering and Engineering Mechanics
New York, NY 10027

ABSTRACT

A pronounced particle size effect was identified for different types of glasses in the accelerated ASTM C 1260 test. The pessimum size was found to depend on both the chemical reactivity of the reactive aggregate and the transport property of the cement matrix, which led to a hypothesis that gel formation and gel permeation are the chemical and physical processes that control the pessimum size effect. A chemo-physico-mechanical model is presented to describe the observed behavior.

INTRODUCTION

Ever since its identification in the early 1940's (1), ASR has been studied mostly, if not exclusively, from the viewpoint of reaction chemistry. Since ASR is chemical in nature, reaction chemistry is doubtlessly an indispensable tool for its study and has indeed laid the foundation of the current theory of ASR mechanisms. However, ASR is such a complex process that a chemical approach alone seems to be incapable of satisfactorily explaining all of its important aspects. Moreover, due to the limited understanding of the nature of ASR, few models, if any, exist that can relate ASR expansion to material properties of individual phases of the concrete composites. Such a model is clearly important for the theoretical prediction as well as practical control of ASR.

One of the important phenomena associated with ASR is the pessimum effect, with respect to both aggregate content and aggregate size. The pessimum content effects reported in the literature (2, 3) for various reactive aggregates were obtained from standard mortar bar tests such as ASTM C 227, with a fixed alkali supply in the cement paste. The pessimum content has been explained with a mechanism based on the Na_2O/SiO_2 ratio of the alkali silicate gel products (4). This hypothesis was later supported by Glasser and Kataoka's experimental finding (5) that the silicate gel of an intermediate Na_2O/SiO_2 molar ratio exhibited maximum expansion. Zhang (6) related further the pessimum particle size effect of silica glass to an effective Na_2O/SiO_2 ratio, assuming that the accessible SiO_2 extends a certain distance below the original particle surface. This hypothesis, however, was challenged by recent test results obtained from accelerated mortar bar tests (such as ASTM C 1260) that unexpectedly exhibited a pessimum content (7), since the supply of alkali in a 1 N NaOH solution was basically unlimited, and no pessimum effect was supposed to exist in such an accelerated test. This means, a better theory is needed to explain the experimental evidence.

During the course of a research project to study the suitability of crushed waste glass in concrete masonry blocks as well as other concrete products, fundamental aspects of ASR in concrete with glass as aggregate were extensively investigated, using ASTM C 1260, an accelerated mortar bar test method. Important variables included glass content, aggregate particle size, glass type, and glass chemical composition. While major experimental results were published elsewhere (8, 9, 10, 11), this paper summarizes studies on chemical, physical, and mechanical factors related to ASR. Results showed that the pessimum size depends strongly on the reactivity of the glass. Furthermore, it seems to depend also on the permeability of the cement matrix, which was modulated by the introduction of air-entraining admixtures (AEA). Mechanical confinement, for example, by fiber reinforcement, produces some reduction in ASR expansion, and strengthening in some cases. The experimental evidence led further to the hypothesis that gel formation and gel permeation are the chemical and physical processes that control the pessimum size effect. As a result, a micromechanical model was developed within a general physico-chemical framework.

EXPERIMENTAL WORK
Materials

Three types of glass were used as reactive aggregate: soda-lime glass, Pyrex glass, and fused silica. Soda-lime glass is typically used in glass containers and other consumer glassware and its sample was obtained by crushing beverage bottles after cleaning. The Pyrex glass was type 7740 borosilicate glass supplied by Schott Scientific Glass Inc., and the fused silica was obtained from Cerac Inc. Table 1 lists the chemical compositions of various glass types.

A local Long Island sand was used in the initial experiments, but proved to be slightly reactive, as the expansions of mortar bars made with it in the ASTM C 1260 test were approximately 0.1%. For the subsequent tests, a non-reactive manufactured sand from Blue Circle Co. of Hamburg, NJ, was used. The cement used in all tests was a type I low-alkali cement containing 0.23% alkalis (Na_2O + K_2O) and manufactured by Holnam Inc. An air entraining admixture (AEA), trademarked as Micro-Air from Master Builders, was added to the mix as a means of air entrainment.

Both polypropylene fibers and steel fibers were added to the mix at the volume ratio of 1.5%. The steel fibers were low carbon steel fibers with hooked ends (SF1), 1.2 inches long and 0.02 inches in equivalent diameter, supplied by the Bekaert Corporation, and low carbon steel fibers with deformed ends (SF2), 15/16 inches long and 0.02 inches in equivalent diameter, supplied by Mitchell Fibercon, Inc. The polypropylene fibers were 0.75 inches long with a diameter of 236 μm and supplied by the Fibermesh Company.

Table I. Chemical compositions of clear soda-lime glass, Pyrex glass, and fused silica (%)

	Clear glass	Pyrex glass	Fused silica
SiO_2	73.5	81	99.97
Al_2O_3	1.9	2	-
$Na_2O + K_2O$	13.6	4	-
$CaO + MgO$	10.8	-	-
SO_3	0.20	-	-
Fe_2O_3	0.04	3.72	-
B_2O_3	-	12 - 13	-

Test Method

ASTM C 1260 test method was used, and the expansions of mortar bars were taken as a measure of the aggregate reactivity. Mortar bars made with graded aggregate, water-cement ratio of 0.47 and aggregate-cement ratio of 2.25 were stored in a 1N NaOH solution at 80 °C for daily length measurements with a digital dial indicator at an accuracy of 0.002 mm.

For each data point three mortar bars were tested. A thorough statistical analysis (11) led to the conclusion that results obtained as the average of three samples are accurate within approximately 10%, at a confidence level of 90%.

RESULTS AND DISCUSSSION
Effect of Particle Size

In order to study the effect of particle size on the reactivity of glass aggregate, 10% of the total graded sand aggregate (Long Island sand) by weight was replaced with clear soda-lime glass aggregate of one particular size at a time. The mortar bar expansions are plotted in Fig. 1, where the replaced particle size is indicated on the horizontal axis in terms of standard sieve numbers. This graph clearly shows that for clear glass the maximum expansion (pessimum) occurs for size #16, i.e., particles retained on sieve #16 (1.18 mm to 2.36 mm). Mortar bars containing 10% of size #50 glass aggregate exhibit approximately the same expansion as the reference bars which contain no glass aggregate, while mortar bars containing 10% of either size #100 or pan (particles passing sieve #100) show less expansion than the reference bars. The intrinsic shape of the distribution curve barely changes during the 14 day test period.

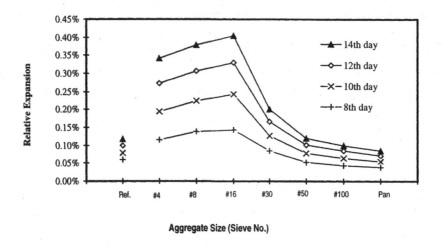

Figure 1 Particle size effect on ASR expansion (10% clear soda-lime glass)

There are two interesting points on the expansion spectrum curve to observe: the one with maximum expansion and the one with zero abnormal expansion. The absence of abnormal expansion for very small glass particles (finer than #100 sieve or 150 μm for soda-lime glass) can be attributed somewhat to the pozzolanic reaction, although such a particle size is much larger than what is believed to be the ideal size (smaller than 45 μm) for pozzolanic reaction to occur. Later experiments (11) confirmed that finely ground glass powder exhibits pozzolanic property. Samples made with a partial substitution of glass powder (minus #400) for Portland cement showed increased strength and less ASR expansion. Yet, the decreased expansion beyond the pessimum size is difficult to explain. Since ASR is usually considered to be of a topochemical type, the expansion is expected to increase for smaller glass particle sizes due to the surface area effect on the reaction rate. Furthermore, as mentioned before, to explain the pessimum effect with the Na_2O/SiO_2 ratio seems to be invalid here, since the supply of alkali is essentially unlimited in the case of such an accelerated test.

Effect of Glass Type

To further study the particle size effect, a test was performed with mortar bars containing non-reactive sand (Blue Circle sand), 10% of which was replaced by either Pyrex glass or fused silica of specific sizes. The results are shown in Fig. 2, together with those for clear soda-lime glass. This graph demonstrates that a pessimum size exists for all three glass types tested: size #16 for clear soda-lime glass, #200 for Pyrex glass, and #400 for fused silica. In terms of absolute reactivity, fused silica is the most reactive among the three, followed by Pyrex and soda-lime glass.

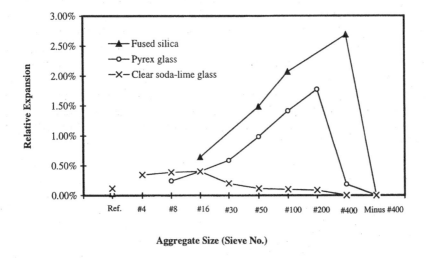

Figure 2 Effect of glass type on ASR expansion

The results of Fig. 2 indicate that the pessimum size depends strongly on the reactivity of the aggregate. As the reactivity increases, the pessimum size is shifted to the smaller size range.

Since the expansions of the three types of glass tested cover quite a large range, it can be postulated that a pessimum size exists for all reactive aggregates. However, the pessimum size of some highly reactive aggregates may not be detectable in the size range routinely used in laboratory tests. This explains why Diamond and Thaulow (12) observed increasing expansions for particle sizes as fine as 20 μm. It is also interesting to note the relationship between pozzolanic reactivity, as indicated by the strengthening and ASR suppressing effectiveness, and the ASR reactivity. It was shown (11) that the higher the ASR reactivity of the glass, the higher its pozzolanic reactivity. Thus, the same material can be either ASR reactive or pozzolanic, with the transition depending on the reactivity and the particle size of the aggregate. This explains the paradox described by Diamond (13) about silica fume: with coarse particles or undispersed agglomerates it can induce ASR rather than mitigate it.

Effect of Air Entrainment

It is a well-known fact that the ASR gel products can fill up pores and cracks within concrete. The micro-pore system of the cement paste can be considered to serve as a reservoir to accommodate the ASR gel and thus relieve the ASR induced stress. Jensen (14) showed that the ASR expansion could be reduced by about 40% when about 4% air voids were introduced. However, the exact role of transport properties, permeability for example, in the ASR expansion process remains unknown. To clarify this issue, a test was performed where an AEA (1 ml / 1 kg cement) was added to mortar bars containing 10% Pyrex glass of one particular size at a time. This dosage is recommended by the manufacturer to generate a stable air content as high as 6%. Fig. 3 shows the resulting expansion spectrum curves. As can be seen, the AEA reduced the maximum expansion at 14 days by approximately 50%. This is consistent with previous results (14). However, there is a shift in the pessimum size with time. Without AEA, the pessimum size

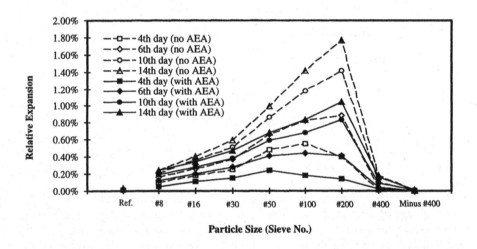

Figure 3 Effect of air entrainment on ASR expansion

Effect of Fiber Confinement

To study the effectiveness of mechanical confinement in restraining the ASR expansion, either 1.5% (by volume) polypropylene fibers or steel fibers were added to the mortar samples containing 10% clear glass, which were then tested for ASR expansion and flexural strength. The results are shown in Fig. 4, where the expansion ratio is defined by the ratio of the 14th day expansion to that of the reference, and the strength ratio is defined by the ratio of the flexural strength with ASR to that without ASR. It can be seen that the addition of fibers reduces the expansion by up to 60%, while decreasing the strength loss due to ASR. Steel fibers are more effective in restraining the ASR expansion and reducing the damage due to ASR. The strength is even increased by up to 30% when steel fiber reinforced mortar bars are experiencing ASR expansion. This fact is attributed to a self-stressing mechanism of fibers introduced by ASR. It is to be noted that this effect can only be achieved with proper bond or anchorage between fibers and cement paste.

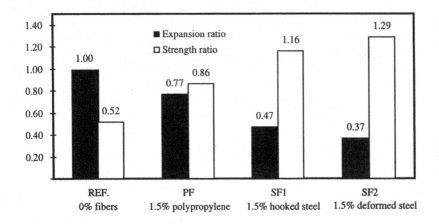

Figure 4 Effect of fiber confinement on the mortar bar expansion and flexural strength

OUTLINE OF A CHEMO-PHYSICO-MECHANICAL MODEL

The relatively simple chemistry and homogeneous nature of glass renders itself as an ideal aggregate material to study the important factors that influence ASR and to establish a general theory that can explain this behavior. Such a theory will be equally applicable to concrete with natural reactive aggregates.

Experimental results showed that the pessimum size depends strongly on the reactivity of the aggregate and to some extent on the transport property of the cement matrix. It can be assumed that the air entrainment increases the permeability, one of the transport properties of the cement paste, as well. This result suggests that some mechanism other than chemical might be involved in the

ASR expansion process and contribute to the pessimum size effect. ASR is indeed a chemo-physical process that results in mechanical damage. If we consider concrete or mortar as a three-phase composite consisting of cement or mortar matrix, interfacial transition zone (ITZ), and aggregate, and further assume the reactive aggregate to be relatively dense, as in the case of glass aggregate, the reaction products between the aggregate and cement pore solution, ASR gel, will form at first on the surface of the aggregate, although some gel will probably form within the aggregate if porous. The swelling pressure of ASR gel can be considered to be osmotic in character (15, 16) or result from electrical double-layer repulsion (17). But under elastic confinement of the cement matrix and aggregate the ASR gel generates hydrostatic pressure. Viscous in nature, it permeates into the surrounding porous ITZ and matrix and thus relieves the pressure caused by the gel expansion. The observed expansion is the result of the elastic and inelastic deformations caused by the net hydrostatic pressure, which depends on the elastic moduli of matrix and aggregate, the viscosity of the ASR gel, the permeability of the matrix, and the aggregate size and content. If the reaction proceeds and the internal pressure exceeds a critical value, cracks form around the reacting aggregate particles. This microcracking decreases the effective elastic moduli and increases the material's effective permeability, and marked expansion will be the result. This process is illustrated in Fig. 5.

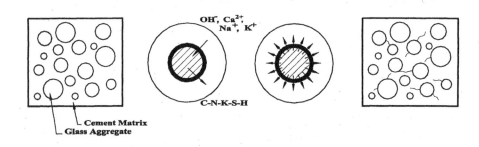

Figure 5 Schematic description of a chemo-physico-mechanical ASR model

Thus, the gel formation and gel permeation are the chemical and physical processes that both affect the internal pressure, which leads to the expansion and microcracking of concrete. The gel formation builds up the internal pressure, while the gel permeation tends to relieve it. The net pressure thus reflects the difference between gel formation and permeation rates. The two rates are different functions of the aggregate surface area, such that there exists a certain particle size at which the two processes balance each other. In other words, there exists a size at which maximum expansion occurs. This pessimum size will be a function of the reactivity of the aggregate and the transport property of the matrix. Fig. 6 schematically shows how these two processes are related to the pessimum size effect. In the left portion of the curve, the process is reaction dominant, while on the right it is transport dominant. The pessimum size will be shifted to the right as the reactivity of the aggregate increases and to the left as the permeability of the matrix increases. This analysis is consistent with the experimental evidence at early ages, when the expansions are mainly elastic.

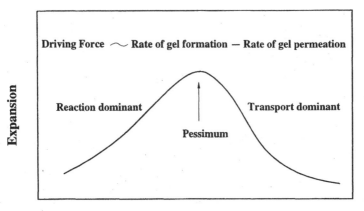

Figure 6 Illustration of the pessimum size effect as a result of two processes

Although still simplified, the hypothesis outlined above makes it possible to develop a mathematical model for the prediction of ASR expansion within the framework of chemical reaction kinetics, transport law, micromechanics of multiphase composites, and constitutive equations of continua. The theoretical formulation and numerical simulation are beyond the theme of this paper and can be found elsewhere (11, 18)). This model takes into account all important variables, not only chemical, but also physical and mechanical. It relates both the swelling ability and viscosity of the gel to the ASR expansion and identifies the importance of the physical property of the gel.

CONCLUSIONS

1. A pessimum size effect exists for clear soda-lime glass, Pyrex glass, and fused silica, even in the accelerated ASTM C 1260 test.
2. The pessimum size depends strongly on the reactivity of the glass. The more reactive, the smaller the pessimum size.
3. Air entrainment is effective in reducing ASR expansion due to the accommodation mechanism provided by the micro-air-void system.
4. Mechanical confinement provided by fiber reinforcement can restrain the expansion and reduce the damage due to ASR. It might even produce a strengthening effect under some conditions due to a self-stressing mechanism.
5. Gel formation and gel permeation are the chemical and physical processes that contribute to the pessimum size effect.

REFERENCES

1. Stanton, T. E. "Expansion of Concrete through Reaction between Cement and Aggregate", Proceedings of the American Society of Civil Engineers, Vol. 66, pp. 1781-1811, 1940.
2. Vivian, H. E. "The Effects on Mortar Expansion of Reactive Component in the Aggregate", Studies in Cement-Aggregate Reactions, Part 10, CSIRO Bulletin, No. 256, pp. 13-20, 1947.
3. Hobbs, D. W. "The Alkali-Silica Reaction - A Model for Predicting Expansion in Mortar", Magazine of Concrete Research, Vol. 33, pp. 208-219, 1981.
4. Hobbs, D. W. "Expansion of Concrete due to the Alkali-Silica Reaction: an Explanation", Magazine of Concrete Research, Vol. 30, pp. 15-220, 1978.
5. Dent Glasser, L. S. and Kataoka, N. "The Chemistry of 'Alkali-Aggregate' Reaction", Cement and Concrete Research, Vol. 11, pp. 1-9, 1981.
6. Zhang, X., Groves, G. W., "The Alkali-Silica Reaction in OPC/Silica Glass Mortar with Particular Reference to Pessimum Effects", Advances in Cement Research, Vol. 3, pp. 9-13, 1990.
7. Shayan, A., "The 'Pessimum' Effect in an Accelerated Mortar Bar Test Using 1 M NaOH Solution at 80 °C", Cement & Concrete Composites, Vol. 14, pp. 249-255, 1992.
8. Meyer, C., Baxter, S., Jin, W., "Potential of Waste Glass for Masonry Blocks", Materials for the New Millenium – Proceedings of ASCE Forth Materials Engineering Conference, Vol. 1, pp.666-673, Washington, DC, November, 1996.
9. Meyer, C., Baxter, S., Jin, W., "Alkali-Silica reaction in Concrete with Glass as Aggregate", Materials for the New Millenium – Proceedings of

ASCE Forth Materials Engineering Conference, Vol. 2, pp.1388-1395, Washington, DC, November, 1996.

10. Meyer, C., Baxter, S., Jin, W., "Glascrete – Concrete with Crushed Waste Glass Aggregate", Proceedings of the CSCE/CGC/JSCE International Conference on Engineering Materials, Vol 2, pp. 735-742, June 8-1, Ottawa, Canada, 1997.

11. Jin, W. "Alkali-Silica Reaction in Concrete with Glass Aggregate – A Chemo-Physico-Mechanical Approach", Ph. D Dissertation, Columbia University, 1998.

12. Diamond, S. and Thaulow N. A. "A Study of Expansion due to Alkali-silica Reaction as Conditioned by the Grain Size of the Reactive Aggregate", Cement and Concrete Research, Vol. 4, pp. 591-607, 1974.

13. Diamond, S. "Alkali-Silica Reaction: Some Paradox", Proceedings of the 10th International Conference on Alkali-Aggregate Reaction in Concrete", pp. 3-14, Melbourne, Australia, August 1996.

14. Jensen, A. D., Chatterji, S., Christensen, P. and Thaulow, N. "Studies of Alkali-Silica Reaction - Part II: Effect of Air-Entrainment on Expansion Cement and Concrete Research", Vol. 14, pp. 311-314. 1984.

15. Powers, T. C. and Steinour, H. H. "An Interpretation of Some Published Researchers on Alkali-Aggregate Reaction. Part 1 - The Chemical Reactions and Mechanisms of Expansion." Journal of the American Concrete Institute, Vol. 26, pp. 497-516, 1955.

16. Powers, T. C. and Steinour, H. H. "An Interpretation of Some Published Researchers on Alkali-Aggregate Reaction. Part 2 - A Hypothesis Concerning Safe and Unsafe Reactions with Reactive Silica in Concrete." Journal of the American Concrete Institute, Vol. 26, pp. 785-811, 1955.

17. Monteiro, P. J. M., Prezzi, M., Wang, K., and Ghio, V. "Durability, Tomography, and Repair of Reinforced Concrete", Report No. UCB/SEMM-95/11, University of California at Berkeley, September, 1995.

18. Jin, W., Suwito, A., Meyer, C., and Xi, Y. "A Mathematical Model for Expansion of Concrete due to Alkali-Silica Reaction", to be presented on the 12th ASCE Engineering Mechanics Conference, San Diego, May, 1998.

DELAYED ETTRINGITE FORMATION AND THE MODE OF CONCRETE FAILURE

Sadananda Sahu, Boyd A. Clark and Richard J. Lee
RJ Lee Group, Inc.
350 Hochberg Road
Monroeville, PA 15146

ABSTRACT

Delayed ettringite formation (DEF) may result in the failure of both air-entrained and non-air-entrained concretes. The mode of failure may differ, as do some of the observed microstructural features.

DEF in the non-air-entrained steam cured concrete examined is exemplified by ettringite deposits at the paste/aggregate interfaces, within voids, in fine paste pores, and cracks. These deposits lead to internal stress development, cracking and failure of the concrete structure.

On the other hand DEF found in an air-entrained steam cured concrete subjected to repeated freezing and thawing was marked by ettringite deposited mainly in the air voids and in fine paste pores, and only rarely at the paste/aggregate interfaces. The study indicates that this mode of DEF creates minimal internal stresses and cracking. Failure appeared to be due to subsequent freeze-thaw deterioration, as most of the air voids were filled with ettringite. Thus it appears that when freeze-thaw mechanisms are operative, deposition of ettringite in air voids reduces the effective air void content and can induce freeze-thaw deterioration.

Comparison of failure mode and microstructure in these two concretes suggest that neither of the currently popular mechanisms of DEF-related expansion is fully satisfactory.

INTRODUCTION

In recent years, a number of investigative reports of distressed concrete cured at elevated temperature show that the failure is due to DEF [1-7].

There are many reports related to distressed concrete of railroad ties and one analysis shows that this was associated with alkali-silica reaction (ASR)[8]. According to these authors ASR is the primary cause of cracking and DEF is a secondary phenomenon. An experimental study examining the relationship between ASR and DEF showed that a limestone mortar did not expand significantly after curing at 95 ^{0}C [9]. However, when reactive silica was added to the mix marked expansion occurred initially from ASR, and was later exacerbated by the effect of DEF. The expansion from DEF was attributed largely to the deposition of ettringite in ASR-induced cracks. The cracking of concrete by freezing and thawing may induce DEF [5]. However, recent studies show that DEF exists in structural elements that are steam cured and exposed to extreme environment. There are cases where ASR symptoms are not observed and DEF is the primary cause of failure [4]. A recent laboratory study shows that a heat-treated cement paste shows significant expansion after an induction period of 2½ years [10]. In that experiment a type III cement having 4.2% SO_3 and about 1% alkali was used, the authors conclude that the expansion of various cement pastes, mortars and concrete is greatly accelerated in the presence of aggregate[11]. These observations lead to the conclusion that DEF can exist alone but is enhanced in association with other forms of concrete deterioration processes.

A characteristic of concrete suffering from DEF in non-air-entrained, heat cured concrete is development of gaps around aggregate. These gaps are very often filled with ettringite. Ettringite can also be deposited in voids, cracks and fine pores in the paste. The occurrence of a band of ettringite around the periphery of aggregate is very often regarded as a diagnostic for this mode of concrete distress [6].

The mechanism of DEF and its relation to expansion of the concrete is only partly understood. There are two different proposed hypotheses: the uniform paste expansion hypothesis [12-14] and crystal pressure growth hypothesis[1-4, 15]. According to the first hypothesis, an elevated heat treatment and subsequent curing forms a microcrystalline ettringite intermixed with innerproduct calcium silicate hydrate[13]. The post-curing formation of this microcrystalline ettringite leads to homogenous paste expansion and gaps at the paste/aggregate interface. Subsequently, ettringite recrystallizes in the pastes/aggregate gaps, cracks, and other void spaces. The second hypothesis postulates that expansion and cracking associated with DEF is the direct result of ettringite crystal growth pressure during the formation in cracks, voids and gaps.

The present study describes the microstructural features of single examples of air-entrained and non-air-entrained steam cured concrete. An interpretation of these observed features is offered with respect to the DEF-induced deterioration mechanisms found in these concretes.

EXPERIMENTAL

A sample of severely cracked precast concrete exposed to the environment for approximately 12 years was examined as the steam cured 'non-air-entrained' concrete.

Another sample from an air-entrained precast concrete structure exposed to conditions of frequent freezing and thawing was examined. This concrete was in service for 4 years and exhibits a pattern of gross cracking. It was designated as 'air-entrained' concrete.

Both concretes were made using type III cement and both were subjected to elevated temperature curing.

Microstructures of the specimens were studied using a RJ Lee Instruments Ltd., Personal Scanning Electron Microscope (PSEM). Back-scattered electron (BSE) images were acquired for each specimen at 20 keV.

The specimens of 50x50 mm were made in order to scan large areas of each sample. Samples were epoxy impregnated and polished with increasingly finer diamond paste to 1/4 micrometer (μm). After polishing the samples were coated with a thin layer of carbon.

MICROSTRUCTURE OF NON-AIR-ENTRAINED CONCRETE

Many investigators have described the microstructural features of concrete, the most recent by Diamond [4]. There is general agreement on the microstructural features of concrete suffering from DEF, observed by various researchers. The typical features observed in the present investigation are described below.

Figure 1, a montage of 25 images at 200X, shows the typical network crack pattern developed in the non-air-entrained concrete. Images were collected contiguously and stitched together to form the montage in an automatic fashion. The viewing area is 2.2x2.2 mm. This enables a larger area of the sample to be imaged at a particular location. A network of cracks is observed in the image. Some are filled with ettringite and some are empty. There are many cracks and/or gaps partially rimming around aggregates are filled with ettringite. The aggregate at the bottom right hand corner is completely surrounded by an uniform layer of ettringite. Cracks and/or gaps rimming around aggregates and passing through the paste constitute part of same network. A few entrapped air voids completely or partially filled with ettringite are observed. Ettringite deposits in fine pores within the paste are also seen.

Figure 2 shows a typical ettringite band along the paste/aggregate interface. The width of the ettringite band is not uniform along the periphery of the aggregate. At some portion of the periphery it is non-existent and calcium hydroxide (CH) is still intact at the interface. The average width of the ettringite band (where it exists) is about 10 μm.

Figure 3 is an X-ray map showing the distribution of elements along the paste/aggregate interface of another aggregate particle in the same concrete. The sulfur distribution shows a narrow band along the periphery of the aggregate.

Figure 4 (left-hand side, the low magnification area of the composite image) shows a typical crack pattern associated with this type of concrete distress. Some cracks rimming aggregates are also visible. The right-hand image, at higher magnification shows a cracked hydrated phenograin. Ettringite deposits are observed in the cracked inner calcium silicate hydrate. Narrow bands of ettringite

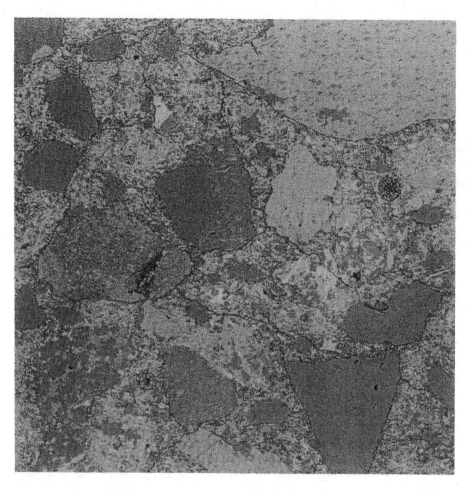

Figure 1. Montage of 25 images at 200X showing the characteristic features of the non-air-entrained concrete.

deposits also partially cover the aggregate particles. Ettringite deposits are also observed in cracks running through the paste and in fine paste pores.

Figure 5 shows ettringite deposits in entrapped air voids in the vicinity of an aggregate. Small inclusions of ettringite are also observed in the void space of former cement grains and within fine paste pores. No ettringite deposits are observed at the paste/aggregate interface. It seems when relatively large void space is available in the neighborhood of an aggregate ettringite is preferentially deposited in those void spaces. Of note is that although significant ettringite has

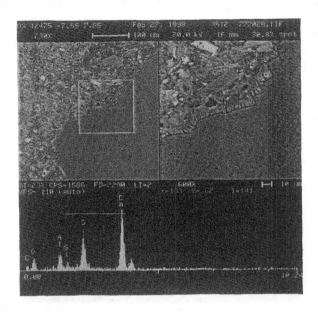

Figure 2. Characteristic band of ettringite around a coarse aggregate.

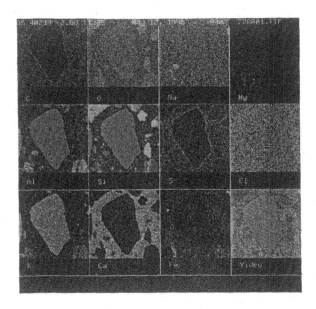

Figure 3. X-ray map showing sulfur deposition along the paste/aggregate interface.

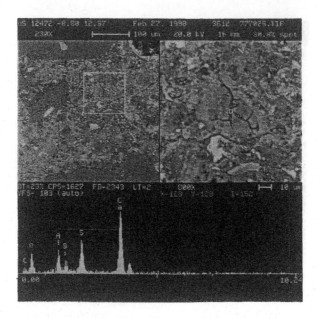

Figure 4. Cracked hydrated phenograin showing ettringite deposits.

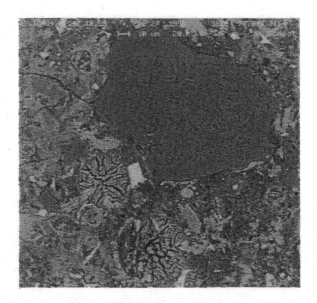

Figure 5. Ettringite deposits in air voids in the vicinity of an aggregate.

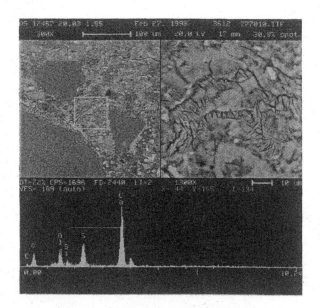

Figure 6. Ettringite deposits in the cracks in the paste.

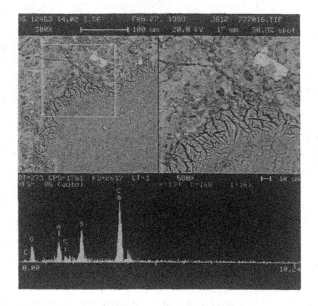

Figure 7. Ettringite deposits at the paste/aggregate interface.

been deposited in voids, no local deterioration of the paste/aggregate interface is visible, indicating there has been no paste expansion.

Figure 6 shows ettringite deposits at paste/aggregate interfaces and deposits filling cracks in the paste. Some empty cracks are still observable. The deposits filling cracks have formed parallel bands. CH deposits are observed at the paste/aggregate interface and within the paste.

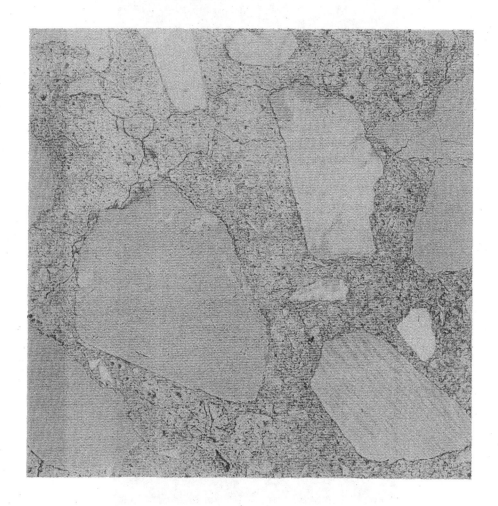

Figure 8. Montage of 25 images at 500X showing the characteristic features of distressed air-entrained concrete.

Figure 7 shows ettringite deposits at a paste/aggregate interface. The width of the deposit varies between 10 and 20 µm (left-hand side of the image). There are several branching cracks filled with ettringite.

MICROSTRUCTURE OF AIR-ENTRAINED CONCRETE

Figure 8 is a montage of 25 images at 500X, a 0.9 X 0.9 mm viewing area. Most of the entrained air voids are completely filled with ettringite. Ettringite deposits are also found in void spaces of former cement grains. A network of empty cracks is observed. There are no gaps or ettringite bands observed at the paste/aggregate interface.

Figure 9 shows examples of air voids completely filled with ettringite. Ettringite deposits are also observed in fine paste pores and voids in former cement grains. Fine cracks radiating from the ettringite-filled air voids are also observed.

Ettringite deposits are also observed at a paste/aggregate interface, Figure 10. Though this pattern of deposit is common to non-air-entrained concrete, it is observed infrequently in the air-entrained sample. Ettringite deposition has formed as a layer between the aggregate and CH, deposited during the pre-steam curing process.

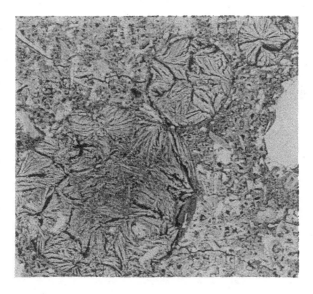

Figure 9. Air voids completely filled with ettringite.

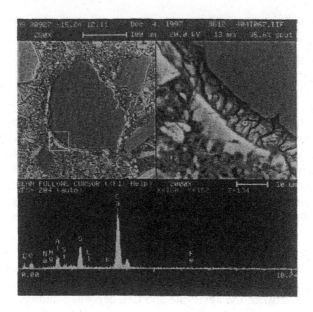

Figure 10. Deposit of ettringite band at the paste/aggregate interface.

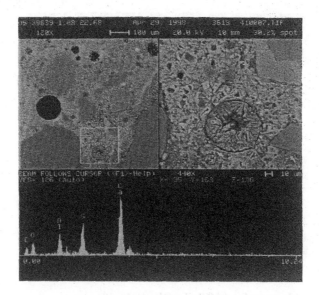

Figure 11. Carbonated zone of the concrete showing empty air voids.

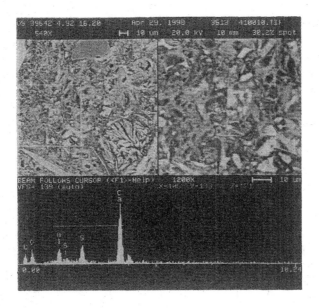

Figure 12. Ettringite deposits in the fine paste pores.

Figure 11 shows the carbonated zone found at the sample surface. The air voids in this carbonated zone are not filled with ettringite. This is due to the reduction of pH of the pore solution in this zone and the consequent destabilization of ettringite. This is an expected effect of carbonation. An ettringite filled air void is seen in a less carbonated area, near the bottom of the field. Thus ettringite recrystallizes below the carbonation zone, creating further overloading of ettringite in this zone which accelerate cracking related to DEF.

Figure 12 shows ettringite deposits in fine pores of the paste and in void spaces of former cement grains.

MECHANISM OF DEF EXPANSION

The mechanism of DEF expansion and cracking of concrete is a controversial issue. As previously indicated, the postulated hypotheses can be classified into two schools. According to one hypothesis the effect of DEF in steam-cured concrete is a 'uniform and homogeneous paste expansion' taking place sometime during post steam curing exposure [12-14]. The other hypothesis postulates that expansion and cracking are the consequence of pressure exerted from 'ettringite crystal growth' [1-4, 15].

The mechanism underlying the uniform and homogeneous paste expansion hypothesis is derived from the fact that during the elevated temperature (>70 ^{0}C) treatment, ettringite which has formed in the system decomposes. During the subsequent curing period it is considered that microcrystalline ettringite forms intermixed with C-S-H. This leads to the uniform and homogenous expansion of the paste, resulting in the formation of gaps around aggregate and cracks through the paste. Subsequently, ettringite is considered as recrystallizing in these gaps, cracks, and in any available void spaces. Ettringite may completely fill the gaps around aggregate or the gaps maybe partially filled. Formation of such secondary ettringite is considered to be non-expansive[14]. The ettringite bands seen at paste/aggregate interfaces are considered to be a result of expansion and not a cause of it [13, 14]. The width of the gaps around aggregate particles is said to be proportional to the aggregate size [12, 16, 17]. Skalny *et al* in a review described the details supporting the uniform and homogenous paste expansion hypothesis [14].

Diamond offers a different hypothesis of the expansion and cracking caused by DEF and several facts are cited when arguing against uniform and homogenous paste expansion [4].

1.   Rimming cracks do not go all the way around aggregate, but only part way around.

2.   Some aggregate particles have rim cracks and others do not.

3.   The rim cracks are portions of a continuous crack network running across the paste, as well as around some aggregate grains. According to Diamond the above phenomena should not occur if a concrete paste matrix has gone through uniform and homogenous paste expansion. Diamond also argues that he observes similar crack patterns and bands of ettringite in non-steam cured concrete. (See Figure 2 of reference #4).

4.   Cracks are not normally observed to radiate from ettringite-filled spherical voids due to the geometry, the stress intensity factor around such voids being too low to induce cracking despite the stresses induced by the ettringite.

According to the crystal growth hypothesis, the expansion and cracking of the paste matrix is the direct result of pressure exerted by ettringite crystal growth. Ettringite deposition in rims surrounding aggregate, pre-formed microcracks and the propagation of these cracks is the cause of expansion. Other authors do not believe to this hypothesis on the grounds that the degree of supersaturation of ettringite in the pore solution of concrete is insufficient to generate the required crystal growth pressure to achieve net expansion[14]. Diamond countered that a modest degree of supersaturation is enough to form ettringite in crack tips and cause crack propagation [4].

Skalny et al [14] in a recent review argues against this hypothesis, They make several arguments:

1.   During precipitation of a crystalline compound from solution, the pressure exerted on its surrounding (the so-called crystallization pressure) requires the following thermodynamic conditions:
    -    The entire precipitating phase must be under pressure.
    -    The activity product of the precipitating phase in solution must be larger than in the saturated solution (supersaturation).

2.   Gaps are formed completely surrounding aggregate particles and the gap width is proportional to aggregate size.

3.   No cracks are seen radiating from the often completely ettringite filled pores and air voids.

DISCUSSION

In the light of the above hypotheses let us analyze the data sets. In Figure 1, ettringite deposits are observed filling cracks running through the hydrated paste and partially around aggregate. In Figure 2, the width of the ettringite band is not uniform; it varies in the range of 0 - 10 μm. The ettringite band is incomplete and CH is in close contact with the aggregate interface. The non-uniformity of ettringite bandwidth is also observed in Figure 7. The above observations support the hypothesis that ettringite crystal pressure is the cause of crack formation and growth.

In figure 4, it is observed that ettringite deposits are present in the hydrated phenograins. Similar crack patterns are also observed in many other hydrated phenograins, but without ettringite deposits (Figure 5). Since the same crack pattern is observed with and without ettringite deposits, these cracks are most probably formed during drying shrinkage.

Figure 3, an X-ray map, shows a sulfur profile almost completely surrounding an aggregate, while the width of the sulfur band appears uniform. A similar band completely surrounding an aggregate, with a uniform width, is observed in Figure 1. This indicates some aggregates show complete and uniform gaps others do not. A range of phenomena are observed; aggregate particles without any gaps, particles with partial gaps and partially filled with ettringite, particles with a complete gap and partially filled with ettringite, particles with a complete gap completely filled with ettringite. There is no indication that the gap width is proportional to the aggregate size.

The observations are not consistent with the uniform and homogenous paste expansion, but they are indicative of localized paste expansion. Uniform paste expansion assumes an even distribution of sulfate throughout the hydrated paste under a substantially elevated curing temperature. If the curing temperature is only slightly elevated or the sulfate distribution is non-uniform, the resulting paste expansion may only occurs over smaller regions. The result of this localized paste expansion, as in the homogenous hypothesis, may be ettringite formation within C-S-H, the formation of gaps at the paste/aggregate interface, with subsequent ettringite recrystallization in these gaps.

The microstructures reflecting uniform and localized paste expansion will be quite different. In uniform paste expansion gaps at paste/aggregate interfaces may completely surround the aggregate and the widths of these gaps may be proportional to the aggregate size [12]. In localized paste expansion, since only one side of an aggregate may be influenced, gaps may not completely surround an aggregate and many aggregates will not have gaps at all. Whether uniform or localized paste expansion occurs, may depend upon both the sulfate concentration in the paste and the degree of elevation of the curing temperature.

The microstructure predicted for uniform and homogenous paste expansion is certainly not observed in the air-entrained concrete; there are very few gaps found surrounding any aggregate. The expected consequences of ettringite crystal pressure phenomenon is also not observed; there are very few cracks filled with ettringite. In air-entrained concrete ettringite fills air voids and fine paste pores, as seen in Figures 8, 9 and 12. The observed phenomena appear to be due to the ettringite recrystallization process preference for open space in which nucleation and growth can occur. There is no further ettringite formation potential in the near vicinity, so no ettringite deposits in open gaps or cracks. In both the air-entrained and non-air-entrained concretes examined in the present investigation it was observed that most of the air voids are completely filled with ettringite and only a few air voids are partially filled. A few cracks filled with ettringite are observed in the air-entrained sample, but there are many empty cracks observed in the system, as seen in Figure 8. There are instances in which ettringite is also deposited as a narrow band at the paste aggregate interface, Figure 10. In figure 10, it is clearly seen that CH originally deposited has intergrown with ettringite.

The failure mechanism of the air-entrained concrete appears to be due to freeze-thaw exposure. Most of the air voids are completely filled with ettringite and no space is available for the freeze-thaw process. This concrete, which was in service in the cold environment, is susceptible to freeze-thaw deterioration. The ettringite deposits in the paste/aggregate interface indicate that some amount of stress has been definitely exerted by DEF. So the expansion, cracking and failure of the concrete is partially due to DEF, but primarily due to freeze-thaw. If there had been no potential for DEF formation, the concrete would likely have performed successfully. If there had been no freezing-thawing cycles, the concrete would have a marginal in-service impact from DEF.

Observations made on both the air-entrained and non-air-entrained concretes do not support the uniform and homogenous paste expansion hypothesis. Most of the observations appear to reflect local expansion. Most probably local paste expansion opens micro-cracks at the interfaces, which than facilitates ettringite crystal growth and result in additional expansion.

CONCLUSIONS

The microstructures and crack patterns of single samples of an air-entrained and a non-air-entrained concrete have been compared. Our observations support the following conclusions:

- DEF leading to concrete failure can exist alone. However, in the presence of other deterioration processes, expansion and cracking and ultimate failure is accelerated.
- DEF was a primary cause of failure in the non-air-entrained concrete.
- On the other hand the failure of the air-entrained-concrete appears to be due to freeze-thaw damage resulting from filling of air voids with DEF-induced ettringite deposits.
- Microstructural observations do not support the hypothesis of 'uniform and homogenous paste expansion' for DEF, but are also not entirely consistent with 'crystal growth hypothesis' for DEF.
- The observations do show evidence that local paste expansion can lead to gap formation and ettringite recrystallization in some areas, which further enhances expansion.

REFERENCES

[1]D. Heinz and U. Ludwig, "Mechanism of Subsequent Ettringite Formation in Mortars and Concretes After Heat Treatment," pp.189-194 in 8th International Congress on the Chemistry of Cement, Vol. V, Rio de Janeiro, 1986.

[2]D. Heinz, U. Ludwig, and I. Rüdiger, "Delayed Ettringite Formation in Heat Treated Mortars and Concretes", Concrete Precasting Plant and Tech., 11, 56-60, (1989).

[3]D. Heinz and U. Ludwig, "Mechanisms of Secondary Ettringite Formation in Mortars and Concretes Subjected to Heat Treatment," ACI SP100, 2, 2059-71 (1987).

[4]S. Diamond, "Delayed Ettringite Formation-Process and Problems," Cement and Concrete Composites 18, 205-215 (1996).

[5]H.M. Ludwig and J. Stark, "Effects of Low Temperature and Freeze-thaw Cycles on the Stability of Hydration Products," pp. 3-9 in 9th International Congress on Chemistry Cements, Vol. IV, New Delhi, 1992.

[6]C. D. Lawrence, "Delayed Ettringite Formation: An Issue?" pp. 113-154 in Materials Science of Concrete IV, edited by J. Skalny and S. Mindess. American Ceramic Society, Westerville, OH, 1995.

[7]V.J. Marks and W.G. Debberke, "Investigation of PCC Pavement Deterioration," Paper presented at 74th Annual Meeting, Transportation Research Board, Washington, 33pp.1995.

[8]A. Shayan and G.W. Quick, "Microscopic Features of Cracked and Uncracked Concrete Railway Sleepers," ACI Materials J., 89, 348-361 (1992).

[9]S. Diamond and S. Ong, "Combined Effect of Alkali Silica Reaction and Secondary Ettringite Deposition in Steam Cured Mortars," pp.79-90 in Cement Technology, Ceramic Transactions, Vol. 40, edited by E.M. Gartner and H. Uchikawa. American Ceramic Society, Westerville, OH, 1994.

[10]R. Yang, C.D. Lawrence, and J.H. Sharp, "Delayed Ettringite Formation in 4-Year Old Cement Pastes," Cement and Concrete Research, 26 [11] 1649-1659 (1996).

[11]C.D. Lawrence, A.J. Daziel, and D. W. Hobbs, British Cement Association, Crowthorne, Berkshire, U. K., interim technical note 12, May 1990.

[12]V. Johansen, N. Thaulow and J. Skalny, "Chemical Degradation of Concrete", paper presented at 74th Annual Meeting, Transportation Research Board, Washington and published as RH and H Bulletin No. 56, Rambøll, Hannemann and Højland A/S, Virum, Denmark, 1995.

[13]K. Scrivener and H.F.W. Taylor, "Delayed Ettringite Formation: a Microstructural and Microanalytical Study", Advances in Cement Research, 5, 139-145 (1993).

[14]J. Skalny, V. Johansen, N. Thaulow, and A. Palomo, "DEF As a Form of Sulfate Attack," Material De Construction, 46 [244] 5-29 (1996).

[15]R. Yang, C.D. Lawrence, C.J. Lynsdale, and J.H. Sharp, "Delayed Ettringite Formation in Heat Cured Portland Cement Mortars, Cement and Concrete Research (in the press).

[16]V. Johansen, N. Thaulow and J. Skalny, Simultaneous Presence of Alkali-silica Gel and Ettringite in Concrete," Advances in Cement Research, 5 [17] 23-29 1993.

[17]R.J. Lee, "Electron Microscopic Approaches to Concrete Deterioration Studies," presented at the H. F. W. Taylor Symposium on Materials Science of Cement and Concrete, High Tatras, Slovakia, June20-25, 1993.

ALKALI-SILICA REACTION - A PROBLEM OF THE INSUFFICIENT FUNDAMENTAL KNOWLEDGE OF ITS CHEMICAL BASE

W. Wieker, C. Hübert, D. Heidemann and R. Ebert
WITEGA Angewandte Werkstoff-Forschung g. GmbH, AG Baustoffchemie
Rudower Chaussee 5, Haus 1.2
D-12484 Berlin-Adlershof

ABSTRACT

With respect to the alkali-silica reaction the reactions of δ-$Na_2Si_2O_5$ and $K_2Si_2O_5$ with reactive silica were investigated by ^{29}Si MAS NMR measurements in the sense of model reactions. It was shown that δ-$Na_2Si_2O_5$ reacts with silica under formation of the layer silicate Kanemite. In the reaction of $K_2Si_2O_5$ with silica only intermediately $KHSi_2O_5$ is formed which is decomposed to a mixture of lower molecular potassium silicates by a quick further uptake of water. According to this the last reaction is connected with a volume expansion of the reaction product in the first reaction stadium, but after relatively short reaction periods expansion decreases to low values. This is in contrast to the reaction of δ-$Na_2Si_2O_5$ with silica, where we found a steady increase of the expansion because of the formation of the relative stable layer silicate Kanemite. Furtheron the influence of $Ca(OH)_2$ on the expansion in the system δ-$Na_2Si_2O_5$ - SiO_2 - H_2O was studied and a possible reaction mechanism is discussed.

INTRODUCTION

Since Stanton's observation that the formation of an alkali containing gel is the source of a destructive process of concrete constructions (1) several efforts have been made to get a deeper insight into the mechanism of this concrete damaging reaction which is influenced by so many unclear factors in the multicomponent system concrete.

At the last International Conference on Alkali-Aggregate Reaction in Melbourne in 1996 Diamond reported that there were published already between 1939 and 1991 about 1300 papers concerning the alkali-silica respectively alkali-aggregate reaction (2). On the other side he pointed on some of those problems that are not exactly understandable and he called them "paradoxes of the alkali-silica reaction". In spite of these till now not fully understandable facts by empirical investigations it was possible to find ways to avoid to a great extent the damaging of concrete structures by alkali-silica reaction (ASR) (3). But by such an approach there always remain questions with respect to the durability of this or those procedure and what happens if one has to leave the empirically explored area of conditions like reactivity and composition of concrete components, humidity, water/cement- and Na_2O/K_2O-ratios and not at least porosity and admixtures used to prepare the concrete by any reasons. By our opinion these uncertainties are the reasons why we hear again and again (and this with an increasing tendency) of new cases of concrete damaging by ASR. Therefore, in the recent years we tried to get a little deeper insight into the mechanism of the ASR, the chemical compounds possibly formed in this reaction and the conditions of their formation (4). It is widely accepted that the alkali-silica gel is formed by a reaction of alkali hydroxides MeOH (Me=Na, K) with reactive SiO_2 contained in aggregates. The MeOH on the other side comes from alkali sulphates in the cements which reacts with $Ca(OH)_2$ according equation 1.

$$Me_2SO_4 + Ca(OH)_2 \longleftrightarrow 2MeOH + CaSO_4 \qquad (1)$$

A second part is generated from the Me_2O content dissolved in the clinker minerals in form of solid solutions. The way by which MeOH reacts with SiO_2 and the alkali containing silica gel is formed is still uncertain.

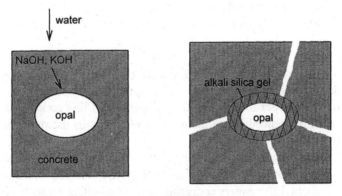

FIGURE 1.
Schematic presentation of the ASR.

According to figure 1 it is suggested that MeOH reacts with the surface of SiO_2 grains, i.e. opal, forming there an alkali silicate that expands in the presence of water resulting in the formation of cracks .

The second proposal for the mechanism of the damage of concrete by ASR came from Johansen et al. (5) who found that the alkali silicate gel can be observed mostly inside those cracks that are primarily present in the aggregates. As one should expect, in this case they observed in ASR damaged concrete that the cracks in the cement matrix have their starting point at these alkali silicate gel containing cracks of the aggregates. But from this observation the following questions arise:

- By which way does the highly viscous ASR gel come into the small cracks of the aggregates, especially when this gel contains Ca^{++}-ions?
- Is the gel usually observed a secondary reaction product of primarily formed alkali silicates?

A first answer of these questions came from Diamond and his co-authors Cong and Kirkpatrick (6) who investigated the reaction of potassium hydroxide and opal by means of ^{29}Si NMR measurements. They found that a part of the four-dimensional network of the opal in which each of the SiO_4-tetrahedra has four bridging oxygen atoms to other Si-atoms (Q^4) (characterised in figure 2 A by a chemical shift δ_{Si}= -112 ppm) is converted to a silicate layer structure in which all SiO_4-tetrahedra share three oxygen atoms with other SiO_4-tetrahedra (Q^3) (characterised in figure 2 A by a chemical shift in the region of -95 till -99 ppm).

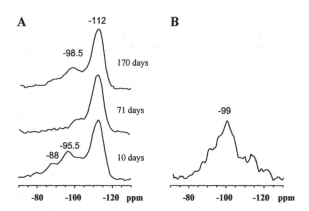

FIGURE 2.

^{29}Si MAS NMR spectra of reaction mixes of Nevada opal with KOH solution at several ages (A) and a ^{29}Si CP MAS NMR spectrum at 10 days (B) (adapted from (6)).

From the results of cross-polarisation (CP) NMR experiments they concluded that Si-OH groups exist in these reaction products with layer silicate structure. The relatively big line width of the corresponding CP NMR signal (see figure 2 B) led to the conclusion that the alkali layer silicates formed were amorphous, so that a characterisation of the reaction product by XRD was impossible.

RESULTS AND DISCUSSION

Starting from the assumption that the alkali-silica gel should be related to crystalline alkali layer silicates we found in the literature the following known compounds (table I). Remarkable of this list is that most of the sodium silicates exist in form of minerals while the potassium compounds can be prepared only artificially. Theoretically all of the compounds of table I could be end or intermediate products of an ASR.

Because Knudsen and Thaulow (7) found that alkali-silica gels from concrete structures mostly have Na_2O/SiO_2 ratios between 0.2-0.5 we decided to investigate at first the Kanemite $NaHSi_2O_5 \cdot 3H_2O$ in the sense of a model system. This compound is easily formed by a reaction of δ-$Na_2Si_2O_5$ with silica gel in the presence of small amounts of water (8) in the temperature range from 20^0 C till 60^0 C according to the following equation:

$$\delta\text{-}Na_2Si_2O_5 + 2SiO_2 + 7 H_2O \rightarrow 2(NaHSi_2O_5 \cdot 3 H_2O) \qquad (2)$$

Table I. Alkali layer silicates

Name	Idealised composition	Name	Idealised composition
Silinaite	$NaLi[Si_2O_5] \cdot 2 H_2O$		
Kanemite	$Na[HSi_2O_5] \cdot 3 H_2O$	(synthetic)	$K[HSi_2O_5]$
Makatite	$Na_2[H_2Si_4O_{10}] \cdot 4 H_2O$		
Octosilicate (synthetic)	$Na_2[H_2Si_8O_{18}] \cdot x H_2O$		
Magadiite	$Na_2[H_2Si_{14}O_{30}] \cdot x H_2O$	K-Magadiite (synthetic)	$K_2[H_2Si_{14}O_{30}] \cdot x H_2O$
Kenyaite	$Na_2[H_2Si_{20}O_{42}] \cdot x H_2O$	K-Kenyaite (synthetic)	$K_2[H_2Si_{20}O_{42}] \cdot x H_2O$

To study this reaction *in situ* by ^{29}Si NMR methods we used an amorphous silica sample which was enriched in ^{29}Si. The δ-Na$_2$Si$_2$O$_5$ contained ^{29}Si in its lower natural abundance. The ^{29}Si MAS NMR spectra of the starting materials, the reaction products at several ages and the corresponding spectrum of pure Kanemite are shown in figure 3.

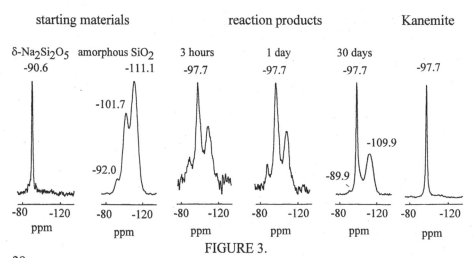

FIGURE 3.

^{29}Si MAS NMR spectra of δ-Na$_2$Si$_2$O$_5$, the used amorphous SiO$_2$ sample enriched in ^{29}Si , the reaction products measured *in situ* at several ages and the corresponding spectrum of pure Kanemite

Caused by the high degree of ^{29}Si enrichment in the SiO$_2$ the measured NMR spectra of the reaction products represent only those products coming from the SiO$_2$ sample. Compared with the spectrum of the starting SiO$_2$ in the spectra of the reaction products already after 3 hours a new signal at -97.7 ppm appears which increases with increasing reaction time. This signal at -97.7 ppm is typical for Kanemite (9). Therefore, these *in situ* investigations show that in the studied reaction the SiO$_2$ is partially converted into Kanemite and that the formation of Kanemite increases with increasing reaction time.

To show that there is really a similarity with the alkali-silica reaction we tried to find out if the reaction between δ-Na$_2$Si$_2$O$_5$ and SiO$_2$ is also connected with an expansion process. Therefore we pressed tablets from 1:2 molar mixtures of δ-Na$_2$Si$_2$O$_5$ and SiO$_2$, treated them at room temperature at nearly 80% relative humidity and measured their expansion in dependence on the reaction time as described in (4). Figure 4 demonstrates clearly that the reaction according to equation 2 produces a strong expansion. Under the same conditions we performed also experiments in which we substituted the amorphous silica by minerals, i.e. opal and siliceous chalk, which are substances with a very well known high reactivity in the alkali-silica reaction. As shown in figure 4 these minerals give a relatively high expansion, too. Crystalline quartz on the other hand does not give any expansion in the reaction with δ-Na$_2$Si$_2$O$_5$.

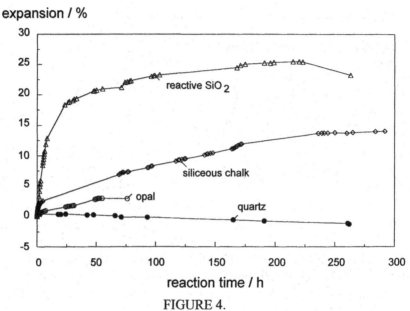

FIGURE 4.
Expansion plot of the reaction product of δ-Na$_2$Si$_2$O$_5$ with amorphous SiO$_2$, siliceous chalk, opal and quartz in dependence on reaction time.

All these results show a considerable similarity to the ASR and provoke questions with respect to the formation of a compound Na$_2$Si$_2$O$_5$ in a real concrete system.

In this connection one should remember on two well known fundamental rules in silicate chemistry (see e.g. (10)):

1. An increasing SiO$_2$/Me$_2$O ratio in silicate solutions leads to an increase in the condensation degree of the silicates formed.

   That means, if for example a concrete pore solution with a definite amount of MeOH (Me=Na, K) dissolves more and more SiO$_2$ the primarily formed monomeric silicates are converted to dimeric, trimeric, long chain, double chain and at least layer silicates.

   This is it what goes on in ASR, too.

2. An increasing concentration of a silicate solution with a definite SiO_2/Me_2O ratio leads to an increase of the condensation degree of the formed silicates. Also this way of increasing the condensation of silicates can occur in cement based systems because the water added to the cement is partially consumed by the hydration and in a normal climate an evaporation of water takes place resulting in an increase of the concentration of the pore solution in concrete.

From these two facts it seems generally possible that alkali layer silicates can be formed in concrete systems. But there remains the question if really only the layer silicates are able to form protonated alkali layer silicates like Kanemite by a reaction with reactive silica. Therefore, we prepared tablets of a mixture of silica gel with the compound $(Na_2SiO_3)_x$ which is built up only by long silicate chains and cured them in a chamber at room temperature in which a relative humidity of 87% was adjusted. The result of this experiment was that a viscous mass was formed which does not give any expansion. This is the reason why we assume that the formation of alkali layer silicates is a necessary condition for expansion which we observe in the alkali-silica reaction. But it is not easy to understand that the formation of alkali layer silicates is possible in the presence of Ca^{++}-ions which are constituent in all hydrating cement containing system, because Ca^{++}-ions should interfere in the condensation mechanism of silicate anions by the formation of very low soluble precipitates already with low molecular silicates. Furthermore, there should be a Ca-Na cation exchange reaction with alkali silicates present in the silicate condensation equilibrium resulting in condensed silicates with mixed cations (Na and Ca).

To get more detailed information in this direction we performed expansion measurements in which we used 1 : 2 : (0; 0.5; 2; 4) molar mixtures of δ-$Na_2Si_2O_5$, SiO_2 and $Ca(OH)_2$. The results in figure 5 show that the expansion is strongly depressed when the amount of $Ca(OH)_2$ in these reaction mixtures increases from 0 to 4. From the fact that in the reaction products by XRD always Kanemite-like structures still can be observed we assume that Ca^{++}-ions enter by an ion exchange process into layer silicate structures which are already formed. Caused by the high positive electrostatic field strength of the Ca^{++}-ions the distance between the negative charged silicate layers is reduced. Under these condition a smaller number of water molecules is able to enter the interlayer space so that the expansion of such layer silicates is reduced. To enhance this interpretation much more investigations are needed. In the already cited paper of Diamond and co-authors (6) it was shown that investigating the reaction of opal with KOH in the presence of $Ca(OH)_2$ in fact layer silicate structures (δ_{Si}= - 97 and -92 ppm) are formed (see figure 6).

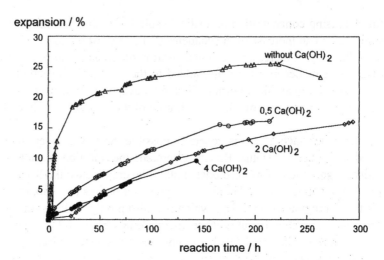

FIGURE 5.
Expansion plot of the reaction products of δ-$Na_2Si_2O_5$, amorphous SiO_2 and different ratios of $Ca(OH)_2$ at nearly 80% relative humidity in dependence on reaction time.

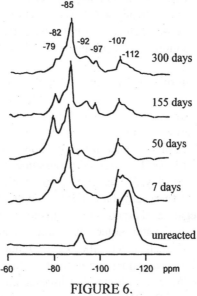

FIGURE 6.
^{29}Si MAS NMR spectra of the reaction products of Beltane opal with KOH solution in the presence of $Ca(OH)_2$ (adapted from (6)).

But the main amount of Si is present in low molecular silicates characterised by signals between -79 and -85 ppm which are very similar to those from C-S-H phases. This could mean that the Ca^{++}-ions interfere in the condensation equilibrium of the alkali silicates as mentioned above resulting only in the formation of insoluble calcium silicates with low molecular anions which are not known to give strong expansion.

Even though the mechanism of the influence of Ca^{++}-ions on the alkali-silica reaction is not quite clear and needs further investigations it is the fact that Ca^{++}-ions reduce the expansion caused by ASR drastically. The highest expansions in these reactions should therefore be observed when the condensation equilibrium of the silicates in the aqueous phase of mortars or concretes is not influenced by Ca^{++}-ions.

How could this be realised in a real concrete system?

First: If the alkali hydroxide concentration is high only a very low $Ca(OH)_2$ concentration is present in the pore solution so that one can expect that there is a slow exchange of alkali ions against Ca^{++}-ions related to the formation velocity of the layer silicates.

Second: The primarily formed diluted low molecular alkali silicate solutions which have a low viscosity easily enter cracks in aggregate particles where they can react with reactive silica and form higher molecular alkali silicates up to layer silicates that are able to give expansions. Because Ca^{++}-ions are mostly relatively strong bonded in the normally dense aggregates there should be only a very small interference between the Ca^{++}-ions and the alkali silicates in these cracks.

The results of Johansen and co-workers (5) who found the alkali silica gels especially in cracks underline this interpretation. A more direct result came from experiments of Scrivener and Monteiro (11) in which they investigated the reactions in the system cement, alkali hydroxide and monolithic opal by scanning electron microscopy (SEM) and energy-dispersive X-ray (EDS) analysis. They found that in "fingers" and cracks a type I gel with relatively high amounts of sodium and potassium but relatively low content of calcium and $(CaO + Na_2O + K_2O)/SiO_2$ molar ratios of 0.5 is formed. This would correspond to a composition 0.56 (Na,K) 0.72 $CaSi_2O_5$ calculated from the data published in (11). This composition is similar to that of the layer silicate structure in Kanemite in which parts of the sodium, potassium and protons are substituted by Ca^{++}-ions. In the interface between the monolithic opal and the hydrated cement a type II gel with a high amount of CaO and lower contents of Na_2O and K_2O and $(CaO + Na_2O + K_2O)/SiO_2$ molar ratio of 0.55 was detected. That would mean that a further exchange of Na^+- and K^+-ions by Ca^{++}-ions took place. According to the authors mentioned above the additional uptake of Ca^{++}-ions going from gel

I to gel II results from a local leaching of the surrounding cement hydration products. Investigations on the structure of such partially Ca-exchanged alkali layer silicates are now in progress.

A further problem we investigated comes from the question if Na_2O- and the K_2O contents in cements have the same influence on the rate and degree of the expansion in alkali-silica reaction.

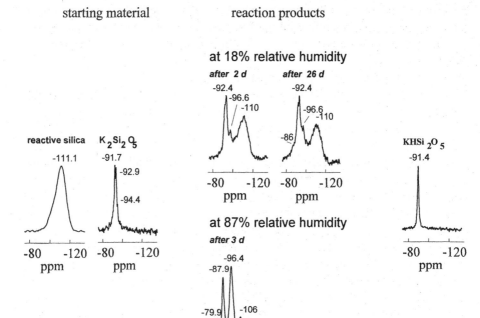

FIGURE 7.

^{29}Si MAS NMR spectra of $K_2Si_2O_5$, the reactive silica, the reaction products at several ages and different relative humidities as well as the corresponding spectrum of pure $KHSi_2O_5$.

To solve this problem we prepared $K_2Si_2O_5$ by a thermal process and mixed this compound in a molar ratio of 1:2 with reactive silica and pressed tablets as we did it in our experiments with $Na_2Si_2O_5$ mentioned above. These tablets we cured in a chamber at relative humidities of 87% and 18%. The starting materials and the reaction products as well as the pure $KHSi_2O_5$ were characterised by ^{29}Si MAS NMR investigations as summarised in figure 7.

The spectra show that we got different reaction products in dependence on the relative humidity. The measured spectra of the reaction products, formed at a low relative humidity of 18%, mainly consist of the signals for the starting materials at -92.4 ppm and -110 ppm for both the reaction products after 2 days and 28 days. Besides these signals a small signal is detectable at -96.6 ppm which must be attributed to Q^3-groups in a layer silicate formed during the reaction. But this layer silicate is not the expected compound $KHSi_2O_5$ (12) because the last is characterised by a chemical shift value of -91.4 ppm (see figure 7). By our NMR investigations the formation of $KHSi_2O_5$ can not be proofed in the reaction products because its NMR signal at -91.4 ppm can not be clearly distinguished from the signals of $K_2Si_2O_5$ between -91.7 ppm till -94.4 ppm. But by XRD it is possible to show that $KHSi_2O_5$ exists in the reaction products formed at the low relative humidity of 18%. The NMR spectrum of the reaction product formed after 3 days at the high relative humidity of 87% (mostly used in the experiments with $Na_2Si_2O_5$) consists of NMR signals at -79.9, -87.9, -96.4 and -106 ppm. These chemical shift values are very similar to those measured for water glass solutions which contain a mixture of low and high molecular silicate species (10). This result is supported by the fact that the reaction mixture is transformed to a plastic mass and with increasing reaction time to a solution. This shows that the $K_2Si_2O_5$-SiO_2 system is much more hygroscopic than $Na_2Si_2O_5$-SiO_2. This fact also influences the expansion behaviour of tablets from mixtures of $K_2Si_2O_5$ and SiO_2. The results of the expansion measurements at nearly 80% relative humidity presented in figure 8 show a fast increase within the first 30 hours reaching values of nearly 40% and after this the expansion decreases rapidly till values lower that 5% at 200 hours.

To demonstrate the large difference between the reaction of $K_2Si_2O_5$ and $Na_2Si_2O_5$ with reactive SiO_2 the expansion plot of the last one is also presented in figure 8. The reason for this different expansion behaviour is obviously connected with the fact that from the $NaHSi_2O_5$ a stable hydrate exists in form of Kanemite $NaHSi_2O_5 \cdot 3H_2O$. From $KHSi_2O_5$ no stable hydrate is known, so that during the reaction the further uptake of water leads to a shift of the condensation equilibrium to lower molecular species via hydrolysis reactions. Such low molecular silicates are not able to produce a high volume expansion. From these results we must conclude that the Na/K ratio plays a role with respect to the extent of the expansion in ASR.

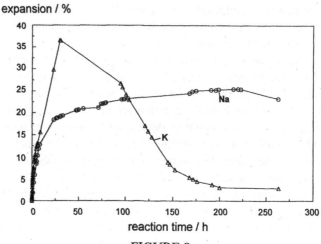

FIGURE 8.

Expansion plot of the reaction product of $K_2Si_2O_5$ with reactive SiO_2 (K) compared with that of δ-$Na_2Si_2O_5$ with reactive SiO_2 (Na) at nearly 80% relative humidity in dependence on reaction time.

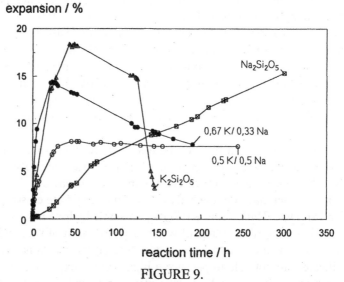

FIGURE 9.

Expansion plot of the reaction products of $Na_2Si_2O_5/K_2Si_2O_5$ mixtures with a less reactive SiO_2 at nearly 80% relative humidity in dependence on reaction time.

To demonstrate this we prepared $Na_2Si_2O_5/K_2Si_2O_5$ mixtures of different molar ratios and added SiO_2 in a molar ratio of $2SiO_2$: $1(xNa_2Si_2O_5+yK_2Si_2O_5)$, $(x+y=1)$. These starting materials were pressed to tablets and cured at a relative humidity of nearly 80%. The expansion of these tablets was determined in dependence on the reaction time. The results presented in figure 9 show that the expansion of the reaction products in the first 120 hours is always higher if one has $K_2Si_2O_5$ in the reaction mixture than in that reaction mixture where only $Na_2Si_2O_5$ and SiO_2 are used. After 150 hours the tablet which contains only δ-$Na_2Si_2O_5$ and SiO_2 gives always the highest expansion. With respect to the ASR this would mean that potassium oxide contents in concretes are less harmful than those of sodium oxide, because there is the possibility that in the first week of the hardening of concrete a compensation of stress caused by the expansion is possible.

CONCLUSIONS

These results show that we have to consider in addition to the several still unresolved problems in the discussion with respect to the ASR a further factor, the Na/K ratio in the cements and possibly in reactive aggregates like graywacke or biotite.

From this we think that further investigations are necessary to get more detailed results especially on the influence of the water vapour pressure and the influence of cations with higher charges on these reactions.

ACKNOWLEDGEMENT

For the financial support of our investigations we thank the Deutsche Forschungsgemeinschaft.

REFERENCES

(1) T. E. Stanton, "Expansion of Concrete through Reaction between Cement and Aggregate", *Proceedings of the American Society of Civil Engineering,* **66** 1781-1811 (1940).

(2) S. Diamond, "Alkali-Silica Reaction : Some Paradoxes", *Proceedings of the 10th International Conference on Alkali-Aggregate-Reaction,* A. Shayan (ed.), pp. 3-14, Melbourne 1994.

(3) M. D. A. Thomas and B. A. Blackwell, "Summery of BRE Research on the Effect of Fly-Ash on Alkali-Silica Reaction in Concrete", *Proceedings of the 10th International Conference on Alkali-Aggregate-Reaction,* A. Shayan (ed.), pp. 554-569, Melbourne 1994.

(4) W. Wieker, C. Hübert and R. Ebert, "Contribution to the Chemical Reaction Mechanism of the Alkali-Aggregate Reaction", *Proceedings of the 10th International Conference on Alkali-Aggregate-Reaction*, A. Shayan (ed.), pp. 919-926, Melbourne 1994.

(5) V. Johansen, N. Thaulow and J. Skalny, "Simultaneous Presence of Alkali-Silica Gel and Ettringite in Concrete", *Advances in Cement Research*, **5** [17] 23-29 (1993).

(6) X.-D. Cong, R. J. Kirkpatrick and S. Diamond, "^{29}Si MAS NMR Spectroscopic Investigation of Alkali Silica Reaction Product Gels", *Cement and Concrete Research*, **23** [4] 811-823 (1993).

(7) T. Knudsen and N. Thaulow, "Quantitative Microanalysis of Alkali-Silica Gel in Concrete", *Cement and Concrete Research*, **5** 443-454 (1975).

(8) A. Tapper, R. Adrian and G. Schimmel, "Procedure of Preparation of a Crystalline Sodium Layer Silicate with Kanemite Structure", patent EP 0627383, April 30, 1994.

(9) W. Wieker, D. Heidemann, R. Ebert and A. Tapper, "On the Chemistry of Kanemite [NaHSi$_2$O$_5$ · 3 H$_2$O]", *Zeitschrift für anorganische und allgemeine Chemie*, **621** 1779-1784 (1995).

(10) R. K. Harris, E. K. F. Bahlmann, K. Metcalfe and E. G. Smith, "Quantitative Silicon-29 NMR Investigations of Highly Concentrated High-Ratio Sodium Silicate Solutions", *Magnetic Resonance in Chemistry*, **31** 743-747 (1993).

(11) K. I. Scrivener and P. J. M. Monteiro, "The Alkali-Silica Reaction in a Monolithic Opal", *Journal of the American Ceramic Society*, **77** [11] 2849-2856 (1994).

(12) Z. Q. Deng, J. F. Lambert and J. J. Fripiat, "A Puckered Layered Silicate, KHSi$_2$O$_5$: Hydrolysis Products and Alkylammonium Intercalated Derivates", *Chemistry of Materials*, **1** [3] 375-380 (1989).

CRACKING OF MORTARS SUBJECTED TO EXTERNAL SULFATE ATTACK

Manu Santhanam and Menashi D. Cohen
School of Civil Engineering
Purdue University
W. Lafayette, IN 47907

ABSTRACT

Microstructural investigations were conducted on mortars with 0.50 water-cementitious materials ratio (w/(c+m)) using backscatter scanning electron imaging technique and energy dispersive X-ray analysis. The mortars were subjected to external attack by 8.90 mass percent sodium sulfate solution (equivalent SO_3 = 50,000 ppm). The effect of partial replacement of portland cement with slag, silica fume, or metakaolin on sulfate attack was also investigated. The mode of sulfate attack on the specimen was found to be divided into three distinct zones : Zone 1 (thickness: 300 - 500 µm) - the "surface zone" which was highly cracked, Zone 2 (thickness: 800 - 1000 µm) - the middle zone, which is practically crack-free; this zone, referred to as "deposition zone", showed deposition of gypsum in pores and around aggregates (at the paste-aggregate interface) and also presence of ettringite within the mass of the cement paste, and Zone 3 (thickness: 1000 -1500 µm) - the interior zone or "cracking zone" which was found to be highly cracked but chemically unaltered. Partial substitution of portland cement with slag, silica fume, or metakaolin did not appear to alter the mode of attack but decreased the intensity. For instance, the degree of cracking in the interior zone (Zone 3) was less. This may have been due to the formation of smaller amounts of ettringite and gypsum in Zone 2. These observations are used to explain a cracking pattern observed in mortar specimens subjected to external sodium sulfate attack.

INTRODUCTION

Sulfate attack from an external source on portland-cement mortar has been studied for many years. A clearer understanding of sulfate attack mechanism has been made possible because of the availability of advanced experimental techniques in microstructural investigations, such as backscatter electron imaging [1-4]. Attack by sodium sulfate solution, as described by researchers [5-8], can cause failure of concrete by (a) expansive pressure generated due to the formation of ettringite and (b) softening of the structure as a result of gypsum formation.

Though the chemical nature of the attack has been well documented, the actual mechanism of cracking has not been explained in detail.

The objective of this paper is to present and discuss a pattern of cracking observed in mortar specimens subjected to external sulfate attack (8.90 mass % sodium sulfate solution.

MATERIALS AND MIXTURES

ASTM Type I-II portland cement was used in this investigation. A commercial ground-granulated iron blast-furnace slag, silica fume and laboratory synthesized metakaolin were used as materials that partially replaced the portland cement. The composition and properties of the materials are presented in Table I. Fine aggregate used was a #20-#30 standard Ottawa sand.

A total of six mixtures were prepared. The design factor selected was a constant ratio of absolute volume of paste to absolute volume of aggregate, in order to overcome the increase in volume of paste as a result of portland cement (PC) substitution. The mixture proportions have been presented in Table II. Mixing procedure was in accordance with ASTM C305.

Dimensions of specimens were 5mm x 14mm x 80 mm. After demolding, the specimens were cured in saturated lime-water solution for six days. Then some of the specimens were transferred to a bath containing 8.90 mass % sodium sulfate solution (50,000 ppm SO_3). The specimens were removed after 140 days of immersion in the solution, fractured in the center, and prepared for backscatter electron imaging. The area of the fracture surface under observation was 5mm x 14mm. Companion specimens were used for thermal analysis by Differential Scanning Calorimetry (DSC). For the preparation of samples for DSC, material was removed from the surface of the specimen to an approximate depth of about 1mm, washed with acetone, ground to a powder, and passed through an ASTM 45 μm sieve.

RESULTS

The results of DSC measurements are presented in Figure 1. The figure shows % ettringite and gypsum by mass of the sample measured at 140 days, against the mixture code (M1 through M6). The pattern shows a reduction in the level of ettringite with the addition of slag, silica fume, and metakaolin, compared to the control mixture. For instance, the mixture M3 (with 70% slag substitution) shows about 4% ettringite as compared to M1 (100% PC) which shows more than 10% ettringite formation. The gypsum content is also reduced as a result of portland cement substitution. For instance, the gypsum content for M3 is only 1% as compared to about 8% for M1.

The backscatter images are presented in Figures 2-8*. Figure 2 shows the image for PC mortar (M1) specimen in saturated lime-water solution. This specimen did not show any cracking, and energy dispersive X-ray analysis confirmed the formation of normal hydration products.

Figure 3 shows PC mortar (M1) in sodium sulfate solution. Three distinct zones can be observed (these have been marked 1, 2 and 3). Zone 1 (shown in detail in Figure 4), is termed the "surface zone". It appears to be highly cracked.

---

*In all the SEM micrographs, the dark region next to the specimen is the polymer in which the specimen was embedded. The surface in contact with the solution is the interface between the polymer and the specimen.

Table I. Compositions in % and properties of materials used[*]

Item Reported	PCL	GGBFS	SF	MK
SiO_2	20.87	37.72	98.39	54.44
Al_2O_3	4.13	7.95	0.75	44.58
Fe_2O_3	3.21	0.44	0.57	0.91
CaO	62.99	39.34	0.03	0.07
MgO	1.99	11.22		
K_2O	0.55	0.37	0.26	
Na_2O	0.07	0.27		
TiO_2	0.21	0.44		
SO_3	2.70	0.16		
P_2O_5	0.09			
MnO	0.03	0.66		
SrO	0.11			
L.O.I.	1.00			
Blaine (m^2/kg)	373	561	2400	1250
Specific Gravity	3.15	2.77	2.20	2.60
Bogue Calculations for PCL				
C_3S	56.14			
C_2S	17.54			
C_3A	6.33			
C_4AF	9.76			

Table II. Mixture Compositions (w/(c+m) = 0.5 for all mixtures)

Mixture Code	PCL (g)	GGBFS (g)	SF (g)	MK (g)	Sand (g)	Proportions
M1	378.0	0	0	0	756.0	Control (100% PCL)
M2	264.6	113.4	0	0	763.3	30% Slag + 70% PCL
M3	113.4	264.6	0	0	773.0	70% Slag + 30% PCL
M4	189.0	162.5	26.5	0	775.3	43% Slag + 7% SF + 50% PCL
M5	351.5	0	26.5	0	764.8	7% SF + 93% PCL
M6	283.5	0	0	94.5	771.3	25% MK + 75% PCL

[*] PCL – Portland Cement, GGBFS – Ground Granulated Blast Furnace Slag, SF – Silica Fume, MK – Metakaolin (Compositions of Metakaolin and Silica Fume were determined by quantitative X – ray analysis)

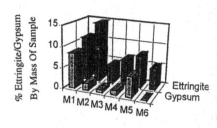

Figure 1. DSC patterns observed

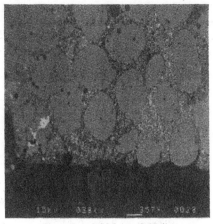

Figure 2.   PC Mortar (M1) in

Saturated Lime Water solution

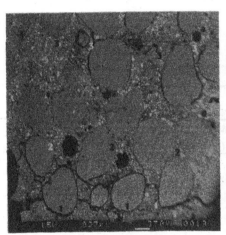

Figure 3.   PC Mortar (M1)

in Sodium Sulfate solution

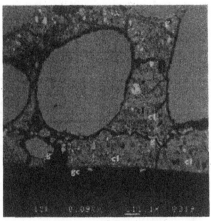

Figure 4.   Zone 1 from Fig. 3

showing gypsum and calcite on surface

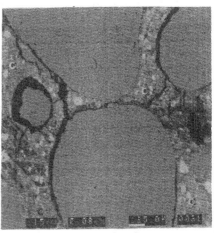

Figure 5.   Zone 2 (deposition zone) from Fig. 3

Figure 6.   Zone 3 (cracking zone) from Fig. 3

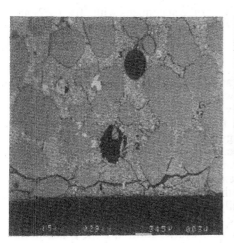

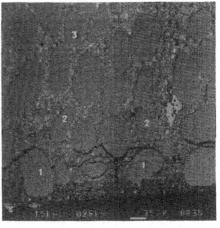

Figure 7.   30% slag replacement mixture –

M2 (showing the same pattern as M1)

in sodium sulfate solution

Figure 8.   70% slag replacement mixture –

M3 (showing the same pattern but lesser

cracks in Zone 3) in sodium sulfate solution

Cracks were seen in the paste matrix and around the aggregates at the paste-aggregate interface. Large quantities of ettringite were also found in the paste (marked c1) in this zone. The surface of the specimen had gypsum and calcite deposits (marked gc); calcite could have formed due to carbonation of CH or probably due to decalcification of the CSH gel and subsequent carbonation (energy dispersive X-ray analysis yielded low C/S ratios in this zone). Gypsum (marked g) was also found in the cracks and around the aggregates. Zone 1 was found to extend from the surface to a depth of 300 - 500 μm.

Zone 2 (detailed view in Figure 5), is termed the "deposition zone". It was found to be devoid of any major cracks. This zone showed gypsum deposits (marked g) around aggregates and in the pores, and also large amounts of ettringite in the paste (marked c1). This zone had a thickness of about 800 - 1000 μm.

Zone 3 (shown in Figure 6) is termed "interior zone". It could also be called "cracking zone" owing to the multitude of large cracks (as wide as 50 μm) present in this zone. Cracks were found in the paste matrix as well as at the paste-aggregate interface. CSH gel seemed to be unaltered, and energy dispersive X-ray analysis confirmed normal C/S ratios. There were no traces of gypsum, ettringite, or any other signs of chemical attack.

Figures 7 and 8 depict the overall view for the specimens with 30% (M2) and 70% (M3) replacements of PC with slag. These specimens show the same trend as the PC mortar specimen, with the formation of the three distinct zones.

DISCUSSION

The formation of zones 1, 2, and 3 as a result of an external attack by 8.90 mass % sodium sulfate solution, as well as the cracking observed in the chemically unaltered Zone 3, can probably be explained as follows:

1. The sulfate solution reacts with hydrated cement compounds near the surface (Zone 1) to form ettringite and gypsum. Ettringite is formed within the paste and gypsum is formed at the surface, in the pores and at the paste-aggregate interface. Thus, to begin with, Zone 1 can be called the "deposition zone". The formation of ettringite generates an expansive stress which results in the expansion of Zone 1 layer. However, the expansion of Zone 1 is somewhat restrained by Zone 2, which is chemically unaltered at this stage of the attack. This may cause the balance of forces to act in such a way that Zone 1 goes into compression, while Zone 2 goes into tension. Cracks form in Zone 2 due to the tensile strain/stress.

2. As the attack progresses, Zone 1 deteriorates due to the softening action of gypsum. This helps the solution to penetrate further inside the specimen and react with the hydrated cement compounds in the inner region (Zone 2). The reactions can lead to deposition of the newly formed ettringite in the paste, and gypsum in the cracks at the paste-aggregate interface and in the matrix, in Zone 2. Hence, Zone 2 acts as the new "deposition zone". Zone 1, because of its state of disintegration, is no longer the deposition zone, but is termed as the "surface zone". Zone 3 is the zone further inside compared to Zone 2, and it has not yet been attacked by the sulfate solution.

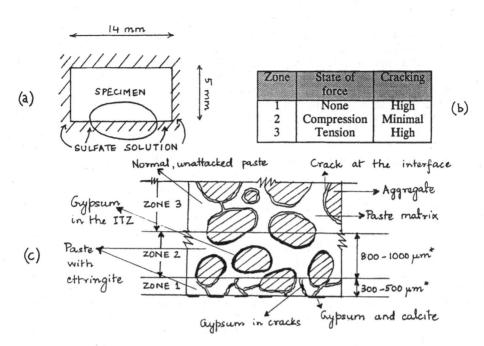

(a)

(b)

Zone	State of force	Cracking
1	None	High
2	Compression	Minimal
3	Tension	High

(c)

Normal, unattacked paste

Crack at the interface

Gypsum in the ITZ

ZONE 3

Aggregate

Paste matrix

Paste with ettringite

ZONE 2

800 - 1000 μm*

ZONE 1

300 - 500 μm*

Gypsum in cracks

Gypsum and calcite

* Thicknesses are time and sulfate concentration dependent

Figure 9.   Schematic representation of the proposed 8.90 mass % sodium-sulfate attack
(a) - The specimen and the surrounding solution.
(b) - Characteristics of the three zones.
(c) - Detailed view of enclosed section from (a).

Zone 1 - 'surface zone', deteriorated; gypsum and ettringite present.
Zone 2 - 'deposition zone', with gypsum and ettringite deposits.
Zone 3 - 'interior zone', chemically unaltered.

The expansion of Zone 2 due to the formation of ettringite is restrained by Zone 3. This cycle repeats the formation of the state of stresses described earlier, with Zone 2 experiencing a compressive force and Zone 3 experiencing a tensile force. This stage is depicted in the SEM micrographs. Zone 3 shows extensive cracking, while Zone 2 is devoid of any major cracks.

The schematic diagram in Figure 9 depicts the stage of attack discussed above. The possible state of forces in the three zones, and the gypsum and ettringite deposits, are shown in the diagram. The relevant images describing the three zones are presented in Figures 4 through 6.

When portland cement is replaced in part by slag, silica fume, or metakaolin, there is a lower availability of Ca^{2+} due to the consumption of $Ca(OH)_2$ in the binder reactions. Thus the rate of attack is slower because the gypsum and ettringite forming reactions in sulfate attack [7] are dependent on the availability of $Ca(OH)_2$. DSC measurements presented in the previous section support this fact, and there is a reduction in the amounts of ettringite and gypsum formed. Hence at the same age, the level of deterioration is lesser when PC is partly replaced by slag and mineral admixtures.

CONCLUSIONS

The following list of conclusions are related to cracking of mortars subjected to external attack by 8.90 mass % sodium sulfate solution :

1. The attack causes damage due to expansions related to ettringite formation, though the level of deterioration is higher when there is more gypsum formed.

2. Three distinct zones are formed as a result of the attack:

Zone 1 - "Surface Zone" 300 - 500 μm thick, highly cracked, with gypsum in the cracks and on the surface, and ettringite within the paste.

Zone 2 - "Deposition Zone" 800 - 1000 μm thick, with minimal cracking, deposits of gypsum in pores and at the paste-aggregate interface, and ettringite within the paste.

Zone 3 - "Interior Zone" which is highly cracked but chemically unaltered, does not show the presence of unusually high amounts of ettringite or any gypsum

3. The replacement of portland cement with slag, silica fume, or metakaolin does not alter this trend, i.e. the three zones are still present. However, it (replacement) causes a reduction in the cracks in Zone 3, as a result of the reduction in the amount of ettringite formed in Zones 1 and 2.

The results yielded in this research pertain to mortar mixtures with a fixed w/(c+m) and exposed to one level of sulfate concentration. The rate of reaction would be faster or slower depending on whether the w/(c+m) is lower or higher, respectively. At different sulfate concentrations, the attack may be governed by different factors. For instance, at lower sulfate concentrations, gypsum may not play an significant.

ACKNOWLEDGMENTS

Financial support from the National Science Foundation - Advanced Cement-Based Materials is thankfully acknowledged. Review comments by Dr. Bryant Mather are appreciated.

REFERENCES

1. David Bonen and M.D. Cohen, "Magnesium Sulfate Attack on Portland Cement Paste - I. Microstructural Analysis", Cement and Concrete Research 22, 1992, pp. 169-180.
2. David Bonen and M.D. Cohen, "Magnesium Sulfate Attack on Portland Cement Paste - II. Chemical and Mineralogical Analyses", Cement and Concrete Research 22, 1992, pp. 707-718.
3. R.S. Gollop and H.F.W. Taylor, "Microstructural and Microanalytical Studies of Sulfate Attack - I. Ordinary Portland Cement Paste", Cement and Concrete Research 22, 1992, pp. 1027-1038.
4. R.S. Gollop and H.F.W. Taylor, "Microstructural and Microanalytical Studies of Sulfate Attack - III. Sulfate Resisting Portland Cement: Reactions with Sodium and Magnesium Sulfate Solutions", Cement and Concrete Research 25, 1995, pp. 1581-1590.
5. F. Akoz, F. Turker, S. Koral and N. Yuzer, "Effects of Sodium Sulfate Concentration on the Sulfate Resistance of Mortars With and Without Silica Fume", Cement and Concrete Research 25, 1995, pp. 1360-1368.
6. P.K. Mehta, "Mechanism of Sulfate Attack on Portland Cement - Another Look", Cement and Concrete Research 13, 1983, pp. 401-406.
7. M.D. Cohen and Arnon Bentur, "Durability of Portland Cement-Silica Fume Pastes", ACI Materials Journal 85, 1988, pp. 148-157.
8. Guiseppe Frigione and Ricardo Sersale, "The Action of Some Aggressive Solutions on Portland, Pozzolanic, and Blast Furnace Slag Cement Mortars", Cement and Concrete Research 19, 1989, pp. 885-893.

INFLUENCE OF THE CEMENT TYPE ON THE RESISTANCE OF CONCRETE TO AN AGRICULTURAL ENVIRONMENT

Jean Pera, Sébastien Rols, Michel Chabannet and Jean Ambroise
Unité de Recherche Génie Civil - Matériaux Institut National des Sciences Appliquées de Lyon - France

ABSTRACT

Concrete in an agricultural environment is subjected to physical and chemical deterioration. The chemical attack is mainly due either to organic acids such as lactic and acetic acids or sulfates such as ammonium sulfate.

This paper deals with the durability of concrete prisms placed in solutions of lactic acid (pH = 4) and ammonium sulfate (200 g/l). Two types of cement and mineral admixtures (metakaolin and silica fume) were investigated. The mass loss, and the erosion or the swelling of specimens were measured. The microstructure was studied by means of scanning electron microscopy associated with energy dispersive X-ray analysis.

The results obtained show that lactic acid reacted with portlandite to produce very soluble calcium lactate. The presence of metakaolin or silica fume at a weight content of 10 % limited this phenomenon. Ammonium sulfate reacted with portlandite to produce expansive gypsum. The addition of pozzolans limited the size of gypsum crystals and provided better durability, according to the type of cement used.

INTRODUCTION

Concrete is a common building material used for structures in the agricultural sector. It is subject to aggressive environmental conditions [1]. The agricultural environment is very complex and can be described by chemical and physical parameters. The cause of the complex environment is the variety of biological activities, which occur within and outside buildings mainly in the organic materials. The organic materials consist of food and manure for and from the animals [2, 3]. Lactic and acetic acids are formed in meal-water mixtures, resulting in pH possibly below 4.5.

Other sources of aggressive substances are effluents from agricultural industries [4]. Latic and butyric acids occur in sour milk and butter and can give rise to pH values as low as 4.0 in the effluents from drairies and cheese factories. These acids, together with acetic and other acids, also arise during the production of silage from green plant material. Ammonium sulfate can be present in liquid fertilizers up to a concentration of 20 %.

Lactic acid reacts very aggressively with concrete because its reaction with calcium hydroxide $(Ca(OH)_2)$ produces very soluble calcium salts [5] which are removed by leaching. Neither pozzolanic nor blast-furnace slags cements offer any advantage over normal portland cement against attack by lactic acid [4]. An investigation of the effects of Na^+, Mg^{2+}, NH_4^+ sulfate solutions on pozzolanic cement showed the most aggressive solution is ammonium sulfate $((NH_4)_2 SO_4)$[6]. A cation-exchange reaction takes place between Ca^{2+} and NH_4^+ leading to the formation of swelling secondary gypsum and ammonium gas emission:

$$(NH_4)_2 SO_4 + Ca(OH)_2 \rightarrow Ca SO_4, 2H_2O + 2NH_3$$

This gypsum may react with the remaining C_3A to precipitate delayed ettringite.

The reduction of the $Ca(OH)_2$ content of concrete seems to be a good solution for improving its durability when subjected to lactic acid or ammonium sulfate solutions. Therefore, the effect of pozzolanic materials such as silica fume and metakaolin was investigated in the present study.

Silica fume is both a very effective filler and a reactive pozzolan. The potential benefits of using silica fume either as a cement replacement material or as an addition to improve concrete properties have been reviewed by different authors [7-8]. Metakaolin is a very reactive pozzolan and the calcium hydroxide content of concrete can be virtually eliminated by its incorporation [9]. A range of concrete property improvements has been recorded, including improved sulfate resistance [10] and resistance to acids [11].

This experimental study investigate the durability of concretes containing either silica fume or metakaolin, and stored in solutions of lactic acid (pH = 4) and ammonium sulfate (200 g/l).

EXPERIMENTAL PROCEDURES

Because C_3A is the cement component most vulnerable to sulfate attack, two types of ordinary portland cement (CEM I 52.5, according to the European Standard EN 197-1) were chosen: OPC 1 with a C_3A content of 7.5 %, and OPC 2 with a C_3A content of 10.7 %. The chemical and Bogue potential compositions of these cements are shown in Table 1. OPC 2 also contained more C_3S than OPC 1, and its hydration led to higher production of calcium hydroxide.

The chemical compositions of silica fume (SF) and metakaolin (MK) are given in Table 1. The BET specific surface areas of SF and MK were 23.4 and 18.7 m^2/g, respectively. MK contained some quartz and muscovite. The metakaolinite content was estimated at 80 %.

Siliceous sand (0/4) and coarse aggregates (4/14) were used to prepare the concrete. The fineness modulus of the sand was 2.7.

A melamine sulfonate was used as superplasticizer in order to get good workability without using too much water. The dry matter content was 30 %.

Table 1. Chemical compositions of cements and mineral additives (w_t %)

Minerals	OPC 1	OPC 2	Silica fume SF	Metakaolin MK
LOI	1.7	2.1	2.0	2.1
SiO_2	19.9	18.8	96.8	54.
Al_2O_3	4.7	5.6	0.1	39.3
Fe_2O_3	2.8	2.5	0.1	1.5
CaO	64.6	63.4	0.2	0.0
MgO	1.5	2.0	0.0	0.1
K_2O	0.6	1.0	0.5	1.3
Na_2O	0.1	0.1	0.0	0.0
SO_3	3.6	3.7	0.0	0.0
TiO_2	0.3	0.3	0.0	1.5
Bogue composition				
C_3S	61.0	63.5		
C_2S	15.0	6.1		
C_3A	7.5	10.7		
C_4AF	10.0	7.4		

Test specimens

Concrete prisms (70 x 70 x 280 mm) were cast according to the mixture proportions given in Table 2. Specimens were vibrated at 50 Hz for 10 sec. and kept in molds for 24 hrs. Then, they were stored in lime saturated water at 20°C ± 2°C for 27 days. At 28 days, some specimens were measured for strength as described in the French standard NFP 18-406. The remaining prisms were subjected to the corrosion tests which lasted more than 6 months.

Table 2. .Mixture proportions of concretes

Concrete designation	1	2	3	4	5	6
Cement type	OPC 1	OPC 1	OPC 1	OPC 2	OPC 2	OPC 2
Mineral addition	-	SF	MK	-	SF	MK
Cement content (kg/m³)	350	330	330	350	330	330
Mineral addition content (kg/m³)	-	35	35	-	35	35
Sand (kg/m³)	780	765	765	780	765	765
Gravel (kg/m³)	1050	1050	1050	1050	1050	1050
Water (kg/m³)	180	197	186	182	200	190
Superplasticizer (kg/m³)	7	9	9	7	9	9

In each concrete, the quantity of superplasticizer was adjusted to get a slump value between 180 and 200 mm. SF (or MK) was introduced partly as cement replacement (pozzolanic activity) and partly as sand substitute (filler action). Due to their respective costs, these mineral admixtures were used at a content of 35 kg/m³, which means 10 % of the cement content of the control concrete (1 and 4).

For the microstructural investigation, specimens of pastes corresponding to the different concretes were prepared. Minicylinders (h = 40 mm, ∅ = 20 mm) were cast and subjected to the same corrosion tests as concretes. Microstructural investigation of concretes was also carried out.

Aggressive solutions and experimental procedure

Lactic acid (pH = 4) and ammonium sulfate (200 g/l, i.e. 20 %) solutions were prepared as aggressive solutions. Lime-water (1,48 g/l) was used as a reference. The ratio between the volume of the samples and the surrounding solution was 0.08.

After demolding, the specimens were stored in lime-saturated water until 28 days of age. Then, they were dried at 20°C for 7 days and immersed in the aggressive solutions. They were subjected to 6 cycles of corrosion, each one consisting of 4 weeks of immersion in aggressive solutions followed by one week of drying at 20°C. The cyclical exposure procedure with alternate wetting and drying accelerated the attack.

After each cycle of corrosion, the weight and length changes were measured. The microstructure of pastes was investigated by means of scanning electron microscopy (JEOL 35 CF) associated with energy dispersive X-ray analysis (TRACOR system).

RESULTS AND DISCUSSION

Properties of concretes before corrosion

The initial properties of the concretes are summarized in Table 3. Concretes cast with OPC 1 showed better strengths than those using OPC 2. The presence of MK enhanced the strength of concrete: + 17 % for OPC 1 and + 28 % for OPC 2. The use of SF led to lower performances, due to a lower pozzolanic activity.

Table 3. Properties of concretes before corrosion

Concrete designation	1	2	3	4	5	6
Flow (mm)	200	175	185	175	180	185
Mean density (kg/m³)	2360	2310	2340	2340	2290	2330
Mean compressive strength at 28 days (MPa)	64.4	67.9	75.3	53.2	58.1	67.9

Mass loss of concretes after corrosion cycles

The mass loss of the different concretes after 6 cycles is shown in Figures 1 to 4.

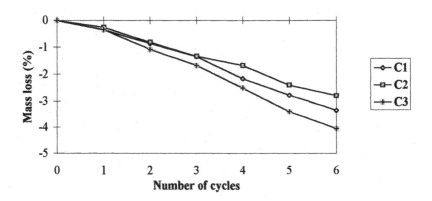

FIG. 1. Mass loss of concretes cast with OPC 1
after 6 cycles in lactic acid solution

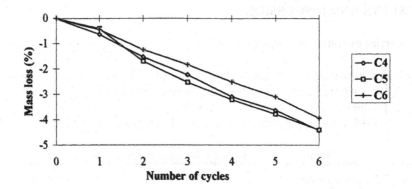

FIG. 2. Mass loss of concretes cast with OPC 2
after 6 cycles in lactic acid solution

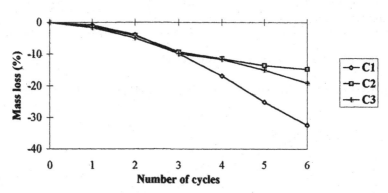

FIG. 3. Mass loss of concretes cast with OPC 1
after 6 cycles in ammonium sulfate solution

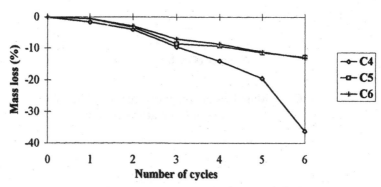

FIG. 4. Mass loss of concretes cast with OPC 2
after 6 cycles in. ammonium sulfate solution

The results show that lactic acid was less aggressive than ammonium sulfate, whatever the binder might be. After 6 cycles of corrosion in lactic acid solution, the mass loss was in the range of 2.8 % to 4.8 %, while it reached 36 % for concrete 4 immersed in ammonium sulfate solution. For concretes 1 and 4 subjected to ammonium sulfate, the mass loss was higher than 30 %. The higher the C_3A content of the cement, the higher the mass loss. The presence of SF or MK decreased the mass loss, which was only 13 % in concretes 5 and 6. SF seemed to behave better than MK, specially when OPC 1 was used.

Swelling of concretes in ammonium sulfate solution

The length variations of concretes exposed to a lactic acid solution were very small. As shown in Figures 5 and 6, the swelling of concretes immersed in ammonium sulfate solution was very significant.

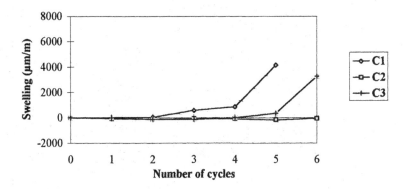

FIG. 5. Swelling of concretes cast with OPC 1 after 6 cycles in ammonium sulfate

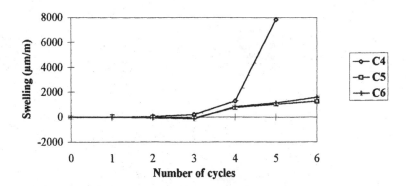

FIG. 6. Swelling of concretes cast with OPC 2 after 6 cycles in ammonium sulfate

Specimens cast with plain OPC behaved well for 3 cycles of corrosion, but were destroyed after 6 cycles. SF was very effective to limit the swelling of concrete, specially with OPC 1. MK was slightly less effective but performed well for 5 cycles of corrosion.

Microstructural analysis

When concrete cast with OPC 2 was subjected to lactic acid, the calcium hydroxide initially present at aggregate interfaces was transformed into calcium lactate and some debonding between paste and aggregates occurred, as shown in Figure 7. A certain decalcification of C-S-H also appeared (Figure 8).

FIG. 7. Formation of calcium lactate (Concrete 4)

FIG. 8. Decalcification of C-S-H (Concrete 4)

Such decalcification of C-S-H was also present in concrete 6 containing MK (Figure 9) but the paste-aggregate interface was preserved (Figure 10).

FIG. 9. Decalcification of C-S-H-
(Concrete 6)

FIG. 10. Paste-aggregate interface
(Concrete 6)

In concretes cast with plain portland cements (Concretes 1 and 4) subjected to ammonium sulfate solution, massive crystallyzations of gypsum occurred (Figures 11 and 12).

FIG. 11. Massive gypsum present in Concrete 1

FIG. 12. Massive gypsum present in Concrete 4

The presence of either MK or SF reduced the size of gypsum crystals (Figures 13 and 14).

When OPC and SF or MK were used (Concretes 5 and 6), many gypsum crystals were present at interfaces (Figures 15-16). This gypsum replaced calcium hydroxide due to the cation-exchange reaction ($Ca^{2+} \rightarrow NH_4^+$). But the size of these crystals was small enough to prevent any expansion.

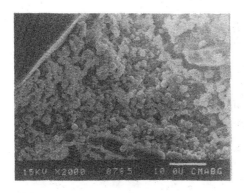

FIG. 13. Small crystals of gypsum in
Concrete 2

FIG. 14. Small crystals of gypsum in
Concrete 3.

FIG. 15. Presence of gypsum at interfaces in
Concrete 5.

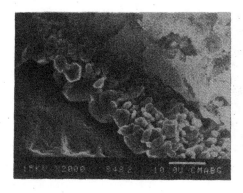

FIG. 16. Presence of gypsum at interfaces in
Concrete 6.

CONCLUSION

From this limited investigation, the following conclusions can be drawn:
1) MK seems to be more pozzolanic than SF but is less effective wich regard to the resistance of concrete to ammonium sulfate solution (20 %);
2) Ammonium sulfate solution (20 %) is a more severe environment than lactic acid at pH=4;
3) Normal portland cements are not durable in ammonium sulfate solution; massive crystals of gypsum precipitate and replace calcium hydroxide, which leads to major swelling and final destruction of concrete;

4) SF is very efficient in reducing and even suppressing the swelling of concrete in ammonium sulfate;
5) The presence of SF or MK limits the size of gypsum crystals and therefore reduces the swelling of concrete.

REFERENCES

1. B. Svennerstedt. Durability of concrete material in agricultural environment. Proceedings of the 6th International Conference on Durability of Building Materials and Components. Omiya, Japan, 26-29 Oct. 1993, Vol. 1, pp. 657-666, (Ed. S. Nagataki, T. Nireki, and F. Tomosawa), (1993).
2. N. De Belie, H.J. Verselder, B. De Blaere, D. Van Nieuwenburg, and R. Verschoore Influence of the cement type on the resistance of concrete to feed acids. Cement and Concrete Research, Vol. 26, N° 11, pp. 1717-1725, (1996).
3. N. De Belie, V De Coster, and V. Van Nieuwenburg. Use of fly ash or silica fume to increase the resistance of concrete to feed acids. Magazine of Concrete Research, Vol. 49, N° 181, pp. 337-344, (1997).
4. M. Eglinton. Resistance of concrete to destructive agencies. In: Lea's Chemistry of Cement and Concrete. Fourth Edition. Edited by P.C. Hewlett, pp. 299-342, (1998).
5. A. Kleinlogel. L'influence des divers éléments physico-chimiques sur les bétons. Ed. Dunod, Paris , (in French), (1960).
6. F.M. Killinckale. The effect of $MgSO_4$ and HCl solutions on the strength and durability of pozzolan cement mortar. Cement and Concrete Research, Vol. 27, N° 12, pp. 1911-1918, (1997).
7. V.M. Malhotra and G.G. Carette. Silica fume concrete properties, applications, and limitations. Concrete International, Vol. 5, N° 5, pp. 40-46, (1983).
8. T.A. Durning and M.C. Hicks. Using microsilica to increase concrete's resistance to aggressive chemicals. Concrete International, Vol. 13, N° 3, pp. 42-48, (1991).
9. J. Ambroise, S. Maximilien, and J. Péra. Properties of metakaolin blended cements. Journal of Advanced Cement-Based Materials, Vol. 1, N° 4, pp. 161-168, (1994).
10. B. Sigh and G.J. Osborne. Hydration and durability of OPC/metakaolin blended concrete. London: Building Research Establishment, BRE, Client Report CR 291/94, (1994).
11. J.A. Kostuch, G.V. Walters, and T.R. Jones. High performance concretes incorporating metakaolin - a review. Proceedings of Concrete 2000 - Economic and durable construction through excellence. Dundee (U.K.), 7-9 september 1993, Vol. 2, pp. 1799-1811, (Ed. R.K. Dhir, and M.R. Jones), (1993).

ADMIXTURES

ISSUES RELATED TO SILICA FUME DISPERSION IN CONCRETE

R.D. Hooton, R.F. Bleszynski and A. Boddy
35 St. George Street
Department of Civil Engineering
University of Toronto
Toronto, Ontario, Canada M5S 1A4

ABSTRACT

Silica fume (micro silica) has been widely used in blended cements and as a partial cement replacement in concrete. When properly proportioned, it provides (i) enhanced workability without segregation, (ii) improved physical properties such as strength, especially at early ages, (iii) enhanced durability through improved resistance to fluid penetration and through chemical modifications (such as alkalis from pore solution). Due to silica fume's low as-produced bulk density it is often densified or compacted for ease of transport and handling. Some of these densified agglomerates may not breakdown when later interground to form blended cement, and/or mixed in concrete. Regardless, silica fume's extreme fineness and tendency to flocculate requires the use of superplasticizing admixtures to ensure proper dispersion within the concrete. If not dispersed, the improvement to concrete properties may be less dramatic, especially with respect to durability, and performance may also be variable. In addition, undispersed clumps of silica fume could potentially act as alkali-reactive, or frost susceptible, aggregates. In this paper, the published work in this area will be reviewed and new microscopic, physical, and durability data are presented.

BACKGROUND

This introduction of silica fume has had a dramatic impact on commercial concrete technology in North America, Europe, and around the world. When adequately dispersed with superplasticizing admixtures, flowable, non-segregating concretes with low water to cementitious materials ratio (w/cm) can be produced with develop high early strengths and offer enhanced durability. The main uses of silica fume concrete in North America have been to achieve high strengths (e.g.: columns in tall buildings), chloride penetration resistance (e.g.: parking garages and bridge overlays), and chemical resistance (e.g.: dairy and chemical plant floors). In Iceland, 7.5% silica fume blended cements have been used successfully since 1979 to prevent deleterious alkali-silica reactivity (ASR) in concrete (1,2). However, in spite of the good Icelandic experience, not all studies have shown silica fume to be effective at the typical 5 to 10 percent by mass replacement levels. (3,4,5,6,7).

Although, not all of these references provide an explanation, improper dispersion of silica fume is implicated in some either due to use of a highly densified or pelletized form, or the lack of use of a superplasticizer. While superplasticizers will aid in dispersion, Lagerblad and Utkin (8) found that concrete mixing sequences were far more important (pre-mixing silica fume with coarse aggregate proves to be beneficial). Dispersants helped but were not enough on their own to disperse silica fume which had been densified to greater than 700 kg/m^3. In this paper, some of the literature on silica fume dispersion and variable performance is reviewed and several pilot-experimental studies are detailed.

FORMS OF SILICA FUME

Silica fumes derived from silicon or 75% silicon alloy production and having at least 85% SiO_2 content are the ones typically allowed in standard specifications such as ASTM C1240 or CSA A23.5. Other lower grade silicon alloy fumes have been used for shotcrete and in oil well cementing, but are not typically allowed for use in ready-mixed or precast concrete. For example, lower grade, low SiO_2 fumes are not as useful for controlling ASR (3). Silica fume is typically characterized by high BET surface areas (20-30 m^2/g) and the spherical particles have an average size of 0.1 to 0.2 μm. The high surface area of these amorphous silica particles is responsible for the unusual properties imparted to concrete. It has been calculated that for a level of replacement of 7.5% by mass of cement by silica fume that there may be up to one million silica fume particles per particle of Portland cement (9).

More importantly, from a commercial point of view, the as-produced silica fume has a very low bulk density (130-430 kg/m^3). At the low density end, silica fume is difficult to handle and uneconomical to transport from the silicon alloy furnaces sites to either cement or concrete plants. A bulk tank truck which normally carries 30 tonnes of Portland cement may only carry 6 to 10 tonnes of raw silica fume. Also upon arrival of the truck, the silica fume will have self-densified due to vibration, up to twice its original bulk density.

Therefore to improve transport or handling efficiency, commercial suppliers have developed several methods of densification, some of which are patented (10).
1. air-densification by blowing air up through a silo containing silica fume.
2. compaction in a cone-shaped auger end extrusion system.
3. moist pelletizing in an inclined, rotating drum.
4. creating a slurry with water.

The pelletized fume is not suitable for direct use in concrete but is typically used for producing interground, blended cements. In air densification, final bulk densities can vary widely depending on the length of the process. Regardless of the process (except for slurries), final bulk densities can be as high as 600 to 700 kg/m^3 or more and agglomerated particles can be up to 1 mm in diameter (Figure 1). Such a particle would contain approximately 10^{12} (one trillion), 0.1 μm diameter spheres. Even a 50 μm agglomerated particle could contain 125 million particles.

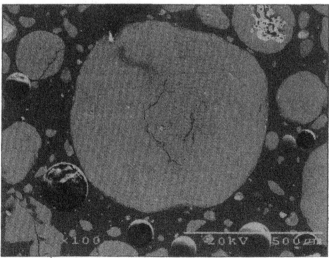

Figure 1: BSE image of large particle (900mm) of densified silica fume.

Two questions arise from use of densified silica fume:

1. Does the densified fume effectively redisperse either when interground in a blended cement or in concrete mixing?
2. If the agglomerates do not breakdown and redisperse in the concrete, what impact do they have on physical properties and durability?

Negative effects, if they exist, could be due to the silica fume particles not being dispersed to the right places i) to act as nucleation sites for hydration, ii) to pozzolanically improve the interfacial transition zone (ITZ) around aggregates, iii) to pozzolanically react to infill and subdivide original pore space, and iv) to remove alkalis or other ions from pore solution in the process. As well, large silica fume agglomerates could potentially act as reactive fine aggregates and induce ASR expansions (7,11,12).

PARTICLE SIZE DISTRIBUTION

St. John (13) studied nine silica fumes of various bulk densities from 128 to 710 kg/m^3 and BET surface areas ranging from 16.9 to 31.0 m^2/g using transmission electron microscopy (TEM) in conjunction with both laser and X-ray particle size analysis. He found that some particles had been fused together during manufacture. Also, using ultrasonic dispersion followed by particle size analysis, he found that the fumes above 500 kg/m^3 bulk density could not be redispersed and were unlikely to do so in a concrete mixer. He also found that the mean particle size varied between 1 and 50 μm and not the 0.1 to 0.2 μm often quoted.

In concrete, large agglomerated particles of densified silica fume of 40-100 μm were found in laboratory concrete (12,14) and in a commercial parking garage, lumps as large as 100-800 μm were observed (11). St John and Freitag (15) state that there are indications that agglomerates larger than 30 μm present in concrete could potentially act as alkali-silica reactive aggregates.

STANDARDS

Neither ASTM nor CSA standards currently address silica fume dispersion. The American Concrete Institute C234 (10) document mentions that "Pelletizing is not a reversible process - the pellets are too hard to break down easily during concrete production", but, "it may be interground with Portland cement clinker to from a blended cement". This document also states that "beyond about the 720 kg/m^3 level, it may become increasingly difficult to disperse densified silica fume particles within concrete".

There is reported to be a Danish Standard, DS 423.36, "Silica Fume Clumps in Hardened Concrete". The authors have not seen this standard, but it is thought to address the large scale clumps that can occur with any form of silica fume if batched incorrectly in a plant or truck mixer (i.e.: large balls of fume up to 100 mm in diameter). An example of such a lump taken from a job site in the USA is shown in Figure 2. Also, Dansk Beton Teknik A/S has developed two "standard" test methods:

1. DBT-LAB-47: Dispersibility of Condensed Silica Fume (CSF)
2. DBT-LAB-46: Silica Fume Lumps in Fresh Concrete

Due to lack of familiarity, the usefulness of these tests is unknown to the authors, but it is interesting to note that there was enough concern to develop the test procedures. However, most, if not all, of these lumps can be avoided by proper batch sequences and mixing time (8).

Figure 2: 50 mm silica fume lump in commercial concrete.

EXPERIMENTAL STUDIES

Laboratory Densification Tests

An undensified, Canadian silica furnace fume (93% SiO_2, BET surface = 23 m^2/g) was obtained. After shipment in bags, the bulk density was found to be 271 kg/m^3. To simulate densification in the laboratory, approximately 2.5 kg of silica fume was placed on a standard Vebe vibrating table in the 8.5 litre Vebe steel bucket (the bucket was fitted with a sealed, steel lid). The fume was vibrated for various periods to obtain a range of bulk densities as shown in Table I.

Table I. Effect of Vibration Time on Bulk Density of Silica Fume

Sample	Vibration Time (minutes)	Bulk Density (kg/m^3)
A	0	271
B	7.5	299
C	60	447
D	780	632

A typical backscattered electron (BSE) image of sample C particles (embedded in epoxy and polished) is shown in Figure 3. These laboratory produced agglomerates do not appear as dense as commercially densified agglomerates. Mortar cubes were made from these four samples according to ASTM C109 using 2.75 parts standard quartz sand and 1.0 part of a high alkali, high C_3A Portland cement replaced 10% by mass with silica fume. Mortar mixtures were made without superplasticizer and were cast to equal flow (w/cm = 0.58 versus 0.485 for the Portland cement control) similar to the procedures outlined in ASTM C1240. As well, another set of mixtures was made using 1.5% by mass of a sodium naphthalated sulfonate condensate type superplasticizer and cast at w/cm = 0.485 (to flows

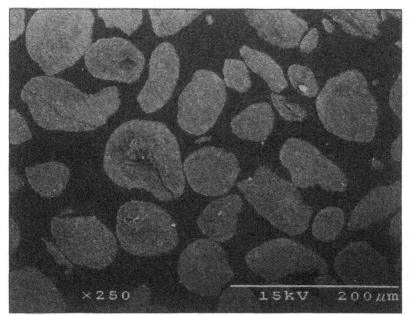

Figure 3: Silica fume densified by vibration to a bulk density of 447 kg/m3.

of 103-109). Table II shows the strength testing results after 7 days moist at 65°C (as per ASTM C1240) as well as age tests where samples were lime water cured at 23°C. At all ages, strength was reduced 15% to 25% with increasing bulk density. Although the superplasticized mixes indicated higher overall compressive strengths, greater losses were observed with increasing bulk density than when compared to the unsuperplasticized ones. This is in contrast with the tests of Cohen and Olek (9) where densified fume gave equivalent strengths to undensified fume in concrete. Polished samples of mortar were examined using BSE imaging but undispersed pieces of silica fume were not obvious and energy-dispersive x-ray (EDX) analysis was difficult due to the siliceous quartz sand used.

In addition, ASTM C1260 accelerated mortar bar tests (at 2 days of age, bars are placed in 1 N NaOH at 80°C for 14 days and expansion is measured) were conducted using a crushed reactive Spratt limestone containing about 5% chalcedonic chert (from near Ottawa, Canada). A high alkali (0.9% Na_2O equivalent) Portland cement (OPC) was replaced with 5% and 10% by mass of either the undensified silica fume (A) or the 632 kg/m^3 densified sample (D) at w/cm = 0.58. In some mixtures, superplasticizers (SP) used to help disperse the silica fume with the w/cm maintained at 0.485.

In Figure 4, it is observed that the OPC mortar bars expanded well in excess of the 14 day expansion limit of 0.10%. Silica fume at 5% level of replacement was insufficient, but 10% silica fume brought 14 day expansions to 0.10% or less. Interestingly, the bulk density of the silica fume had no impact on expansions at either the 5 or 10% replacement levels. Use of SP actually raised expansions slightly, likely due to a contribution of alkalis. The lack of effect of bulk density is good except that these laboratory vibrated particles did not appear as dense as those produced commercially. This indifference of bulk density on expansions is supported by the results of Gudmundsson and Olafsson (1). However, other methods of densification may have different effects.

Table II. ASTM C109 Mortar Cube Strengths with 10% Silica Fume

Fume Density	Compressive Strengths (MPa)			
(kg/m³)	7 days	14 days	21 days	28 days
Without SP (w/cm = 0.58)				
271	36.5	38.6	42.3	45.2
299	35.0	36.1	39.2	41.4
447	33.6	34.3	34.8	36.5
632	30.2	30.7	31.2	32.4
With 1% SP (w/cm = 0.485)				
271	50.8	52.0	54.2	57.6
299	44.6	45.8	47.5	49.9
447	41.6	42.7	44.1	46.4
632	37.6	38.6	39.3	40.1

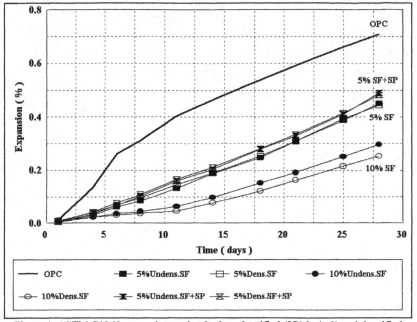

Figure 4: ASTM C1260 expansions using both undensified (271 kg/m3) and densified (632 kg/m3) silica fume.

Microscopy and ASTM C1260 Tests on Pelletized Silica Fume

While normally only used as a raw material for intergrinding into a Canadian blended cement, a sample of pelletized silica fume was obtained (bulk density = 682 kg/m³) from the same source as used in the previous tests. The same high alkali cement and reactive Spratt aggregate was used to cast ASTM C1260 mortar bars. The pelletized silica fume was simply added to the mortar mixer at 5%, 10% and 15% by mass replacement of cement. A BSE image of the pelletized fume (epoxy mounted, cut and polished) is shown in Figure 5. The pellets appear to range in size from 10 to 100 μm diameter.

A sample of the 15% pelletized silica fume mortar was dried at one day of age, cut, polished and examined by BSE imaging. Figure 6 shows a 50 μm diameter, agglomerated particle of silica fume (the spots on the particle are due to electron beam damage). This was confirmed by EDX analysis.

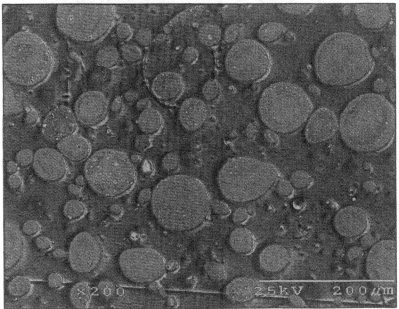

Figure 5: BSE image of pelletized silica fume.

In Figure 7, the mortar bar expansion of the 5% pelletized silica fume is compared to the undensified 5% fume mix and the OPC control. The pelletized fume was not effective in reducing expansion whereas the undensified fume reduced expansions by more than 50%, although not enough to control deleterious expansion. The expansions of all of the replacement levels are shown in Figure 8. Only the 15% pelletized fume mortar expanded less than the threshold value of 0.10% at 14 days as suggested by Davies and Oberholster (16). (This method and limit is currently being considered for inclusion in CSA A23.5 for evaluation of the effectiveness of Supplementary Cementing Materials for control of deleterious ASR). It is also interesting to note that the use of a superplasticizer raised the expansions of the 10% and 15% mixtures. This agrees with Diamond (17) who found that alkali neutralized, sulfonate type superplasticizers raised the OH⁻ level of the pore solution. The next logical step would be to see whether the interground, pelletized silica fume, blended cement performs as efficiently as the same clinker combined with the undensified silica fume. In the tests of Gudmundsson and Olafsson (1) there was no difference in ASTM C227 expansions for an interground blended cement (7.5% of 600 kg/m³ pelletized silica fume) and 7.5% undensified silica fume added to the mortar mixer.

The next logical step would be to see whether the interground, pelletized silica fume, blended cement performs as efficiently as the same clinker combined with the undensified silica fume. In the tests of Gudmundsson and Olafsson (1) there was no difference in ASTM C227 expansions for an interground blended cement (7.5% of 600 kg/m³ pelletized silica fume) and 7.5% undensified silica fume added to the mortar mixer.

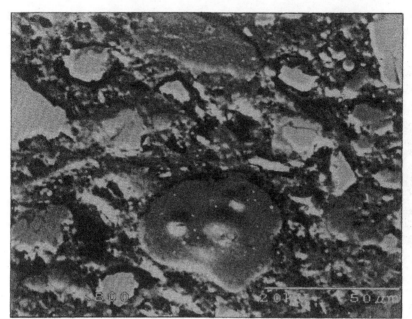

Figure 6: BSE image of C1260 mortar with 15% pelletized silica fume showing 50 mm undispersed silica fume agglomerate (centre near bottom).

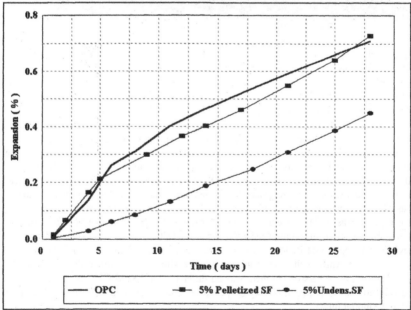

Figure 7: ASTM C1260 expansions of 5% pelletized silica fume, 5% undensified fume and OPC control mortars.

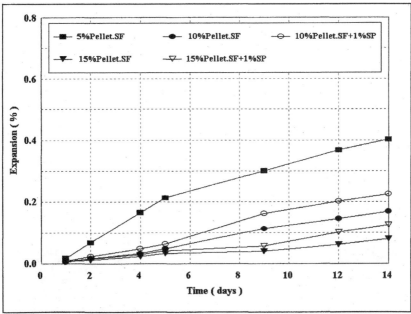

Figure 8: ASTM C1260 expansion with 5%, 10% and 15% pelletized silica fume (not interground).

<u>Effect of Silica Fume Form on Concrete Strength and Durability</u>

Efforts to elucidate the effect of the delivery form of silica fume on concrete strength and durability have resulted in a preliminary study. Three forms of silica fume were considered; an undensified fume (approximately 270 kg/m^3), a densified fume preblended with OPC, and finally a fume in the form a slurry (1:1 SF to water). Concrete mixes were cast with each of these silica fume forms at a w/cm ratio of 0.45 and a level of silica fume replacement of 7% by mass of cementitious materials. Both water reducer and superplasticizer were used to achieve equal workability of the mixes. Specimens were subsequently tested in compressive strength, resistivity, and rapid chloride penetration tests (ASTM 1202). Table III shows a summary of the results of these tests.

Table III. Effect of Silica Fume Form

Silica Fume Form	Compressive Strength		Resistivity (ohm-cm)	Rapid Chloride Penetration	
	7 days (MPa)	28 days (MPa)		coulombs	ASTM Indicator
undensified	51	62	8772	4193	high
blended	43	51	12122	2735	moderate
slurried	61	73	16376	1960	low

The concrete cast with the slurried silica fume outperformed the other silica fume concretes in all tests. These preliminary results suggests that a slurry is the most effective form of silica fume to yield improved strength and durability in concrete. It should be noted that further studies are ongoing and will be reported at a later date.

4. Microscopy Tests to Measure the Efficiency of Intergrinding

Another source of densified silica fume and the interground blended cement produced from it were examined by BSE imaging and EDX. In Figure 9, an epoxy impregnated, polished mount of the densified fume is shown. The large particle is approximately 350 µm in diameter and the density appears to be less on the outer rims. (The dark hole on the right is an air bubble in the epoxy matrix).

In Figure 10 the commercially interground, blended cement is shown. The light grey particles are clinker while the darker particles are silica fume agglomerates (confirmed by EDX). The few white particles are gypsum. Silica fume agglomerates as large as 50 µm were found (left side of micrograph) and generally, the agglomerates appear to have only been ground down to the same size as the clinker grains. With modern milling circuits, using air sweep separators, this is not surprising since fine silica fume particles could be swept out of the mill at the same size as clinker grains.

The effectiveness of concrete mixing action in further dispersing the silica fume agglomerates was not studied. This is the more important issue and needs to be investigated.

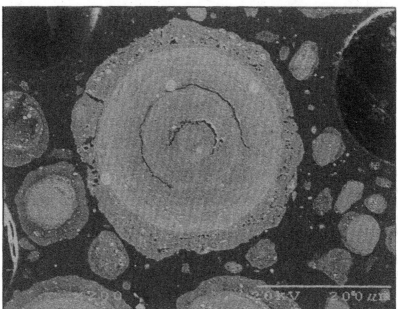

Figure 9: BSE image of densified silica fume particles to be interground in blended cement.

DISCUSSION ON MICROSCOPY FOR SILICA FUME DISPERSION DETERMINATION

Due to their inherently small size (0.1 - 0.2 µm), individual spheres of silica fume are difficult to image by BSE techniques, especially in concrete. While other forms of electron microscopy can be used (TEM), the atomic number grey level separation in BSE images makes it easy to identify SiO_2 grains in a complex matrix of mortar or concrete. Therefore, unless the dispersion of silica fume is very poor (i.e.: large lumps or 10-100 µm agglomerates), it is unlikely that microscopy alone can be used to assess silica fume particles. Physical testing for specific issues of concern (e.g.: alkali silica reactivity) serves as a better performance measure.

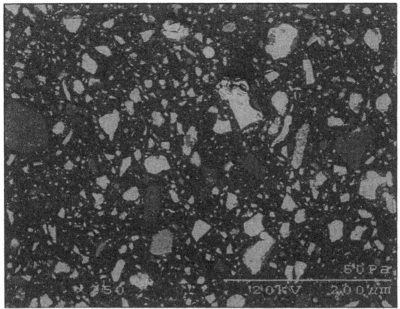

Figure 10: BSE image of unhydrated blended cement showing silica fume agglomerates (dark grey), clinker grains (medium grey) and gypsum particles (white).

SUMMARY

Due to its low bulk density, silica fume is often densified prior to transport, and then used in either blended cements or as a supplementary cementing material. The ability to redisperse agglomerated particles in concrete appears to depend in part on: i) the degree of densification (i.e.: bulk density) of the fume; ii) the efficiency of the intergrinding (in blended cement); iii) concrete mixing. This last point is likely a function of batching sequence as well as mixer efficiency and mixing time. In addition, superplasticizing admixtures may help to disperse very small agglomerates.

There is likely an upper bound on the bulk density of silica fume above which it can not be properly dispersed. This upper bound may be lower than the 720 kg/m^3 mentioned by ACI (10) and may vary with the form of densification employed. Further work is needed to establish these bounds as well as set standards for measuring dispersibility.

The potential adverse effects of undispersed silica fume on both strength and control of alkali-silica expansion have been shown, but more rigorous study is warranted.

Silica fume has been successfully used to enhance both strength and durability in concrete. However, certain delivery forms may hinder adequate dispersion thereby reducing the positive effects imparted to concrete.

ACKNOWLEDGEMENTS

The experimental work of Ms. U. Nytko, Mr. S. Mahini, Mr. J. Botnick, Mr. G. Goy, and Mr. R. Mohamed is gratefully acknowledged.

REFERENCES

1. Gudmundsson, G. and Olafsson, H., "Silica Fume in Concrete - 16 Years of Experience in Iceland", in Alkali-Aggregate Reaction in Concrete (A. Shayan - ed). Proceedings of the 10th International Conference, Melbourne, pp. 469-562, 1996.

2. Olafsson H., "AAR Problems in Iceland - Present State", Proceedings of the 8th International Conference on Alkali-Aggregate Reaction, Kyoto (K. Okada, S. Nishibayashi and M. Kawamura - ed), pp. 65-60, 1989.

3. Duchesne, J. and Berube, M.A., "The Effectiveness of Supplementary Cementing Materials in Suppressing Expansion due to ASR: Another Look at the Reaction Mechanisms, Part 1: Concrete Expansion and Portlandite Depletion", Cement and Concrete Research, Vol. 24, pp. 73-82, 1994.

4. Fournier, B., Bilodeau, A. and Malhotra, V.M., "CANMET/Industry Research Consortium on Alkali-Silica Reactivity (ASR)" Alkali-Aggregate Reaction in Concrete (A. Shayan - ed.), Proceedings of the 10th International Conference, Melbourne, pp. 101-108, 1996.

5. Oberholster, R.E. and Davies, G., "The Effect of Mineral Admixtures on the Alkali-Silica Expansion of Concrete under Outdoor Exposure Conditions", in Concrete Alkali-Aggregate Reactions (P.E. Grattan-Bellew - ed.), Proceedings of the 7th International Conference, Ottawa, Noyes Press, pp. 60-65, 1986.

6. Oberholster, R.E., "Alkali-Aggregate Reaction in South Africa: Some Recent Developments in Research", Proceedings of the 8th International Conference on Alkali-Aggregate Reaction, Kyoto (K. Okada, S. Nishibayashi and M. Kawamura - ed), pp. 77-82, 1989.

7. Pettersson, K., "Effects of Silica Fume on Alkali-Silica Expansion in Mortar Specimens", Cement and Concrete Research, Vol. 22, pp. 15-22, 1992

8. Lagerblad, B. and Utkin, P., "Undispersed Granulated Silica Fume in Concrete - Chemical System and Durability Problems", Materials Research Society Symposium Proceedings, Vol. 370, pp. 89-97, 1995.

9. Cohen, M.D. and Olek, J., "Silica Fume in PCC: The Effects of Form on Engineering Performance", Concrete International, ACI, Vol. 11, No. 11, pp. 43-47, 1989.

10. ACI, "Guide for the Use of Silica Fume in Concrete", American Concrete Institute Committee 234, Report 234R-96, in ACI Manual of Concrete Practice, 1997, Vol. 1, pp. 51, 1996.

11. Marusin, S.L. and Shotwell, L.B., "Alkali-Silica Reaction in Concrete Caused by Densified Silica Fume Lumps - A Case Study", Supplementary Volume, Fifth CANMET/ACI International Conference on Fly Ash, Silica Fume, Slag and Natural Pozzolans in Concrete, Milwaulkee, pp. 45-59, 1995.

12. Shayan, A., Quick, G.W. and Lancucki, C.J., "Morphological, Mineralogical and Chemical Features of Silica-cured Concretes Containing Densified Silica Fume and Various Alkali Levels", Advances in Cement Research, Vol. 5, No. 20, pp. 151-162, 1993.

13. St. John, D.A., "The Dispersion of Silica Fume", Industrial Research Limited Report IRL NO. 244, Lower Hutt, NZ, pp. 29, 1994.

14. Bonen, D. and Diamond, S., "Occurrence of Large Silica Fume-Derived Particles in Hydrated Cement Paste", Cement and Concrete Research, Vol. 22, pp. 1059-1066, 1992.

15. St. John, D.A. and Freitag, S.A., "Fifty Years of Investigation and Control of AAR in New Zealand", Alkali-Aggregate Reaction in Concrete (A. Shayan - ed.), Proceedings of the 10th International Conference, Melbourne, pp. 150-157, 1996.

16. Davies, G. and Oberholster, R.E., "Use of the NBRI Accelerated Test to Evaluate the Effectiveness in Preventing Alkali-Silica Reaction", Cement and Concrete Research, Vol. 17, pp. 97-107, 1987.

17. Diamond, S., "Alkali Silica Reactions - Some Paradoxes", in Alkali-Aggregate Reaction in Concrete (A. Shayan - ed). Proceedings of the 10th International Conference, Melbourne, pp. 3-14, 1996.

ROLE OF HIGH-REACTIVITY METAKAOLIN AND SILICA FUME IN
CONTROLLING THE EFFECTS OF ASR IN CONCRETE

Jan Olek[1] and Wilkins Aquino[2]
[1]School of Civil Engineering, Purdue University, West Lafayette, IN 47907, USA
[2]Dept. of Civil Engineering, University of Illinois, Urbana, IL 61801, USA

ABSTRACT

In this paper we examine the reduction in alkali hydroxide concentrations in
pore solutions expressed from pastes containing two types of mineral admixture,
silica fume (SF) and high-reactivity metakaolin (HRM). In addition, we present
expansion data for a series of mortar bars containing reactive opal aggregate and
the same mineral admixtures, exposed to a high concentration of sodium hydroxide
at 80°C under ASTM C1260 test conditions. The purpose of the study was to
compare the effectiveness of SF and HRM in reducing the risk of ASR reaction in
concretes containing reactive aggregates. Both SF and HRM were found to substantially reduce the alkali hydroxide concentrations of cement paste pore solutions, the SF acting more rapidly and to a greater extent than the HRM. Both
were effective in reducing the expansion of the mortar bars during the standard 14
day ASTM C1260 test period, the HRM being more effective in this regard.
However, it was found that on prolonged periods of exposure both sets of mortar
bars continued to expand, and eventually reached levels of expansion about 20%
lower (HRM) or slightly higher (SF) than the control mortar bars without mineral
admixtures. Nonetheless, it was considered that the reduced 14-day expansion,
and especially, the reduction in pore solution alkali hydroxide concentration in
normal exposure, were strong indications of the potential effectiveness of both
mineral admixtures.

INTRODUCTION

The occurrence of alkali-silica reaction (ASR) has the potential to become an
increasingly serious problem as environmental regulations requiring changes in the
cement burning process result in difficulties in the manufacture of cements with
low alkali content. In addition, the depletion of existing aggregate sources in some
areas of the United States, has led to an increase in the relative availability of potentially alkali reactive aggregate.

The circumstances explained above have led many investigators to search for new alternatives to mitigate the problem of concrete deterioration due to ASR. It is widely recognized (1-4) that the high alkalinity of the concrete pore solution is mainly responsible for the depolymerization of the silicon-oxygen bonds in aggregates containing amorphous silica and the formation of an expansive gel composed of silicon, OH^-, and alkalis.

Considerable amount of research has been done in the past on the use of fly ash, silica fume, and ground blast-furnace slag in mitigation of ASR in concrete (5-10). In some cases, natural pozzolans containing thermally activated kaolinitic clay (often called "metakaolin") were used to control ASR reactions in marine and underground structures (11, 12). In recent years a new mineral admixture, high-reactivity metakaolin (HRM), has been successfully used in production of high-performance concrete (HPC) (13-15). This admixture is produced by calcining purified kaolinite clay at a specified temperature range. However, to date no research has been reported on the effectiveness of HRM in controlling the ASR.

In the present work we investigate and compare the effects of two mineral admixtures, silica fume (SF) and HRM, on chemistry of pore solution of cement pastes and on the expansion of mortars containing reactive aggregate and prepared using the same admixtures. The pore solution analysis was performed on liquid extracted from cement pastes prepared using different combinations of high-alkali Portland cement (PC), HRM, and SF. The expansion of mortar bars due to ASR was studied using the accelerated mortar bar test (ASTM C1260).

EXPERIMENTAL METHODS

Materials

Mortars and pastes were cast using a high-alkali Portland cement (equiv. Na_2O = 1.14%). The mill analysis of this cement included the following: CaO = 61.8%, SiO_2 = 20.3%, Al_2O_3 = 4.4%, and Fe_2O_3 = 3.3%. The SF used in this study was commercially available densified type with reported SiO_2 content of 93%. The HRM was supplied by commercial producer and the reported oxide composition (SiO_2 + Al_2O_3 + Fe_2O_3) was 95.3%.

The control mortar specimens were prepared using nonreactive calcitic limestone aggregate. All ASR mortar specimens were prepared by replacing 5% (by total weight of aggregate) of nonreactive aggregate with Beltane opal. The gradation of the aggregates followed limits specified in the ASTM C1260 test procedure. No high-range water reducer was used in these mixes.

Sample Preparation and Curing

Six types of mortar bars were cast with aggregate:cement ratio fixed at 2.25. All of the mortars were prepared at w/c ratio of 0.55. Two types of the mortar specimens (with and without reactive aggregate) were used as control samples. The other four types of mortars contained 10% (by the weight of cement) of either SF or HRM. For these four series of mortars the actual water/cementitious material (cement + mineral admixture) ratio was 0.495.

In addition, three types of paste specimens were cast to obtain pore solution needed for chemical analysis. The pastes were prepared at the same w/c ratio as the mortar specimens. Details of all mixes are provided in Table I.

Table I. Mix details for mortar and paste specimens

Mix Type	Beltane Opal Content (% total aggregate)	HRM Content (% wt. cement)	SF Content (% wt. cement)
M1	0	0	0
M2	0	10	0
M3	0	0	10
M4	5	0	0
M5	5	10	0
M6	5	0	10
P1	0	0	0
P2	0	10	0
P3	0	0	10

From each mortar mix three bars (280 x 25.4 x 25.4 mm) were cast with stainless-steel inserts to measure the expansion. After storing for 24 hours in a moist room, the samples were removed from the molds, immersed in tap water, and stored at 80°C for additional 24 hours.

The paste specimens for expressing pore solution were cast in plastic jars of 118 cm^3 nominal volume. The jars were sealed using electrical insulating tape and stored in a standard laboratory environment until testing.

Test Program

Pore Solution Analysis: At the designated ages, the hardened paste specimens were removed from the jars and the pore solution was expressed from each sample using a high-pressure die and piston assembly. The expressed pore solutions were stored in 4 ml sealed bottles, prior to analysis.

The chemical analysis of the pore solutions involved the determination of the K^+, Na^+, and OH^- ionic concentrations at successive ages, up to 180 days. The concentrations of the cations (K^+, Na^+) were determined using atomic absorption spectrophotometry. The OH^- ion concentrations were determined by titration with 0.05N HCl.

Mortar Bar Expansion Tests: These tests were carried out according to ASTM C1260-94 (Standard Test Method for Potential Alkali Reactivity of Aggregates, Mortar Bar Method). In carrying out these tests, an initial mortar bar length reading was taken after 24 hours of exposure to water at 80°C (48 hours after casting). The specimens were than immersed in a 1N NaOH solution and stored at 80°C. Two intermediate readings were taken during the first week of exposure, and subsequent readings were taken every week thereafter. Although the standard ASTM C1260-94 procedure stipulates that measurements be taken only up to 14 days, the present specimens were monitored for a period of 7 weeks in order to more fully evaluate the effects of pozzolanic admixtures.

TEST RESULTS AND DATA ANALYSIS

Chemistry of Pore Solutions

Figure 1 shows combined K^+, Na^+ concentration in pore liquid of both plain pastes and pastes containing pozzolanic admixtures at ages up to 180 days.

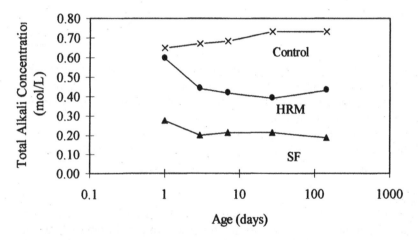

Figure 1. Changes in total alkali content with time in pore solutions expressed from pastes with and without pozzolanic admixtures

The total alkali content in the pore solution of the control samples increased slightly with time for a period of about 4 weeks, and then it stabilized at the level of about 0.73 mol/L. Replacing 10% of cement by SF resulted in very dramatic reduction of total alkali content very early in the hydration process. After 1 day of hydration SF samples showed 58% reduction in the total alkali concentration with respect to the control samples. The process of removal of alkalis from pore solution of pastes containing SF continued for about 3 days and than their concentration stabilized at the level of about 20 mol/L. In contrast to SF, the effect of HRM on the chemistry of pore solution was less pronounced. After one day of hydration only slight reduction in alkali concentration (about 9%) was observed. However, the process of alkali removal from the pore solution continued for about three weeks after which their concentration essentially stabilized at levels much less than that of control samples. In all cases, concentration of OH⁻ ions followed the same trend as concentration of the combined alkalis.

Mortar Bar Expansion

The effects of SF and HRM on the expansion of mortars in the ASTM C1260 test are shown in Figure 2.

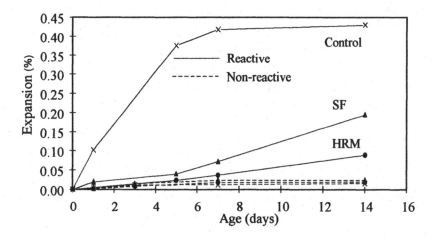

Figure 2. Effect of SF and HRM on the expansion of mortars stored for 14 days in 1N NaOH solution at 80°C

As expected, the expansion of samples prepared using inert aggregate was negligible at all ages. The control samples prepared with reactive aggregate expanded rapidly and reached equilibrium at approximately 7 days, at an expansion of about 0.42%. In contrast, all of the specimens prepared with reactive aggregate and pozzolanic admixtures experienced much reduced levels of early expansion. After 14 days of exposure the expansion of SF-bearing specimens was 56% lower, and the expansion of HRN-bearing specimens was 81% lower than the expansion of control specimens.

Figure 3 presents the expansion behavior of the same bars that provided data for Figure 2 but over a longer period of time (49 days).

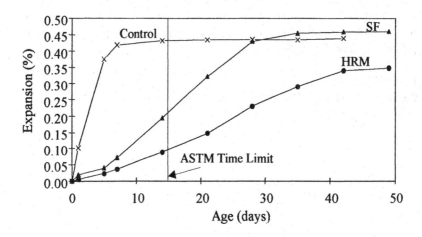

Figure 3. Effect of SF and HRM on the expansion of mortars stored for 49 days in 1N NaOH solution at 80°C

While the levels of expansion of the control (no pozzolanic admixtures) specimens remained essentially unchanged, after about a week the specimens containing SF and HRM continued to expand. The SF bars reached the same level of expansion as control bars after about 28 days of exposure and their expansion leveled-off at a slightly higher level after 5 weeks. The HRM bars continued to expand for about 6 weeks before stabilizing at the level of about 20% lower than the control bars.

SUMMARY AND CONCLUSIONS

The results of these investigations showed that both SF and HRM appear to be effective in prospective mitigation of ASR-induced distress.

It was found that both were effective in reducing the alkali content of paste pore solutions exposed at room temperature to significantly lower levels than that induced by high-alkali cement in control pastes. SF was immediately effective in this regard, and reduced alkali (and OH⁻ ion) concentrations from about 0.7 mol/L to around 0.2mol/L. This low level is not likely to induce significant ASR (16). HRM required a few days to effect its reduction. At equilibrium, the alkali level appeared to be about twice that of SF-bearing pastes, although it was much lower than the level of alkali in pore solutions of the control paste.

Both admixtures were effective in significantly reducing the rate of expansion experienced by mortar bars during 14-day period specified in the ASTM C1260 test. Here the HRM appeared to be more effective than SF, the rate of expansion observed with HRM-bearing mortars being only about half that observed with corresponding SF-bearing mortars.

However, both sets of mortars continued to expand over a period of more than 28 days of high-temperature exposure. The final expansion level attained by HRM-bearing mortars at 48 days was somewhat less than that of the control mortar; that of the SF-bearing mortars was not.

That the expansion in the C1260 test eventually reached high levels does not point to any lack of effectiveness of either admixture in mitigating ASR in concrete at ordinary temperatures. In the C1260 test not only is an 80°C temperature maintained, but more importantly, small specimens are exposed to what is effectively an infinite supply of alkali hydroxide at high concentrations (1N). The mechanism by which such admixtures inhibit ASR is in large part associated with their removal of alkali hydroxide from the pore solutions before the pore solutions can react with coarse reactive aggregate grains. In the presence of an infinite supply of alkali hydroxide, such inhibitory effect is eventually overcome.

ACKNOWLEDGEMENTS

The authors would like to thank K. A. Gruber of Engelhard Corporation for supplying the metakaolin and J. Lovell of Purdue University for technical assistance.

REFERENCES

1. H.E. Vivian, "The Effect of Hydroxyl Ions on the Reaction of Opal," *Australia Journal of Applied Science*, Vol. 2, pp. 108-113 (1951).
2. S. Diamond, "Mechanism of Alkali-Silica Reaction," pp. 83-94, in Alkali-Silica Reaction, *Proc. 8th International Conference*, Kyoto, ICAAR, 1989.
3. S. Diamond, "A Review Of Alkali-Silica Reaction And Expansion Mechanism:1, Alkalis In Cements And Concrete Pore Solutions," *Cement and Concrete Research*, Vol. 5, pp. 329-346 (1975).
4. D.W. Hobbs, "Alkali-Silica Reaction in Concrete," Thomas Telford, London, pp. 123-130 (1988).
5. K. Petterson, "Effects of Silica Fume on Alkali-Silica Expansion in Mortar Specimens," *Cement and Concrete Research*, Vol. 22, pp. 15-22 (1992).
6. M.M. Alasali and V.M. Malhotra, "Role of Concrete Incorporating High Volumes of Fly Ash in Controlling Expansion due to Alkali-Aggregate Reaction," *ACI Materials Journal*, Vol. 88, No. 2, pp.159-163 (1991).

7. H. Chen, J.A. Soles, and V.M. Malhotra, "CANMET Investigations of Supplementary Cementing Materials for Reducing Alkali-Aggregate Reactions," *International Workshop on Alkali-Aggregate Reactions in Concrete*, Halifax, N.S., CANMET, Ottawa, pg. 20 (1990).

8. P.C. Aitcin and M. Regourd, "The Use of Condensed Silica Fume to Control Alkali Silica Reaction - A Field Case Study," *Cement and Concrete Research*, Vol. **15**, pp. 711-719 (1985).

9. R.N. Swamy, "Role and Effectiveness of Mineral Admixtures in Relation to Alkali-Silica Reaction," pp. 96-121, in *The Alkali-Silica Reaction in Concrete*, Edited by R.N. Swamy, Van Nostrand Reinhold, New York, 1992.

10. R.F. Bleszynski and M.D.A. Thomas, "Microstructural Studies of Alkali-Silica Reaction in Fly Ash Concrete Immersed in Alkaline Solutions," *Advanced CementBased Materials*, Vol. **7**, No. 2, pp. 66-78 (1998).

11. K.A. Gruber and S.L. Sarkar, "Exploring the Pozzolanic Activity of High Reactivity Metakaolin," *World Cement*, Vol. **27**, No. 2, pp. 78-80 (1996).

12. F.M. Lea, *The Chemistry of Cement and Concrete*, Third Edition, Chemical Publishing Company, Inc., New York, NY, pp. 414-453 (1971).

13. M.A. Caldarone, K.A. Gruber, and R.G. Burg, "High-Reactivity Metakaolin: A New Generation Mineral Admixture," *Concrete International*, Vol. **16**, No. 11, pp. 37-40 (1994).

14. A. Balogh, "High-Reactivity Metakaolin," *Aberdeen's Concrete Construction*, Vol. **40**, No. 7, pg.4 (1995).

15. S. Wild, J.M. Khatib, and A. Jones, "Relative Strength, Pozzalanic Activity and Cement Hydration in Superplasticized Metakaolin Concrete," *Cement and Concrete Research*, Vol. **26**, No. 10, pp. 1537-1544 (1966).

16. S. Diamond, "Alkali Reactions In Concrete: Pore Solution Effects," *Proc. 6th International Conference on Alkalis in Concrete*, Danish Concrete Association, pp.155-166 (1983).

POZZOLANIC ACTIVITY OF INCINERATOR RESIDUES OF MUNICIPAL SOLID WASTE

E. Fernández, A. Macías and S. Goñi
Instituto de Ciencias de la Construcción "Eduardo Torroja" CSIC
Serrano Galvache s/n, 28033 Madrid. Spain

ABSTRACT

The pozzolanic activity of the different incinerator residues of municipal solid waste has been investigated to find a viable solution for its use as construction material. Two types of incinerator residues have been studied: the bottom ash, that is, the slag like material which is dumped from the grate after combustion, and the fly ash collected by electrostatic precipitation and other pollution-control devices in incinerators. The pozzolanic activity of both materials was determined by putting them in contact with a saturated lime solution at $40 \pm 1°C$ for 1, 7, 28 and 90 days. After these periods of time, the Ca^{2+} concentration in the solution was measured and the fixed lime was calculated by the difference between these values and the Ca^{2+} concentration measured in the saturated lime solution. Results show that the bottom ash from municipal solid waste incineration has a pozzolanic activity similar to a natural pozzolanic opaline rock. Finally, tests were made to determine the mechanical properties and porosity of OPC paste with 20% and 40 % of bottom ash replacement.

INTRODUCTION

Although the incineration of Municipal Solid Waste (MSW) results in a mass reduction of approximately 70 %, the amount of residue remaining to be disposed of after incineration is substantial. Thus, municipal waste incineration residues represent approximately 10 % by volume and 25 % by mass of the solid waste combusted and are comprised primarily of bottom ash and air pollution control residues (fly ash). Bottom ash is generally a combination of partially or completely combusted waste that is discharged from the primary combustion grates and materials that pass through these grates. The fly ash residues are comprised of acid gas scrubber residues and dust particles.

Since the 1980's there has been a significant concern about the management of the residues from the combustion of municipal solid waste. Much of this concern is based on the fact that when the residues are subjected to toxicity tests concentrations of toxic metals in the leachate will sometimes exceed those levels defined as hazardous by these tests. This occurs more often for the fly ash than for the bottom ash. This growing concern has moved different countries to investigate several solidification/stabilization technologies for treating the residues(1-5).

Also, due to the problem of landfill shortage, it was recommended that the properties of MSW incinerator residue be investigated to find a viable solution for its use as a construction material. There are several potential options for using MSW incinerator residues in a beneficial manner. Examples include as aggregate to replace natural gravel in the production of concrete (6, 7), construction blocks (8), in asphalt for road constructions (9, 10), as a raw material for a special cement containing hydraulic materials such as alinite or calcium choroaluminate (11), erosion control, artificial reefs, landfill cover, etc. While there has been significant interest in use of residues very little actual utilization has occurred, partly because there is a need for technically sound criteria for utilizing MSW incinerator residues. With regard to the potential pozzolanicity of these type of residues, controversial results can be found in the literature. Thus, the addition of calcium oxide studied by Forrester and Goodwin (12) for the stabilization of MSW ashes, achieved a reduced permeability, suggesting the pozzolanic behavior of MSW ash. Hamernik and Frantz (13) studied the physical and chemical properties of MSW incinerator residues and concluded that most of the mass-burn fly ashes met the requirements of ASTM C618 for pozzolanic activity. They also reported on the compressive strength properties of MSW fly ash-based concrete (14). The concrete achieved a better strength, with up to a 30 % replacement of cement with MSW fly ash. However, the rate of strength gain was slower at early stages.

On the other hand, Bhatty and Reid (15) concluded that although it is possible to produce some reasonable mortar strengths by incorporating major amounts of sludge slag, the mortars remain much weaker than the required level of strength for normal mortars even after long curing times. This could be due to the finely dispersed and porous nature of the ash which, despite its high silica, alumina and lime contents, lacks pozzolanic activity .

The present research was performed to investigate the pozzolanic activity of various MSW incineration residues. The incineration residues were obtained from a municipal waste incineration facility in Melilla (Spain). The method of incineration used in the Melilla facility is a mass burning, which incinerates all MSW in the as-received state with energy recovery. This mass burning procedure generates two types of incineration residues: the bottom ash (MSWS) and the fly ash (MSWFA).

EXPERIMENTAL

Materials

In Table I the chemical composition of the portland cement (OPC), used to make the specimens for the evaluation of mechanical strength, and of the MSW incinerator residues studied are presented. MSWS 1 and MSWS 2 are bottom ash and MSWFA is the fly ash of the incinerator in Melilla.

Table I. Chemical Composition of OPC, MSWS 1, MSWS 2 and MSWFA (% by weight)

	OPC	MSWFA	MSWS 1	MSWS 2
CaO	65.3	23.4	20.6	23.7
SiO_2	20.1	12.8	46.2	40.3
Al_2O_3	5.3	10.0	9.3	11.5
Fe_2O_3	2.9	1.8	7.2	8.9
MgO	1.3	3.1	1.9	2.1
SO_3	3.1	7.6	1.1	2.0
Na_2O	0.2	9.4	5.8	3.2
K_2O	1.0	4.9	1.6	1.4
Cl^-	---	16.8	0.7	0.5
I.R.	0.8	0.5	1.5	0.7
I.L.	1.5	16.61	7.5	10.4
Si+Fe+Al [*]	---	23.6	62.7	70.7
S.G. (gr/cm^3)	3.5	2.32	2.68	2.72
S.A. (cm^2/gr) (Blaine)	5 090	5 606	3 959	3 920

I.L. = Ignition Loss
I.R. = Insoluble Residue
S.G. = Specific Gravity
S.A. = Surface Area
[*] ASTM C 618 requires minimum value of 70 % for Class F pozzolan and 50 % for Class C pozzolan.

The bottom ash was received in a granulated form and was composed of a grey slag-like material and of glass and ceramic particles, unburned paper and metallic particles in the proportions shown in Table II. After grading to a maximum grain size of 3.36 mm, the coarse fraction of the sample was pre-screened and the metallic, glass and ceramic and paper particles were removed. The sample without these types of particles was named MSWS 2, to differentiate it from the as-received material, called MSWS 1.

Table II. Composition of MSWS (% by weight)

	Glass and Ceramic	Paper	Metallic Particles	Slag-like material	Total
MSWS	36.25	3.8	4.12	55.35	99.52

Samples MSWS 1 and MSWS 2 were ground in a ring mill for 3 minutes. This procedure was not needed for the fly ash residue (MSWFA) which was received as a fine powder. The values of the specific gravity and surface area (Blaine) for the materials are also presented in Table I.

The oxide content Si + Fe + Al has also been included in Table I. The ASTM C 618 standard specification (16) requires a minimum value of 70 percent for class F pozzolan and 50 percent for class C pozzolan. Thus, according to the total oxides of Si, Al and Fe, MSWS2 could be considered as a class F pozzolan and MSWS 1 as a class C pozzolan. Nevertheless, ASTM C 618 sets a maximum limit of 6 percent for ignition loss (12 percent with satisfactory test results) and none of the MSWS residues met this 6 percent limit, and MSWFA failed even the 12 percent limit.

Procedure

Pozzolanic activity test: In order to study the pozzolanic activity of MSW incineration residues an accelerated method was used in which the reaction of lime and residue is followed over time (17). The test consisted of putting the different potentially pozzolanic materials in contact with a saturated lime solution at 40 ± 1°C for 1, 7 and 28 days. At the ends of those times, the CaO and OH⁻ concentrations in solution were measured by titration and compared with those obtained from a saturated calcium hydroxide solution.

Samples of 1 gr. of the MSW incineration residues to be tested were dried in a conventional oven at a temperature of 105 ± 5°C until constant weight and then put in contact with 75 cc of filtered saturated calcium hydroxide solution to proceed according to the test previously described.

Flexural Strength and Porosity measurements: Cement pastes with and without the addition of 20% and 40 % of MSWS 1 were prepared at a w/c of 0.4.

Specimens of 1 x 1 x 6 cm were molded and compacted by vibration. After 1 day at 100 % r.h. the specimens were demolded and kept at 100 % r.h. and 20 ± 1°C for 7, 28, 90 and 180 days before measuring their flexural strength. Each flexural strength value is the average of six measurements. A seventh sample was used for porosity and pore-size distribution measurements by mercury intrusion porosimetry carried out with a Micromeritics Pore Sizer 9310.

RESULTS AND DISCUSSION

Pozzolanic Activity of MSW Incineration Residues

The CaO concentration and the pH value measured in the pozzolanic activity test for the MSW incineration residues are presented in Fig. 1 and Fig. 2, respectively. The solid lines in these figures are the CaO concentration and the pH value measured in the saturated calcium hydroxide solution. If the material tested shows pozzolanic activity and fixes calcium hydroxide, the measured CaO concentration in solution must be below this line. As can be seen in Fig. 1, MSW bottom ash shows pozzolanic activity, since the samples have fixed calcium hydroxide from the beginning of the test and particularly significant amounts are fixed after 28 days. On the contrary, the MSW fly ash hardly shows any reaction with the lime before the 28th day, eventhough the concentration of CaO measured in the solution is higher than that measured in the saturated calcium hydroxide solution at 1 and 7 days. This increase in CaO concentration could be due to the dissolution of calcium containing phases present in the starting fly ash (see Table I).

The CaO concentration in the saturated calcium hydroxide solution, taken as reference in the pozzolanic activity test, is related to its pH value through the solubility product, and consequently any decrease in any of these parameters can be interpreted as an indication of the reaction taking place when the saturated calcium hydroxide solution is put in contact with the samples to be tested. That is why similar to the CaO results presented in Fig. 1, from the analysis of the pH values presented in Fig. 2 it can be concluded that MSW bottom ash can be considered as a pozzolanic material while MSW fly ash can not. This conclusion is in agreement with the degree of pozzolanicity suggested by the sum of oxides of Si + Fe + Al deduced from the chemical composition according to ASTM C 618.

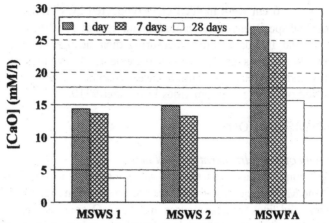

Figure 1. Pozzolanic Activiy: Evolution of CaO concentration over time

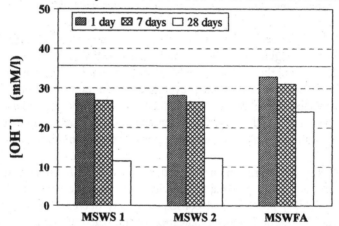

Figure 2. Pozzolanic Activiy: Evolution of OH⁻ concentration over time

The slight difference in the pozzolanic activity for both materials MSWS 1 (MSW bottom ash as-received) and MSWS 2 (MSW bottom ash without ceramic and glass, metallic and paper particles) does not justify the separation process, specially taking into consideration the better behavior shown by the sample in the as-received state. Besides, some authors (18) have pointed out that the total amount of contaminant toxic metals, anions (Cl^- and $SO_4^=$) and unburned carbon remaining in the sample are concentrated in the small size particles and this content increases in the leachate. In consequence removing big particles could indirectly contributed to increase the toxicity of MSW bottom these reasons the only material chosen for the mechanical tests was MSWS 1 sample.

Finally, in Fig. 3 and 4 the pozzolanic activity results obtained for the MSW incineration residues tested are presented jointly with those obtained for other well known pozzolanic materials. From the comparison of the results it can be deduced that MSW bottom ash shows a pozzolanic activity similar to that of a natural pozzolan although the pozzolanic reaction takes place at a low rate initially.

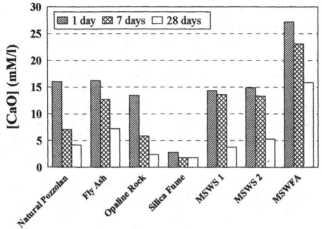

Figure 3. Pozzolanic Activiy: Evolution of CaO concentration over time

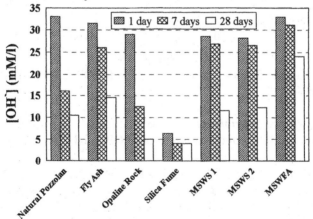

Figure 4. Pozzolanic Activiy: Evolution of OH⁻ concentration over time

Flexural Strength Analysis

The evolution of the flexural strength (F_s) versus time is given in Fig. 5 for the plain OPC paste and OPC with 20% and 40 % MSW bottom ash addition, each value representing the average of 6 measurements. Initially, the highest flexural strength values are measured in the plain OPC sample as can be expected, because a plain cement paste typically gains strength more rapidly at early ages than a paste with part of the cement replaced with a pozzolanic material due to the slower rate of pozzolanic reaction. However, a paste containing a pozzolanic materials may have a final strength higher than a pure cement paste, as is the present case. Thus, for OPC with 20% and 40 % MSW bottom ash addition, the flexural strength increases after 28 days continuously and higher values than that of OPC pastes are measured for periods of time longer than one month, that is, for periods of time at which significant pozzolanic activity was detected (see Fig. 1). Besides, as the cement replacement percentage increased, the flexural strength also increased.

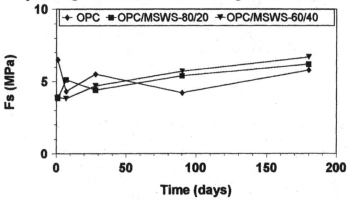

Figure 5. Evolution of flexural strength over time for OPC, OPC/MSWS-80/20 and OPC/MSWS-60/40

Porosity and Pore-size Distribution Analysis

The pore size distribution curves of plain OPC and OPC/MSWS samples at 1, 7, 28, 90 and 180 days appear in Fig. 6.

As shown, at 1 day OPC sample has a porosity value about 50 % lower than the sample with a 20 % cement replacement by bottom ash and 100 % lower than the sample with 40 % cement replacement. This increase of porosity as the cement replacement level increases is mainly related to an increase in the percentage of coarse porosity, that is, of pore sizes bigger than 0.05 μm.

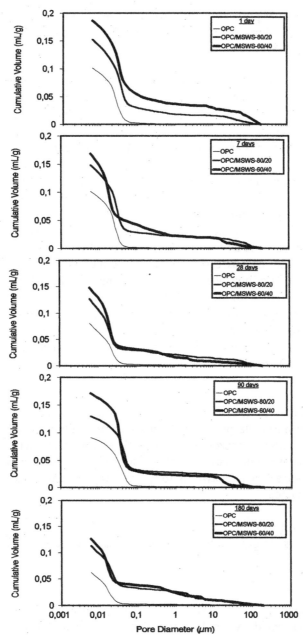

Figure 6. Pore size distribution curves of OPC, OPC/MSWS-80/20 and OPC/MSWS-60/40

The increase in porosity could also be related to the increase in volume reported by some authors (7) (19) and for which two mechanisms have been formulated:

- Chemical reactions on non-ferrous metallics, particularly on aluminum and zinc, can cause an increase in volume. Under alkaline conditions, which occur during cement hydration, metallic aluminum can form hydroxides or lead to emission of hydrogen.
- Ettringite formation is another well-known reaction which may occur.

The increase in porosity with cement replacement by MSW bottom ash could be a serious drawback for its utilization in concrete and needs to be further investigated in the near future.

Finally, from the results presented in Fig. 6, it can be deduced that the total porosity decreased in all the samples over time irrespective of the presence of MSW bottom ash, as was expected due to the high degree of hydration of cement and the extent of pozzolanic reaction with time. Particularly interesting is the remarkable decrease of porosity in OPC/MSWS - 60/40 sample which reached almost the same value of OPC/MSWS -80/20 sample at 180 days, suggesting the pozzolanic activity of MSW bottom ash again and being in agreement with the higher flexural strength values measured for the OPC/MSWS -60/40 sample at this age.

CONCLUSIONS

-Pozzolanic activity test results and flexural strength values measured in cement pastes confirm the pozzolanic activity of MSW bottom ash which is similar to that of a natural pozzolan up to a 40 % cement replacement level.

- The increase in porosity that takes place for MSW bottom ash addition to cement could be a serious drawback to its utilization in concrete and needs to be further investigated.

ACKNOWLEDGMENT

The authors are grateful for the financial assistance from CICYT (project AMB 96-0415).

REFERENCES

[1]S.E. Sawell, T.R. Bridle and T.W. Constable. "Heavy Metal Leachability from Solid Waste Incinerator Ashes". Waste Management and Research.,6 227-238 (1988)

[2]C.C. Wiles. "The U.S. Enviromental Protection Agency Program for Evaluation of Treatment and Utilization Technologies for Municipal Waste Combustion Residues". Second International Conference on Municipal Waste Combustion, Tampa, Florida, April 1991.

[3]T.T. Holmes, D.S: Kosson, C.C. Wiles. "A Comparision of Five Solidication/Stabilization Processes for Treatment of Municipal Waste Combustion Residues: Phisical Testing". Waste Materials in Construction , J.J.J.R. Goumans, H.A. Van der Sloot and Th.G. Aalbers (Editors). Elsevier Science Publisher, 107-118 (1991)

[4]D.S. Kosson, H. Van der Sloot, T. Holmes, and C. Wiles, "Leaching Properties of Untreated and Treatred Residues Tested in the USEPA foe Evaluation of Treatment and Utilization Technologies for Municipal Waste Combustor Residues". Proceedings of WASCON'91, Maastricht, The Netherlands, November 1991.

[5]S. Aver, H.J. Kuzel, H. Pöllmann and F. Sorrentino. "Investigations on MSW Fly Ash Treatment by Reactive Calcium Aluminates and Phases Formed". Cem.Concr.Res, 25 [6] 1347-1358 (1995)

[6]P.J. Wainwright and P. Robery. "Production and Properties of Sintered Incinerator Residues as Aggregate for Concrete". Waste Materials in Construction , J.J.J.R. Goumans, H.A. Van der Sloot and Th.G. Aalbers (Editors). Elsevier Science Publisher, 425-432 (1991)

[7]J. Pera, L. Coutaz, J. Ambroise and M. Chababbet. "Use of Incinerator Bottom Ash in Concrete". Cem.Concr.Res, 27[1] 1-5 (1197)

[8]M.T. Ali and W.F. Chang. "Strength Properties of Cement Stabilized Municipal Solid Waste Incinerator Ash masonry Bricks", ACI Materials Journal, 91 [3] 256-263 (1994)

[9]D.L. Gress, X.Zhang, S. Tarr, I. Pazienza and T.T. Eighmy. "Municipal Solid Waste Combustion Ash as an Aggregate Substitute in Asphaltic Concrete". Waste Materials in Construction , J.J.J.R. Goumans, H.A. Van der Sloot and Th.G. Aalbers (Editors). Elsevier Science Publisher, 161-175 (1991)

[10]D.J. Nonneman, F.A. Hansen and M.H.M. Coppens. "The Use of Incinerator Slag in Asphalt for Road Construction". Waste Materials in Construction , J.J.J.R. Goumans, H.A. Van der Sloot and Th.G. Aalbers (Editors). Elsevier Science Publisher, 569-578 (1991)

[11]Y. Takuma, Y. Tsuchida and S. Uchida. "Characteristics and Hydratation of Cement Produced from Ash from Incinerated Urban Garbage". Proceedings of the 10th International Congress on the Chemistry of Cement, 3 . 3ii118, Gothenburg, Sweden, June 2-6 (1997)

[12]K.E. Forrester and R.W. Goodwin, "Engineering Management of MSW Ashes Field Empirical Observations of Concrete Like Characteristics". Proceedings International Conference on Municipal Solid Wasre Combustion, 1 5B15-5B32 April 11-14 (1989)

[13]J.D. Hamernik and G.C. Frantz. "Phisical and Chemical Properties of Municipal Solid Waste Fly Ash ", ACI Materials Journal, 88 [3] 294-300 (1991)

[14]J.D. Hamernik and G.C. Frantz. "Compressive Stength of Municipal Solid Waste Fly Ash Concrete", ACI Materials Journal, 88 [5] 508-517 (1991)

[15]J.I. Bhatty and K.J. Reid. "Compressive Srength of Municipal Sludge Ash Mortars", ACI Materials Journal, 86 [4] 394-400 (1989)

[16]"Standard Specification For Fly Ash and Raw or Calcined Natural Pozzolan for Use as a Mineral Admixture in Portland Cement Concret (ASTM C 618-85)". Annual Book of ASTM Standars, Vol. 04.02, ASTM, pp. 389-392, Philadelphia, (1987)

[17]M.I. Sanchez de Rojas and M. Frias. "The Pozzolanic Activity of Different Materials, its Influence on the Hydration Heat in Mortars". Cem.Concr.Res. 26 [2] 203-213 (1996)

[18]J.A Stegeman and J. Scheider. "Leaching potential of Municipal Waste Incinerators Bottom Ash as a Function of Particle Size Distribution". Waste Materials in Construction , J.J.J.R. Goumans, H.A. Van der Sloot and Th.G. Aalbers (Editors). Elsevier Science Publisher, 135-143 (1991)

[19]E. Berg and J.A Neal. " Municipal Waste Combustor Ash in Concrete". Proceedings of the 5th International Conference on Municipal Solid Waste Combustor Ash Utilization. Edited by W.H. Chesner and F.J. Roethel. Arlington, November 17-18, pp. 155-170 (1992)

CALCINED PAPER SLUDGE : AN INTERESTING ADMIXTURE FOR THE CEMENT INDUSTRY

Jean Pera, and Jean Ambroise
Unité de Recherche Génie Civil - Matériaux, Institut National des Sciences Appliquées de Lyon - France

ABSTRACT

Processing of waste paper generates fiber residues and sorting wastes (de-inking and sewage sludges). These sludges contain cellulose fibers and mineral fillers such as kaolinite, calcite and talc.
By a controlled-calcination process in the temperature range of 630 to 750°C, it is possible to transform the sludge into a highly pozzolanic material usable in the cement industry. In this process, kaolinite is transformed into metakaolinite, while calcite and talc are preserved.
This paper deals with the influence of the sludge composition and the calcination parameters on the pozzolanic activity of the calcined product. A multiple linear regression analysis shows that the main factors affecting the pozzolanicity of the calcined sludge are: the kaolinite content in the sludge, the quantity of lime produced during calcination and the quantity of particles lower than 10 μm present in the product.
Despite a lower metakaolinite content, calcined sludge exhibits higher pozzolanic activity at early ages than commercially available metakaolins, due to superficial defects occurring during calcination, as shown by thermodesorption data. The higher calcium hydroxide consumption is also explained by the reaction between calcium hydroxide, metakaolinite and calcite which leads to the production of carboaluminate and C-S-H.
Finally, the performances of different concretes containing calcined paper sludge are presented.

INTRODUCTION

The recycling of paper has become an important industry in the past decade. Recycling reduces the requirement for virgin fiber and energy and can lower the environmental impact of paper manufacture. In 1983, the amount of recovered paper utilized by the European paper industry was 14.4 million tonnes, equivalent to 33 % of paper production. Twelve years later in 1995, the amount of recovered paper utilized by the industry had more than doubled to 30.7 million tonnes, equivalent to 43 % of total production [1]. In 1993, 39.2 % of the paper consumed in the United States was recovered [2]. The U.S. pulp and paper industry has set a 50 % paper recovery goal for the year 2000 [3].

The re-use of recovered paper in paper making inevitably results in the generation of process residues that cannot be used in paper production. De-inking sludges and primary clarified sludges are such residues. Generally, they are dewatered together in a screw press to about 50 % solids before they are discharged. On a dry basis, these sludges contain approximately equal amounts of organic (fibers) and inorganic components (clay, calcium carbonate, titanium oxide). With the current and projected high growth rate in de-inking capacity worldwide, the quantity of residues could reach 20 million tonnes by 2005 [4].

The residues most often are disposed of in landfills. However, in the United States, and especially in Europe, landfills are becoming increasingly undesirable and prohibitively expensive. Therefore, it is important to find alternate disposal means for the sludge from recycled paper plants. Recovery of energy, useful materials, or both from the sludge could provide an economical alternative to landfill disposal. Many mills are choosing incineration as a way to handle sludge. Air emissions are not a problem for combustion temperatures above 870°C. The ash produced is not a hazardous material [5]. The volume of ash from the incineration process still is significant, typically around 25 % of total dry solids [6]. Disposal of incinerator ash is a problem that is becoming increasingly important. Few constructive uses for incinerator ash have been developed, and none that are used by the pulp and paper industry. Typically, it is used for land-spreading or as a component of cement [7-9].

The present paper shows that a controlled calcination process can allow the re-use of the resulting ash as pozzolanic material. Such application is possible due to the composition of the inorganic phase present in the sludge. This material contains a significant fraction of clay, calcium carbonate, titanium oxide, and talc. As shown by many authors [10-13], clays once calcined in the temperature range of 700°C to 800°C, develop interesting pozzolanic properties. In this temperature range, calcium carbonate does not decarboxylate and can also react with cement [14]. Calcined clay, calcium hydroxide, and calcium carbonate react together to precipitate carboaluminate which improves the performance of hydrated cement

[12]. Other minerals (i.e, talc and titanium oxide) only act as fillers.

EXPERIMENTAL PROCEDURES

Description of raw paper sludges

Sixteen sludges supplied by different European facilities were investigated. The moisture content of the sludges as received was in the range of 35 to 60 % by weight. The average wet density of sludges was 600 kg/m^3. Table 1 provides chemical analyses of sludges showing their main components. The chief inorganic components were silica, alumina and calcium oxide, coming from coatings and fillers used in the manufacture of paper. The organic carbon content, due the presence of cellulose fibers, was in the range of 11 to 20 %. That explains the high values of loss on ignition observed (> 40 %).

Table 1. Chemical composition of sludges (w$_t$ %)

Designation of sludge	SiO$_2$	Al$_2$O$_3$	TiO$_2$	CaO	MgO	K$_2$O	C$_{org}$	LOI
BEG	21.8	16.4	0.2	6.8	1.2	0.7	16.2	50.6
BEN	21.9	11.2	0.5	14.3	4.1	0.2	17.3	46.0
GE	11.5	8.3	0.2	23.2	0.7	0.3	16.5	55.1
I	12.2	10.0	0.3	14.2	0.4	0.2	19.1	62.4
HD	11.6	8.4	0.3	19.2	0.9	0.2	17.3	58.6
GMD	9.4	7.4	0.2	26.8	0.6	0.2	13.9	54.4
GME	12.7	10.1	0.5	14.0	0.4	0.4	18.9	60.6
DK	12.3	9.1	0.3	27.6	1.0	0.2	11.2	48.7
BSE	13.1	10.2	0.7	17.9	0.7	0.4	16.6	55.6
DES	11.0	8.5	0.6	25.8	0.8	0.2	12.9	52.4
BL	24.1	19.3	1.2	12.0	1.8	0.3	11.6	40.1
KNP	12.0	10.9	0.3	17.5	0.5	0.2	17.3	57.8
HAK	12.1	7.8	0.4	17.5	1.1	0.4	17.6	59.8
GEN	11.1	8.3	0.6	23.6	0.9	0.3	14.2	54.8
PM	12.4	7.8	0.3	11.4	1.0	0.5	20.4	65.6
ESP	13.9	11.1	0.4	20.2	0.8	0.2	14.7	52.8

The mineralogy of the inorganic fraction of sludges was assessed by means of

X-ray diffractometry and Fourier Transform Infra-Red (FT-IR) spectrometry. The results are summarized in Table 2.

The inorganic fraction of dry sludges was in the range of 44 to 75 %. It was mainly composed of kaolinite and calcite. The ratio between kaolinite and calcite varied from 0.36 (DK) to 2.94 (BEG).

In this inorganic fraction, the proportion of kaolinite reached 61 % for BEG and BL, while it was only 25 % in GMD and DK sludges. The calcium carbonate content was as low as 21 % in BEG sludge and reached 70 % in GMD sludge. Some sludges as BEN and BL contained more than 5 % talc. The amounts of muscovite and quartz were generally low.

Table 2. Mineralogy of raw sludges (w_t %)

Sludge	Organic content	Inorganic components					
		Kaolinite	Talc	Muscovite	Quartz	Calcite	Total
BEG	39.7	35.8	3.8	5.8	0.7	12.2	58.3
BEN	35.0	26.3	12.8	2.0	0.6	18.6	60.3
GE	34.0	18.3	2.3	2.9	0.2	41.5	65.2
I	47.7	23.9	1.2	1.4	0.0	25.4	51.9
HD	40.6	19.5	2.7	1.9	0.0	34.3	58.4
GMD	31.6	17.4	1.9	1.3	0.0	47.8	68.4
GME	46.4	22.5	1.3	3.2	0.0	25.0	52.0
DK	23.8	20.9	3.2	2.0	0.0	49.3	75.4
BSE	38.0	22.6	2.2	3.2	0.0	31.9	59.9
DES	28.6	19.7	2.7	1.9	1.0	46.1	71.4
BL	23.9	45.4	5.7	1.5	0.0	21.4	74.0
KNP	39.5	26.2	1.6	1.4	0.0	31.3	60.5
HAK	43.5	16.5	3.6	3.4	0.6	31.2	55.3
GEN	33.5	18.7	2.7	2.3	0.0	42.1	65.8
PM	54.2	15.7	3.0	4.1	1.3	20.3	44.4
ESP	32.3	26.5	2.6	1.7	0.8	36.1	67.7

Calcination of sludges

Sludges were burnt in a laboratory electrical fixed-bed furnace. The volume of the furnace was 125 L and the quantity calcined was 5 kg of wet sludge. The temperature of calcination was 700°C. A previous study has shown that higher temperatures led to significant decarbonation of calcite and the formation of unreactive products (i.e, anorthite and gehlenite) [15].

The temperature was increased to a rate of 250°C/hr and the product was maintained at 700°C for 5 hours. Then, it was cooled, with the furnace being closed. Discharging took place as the furnace recovered to ambient conditions. The cooled burnt sludge was milled for 30 minutes to obtain particles less than 100μm. Laser granulometry and BET surface area analyses were carried out to characterize the physical properties of the calcined products. XRD and FT-IR spectrometry analyses were also done to investigate the dehydroxylation of kaolinite and the preservation of calcite during calcination.

Pozzolanic activity of calcined sludges

The term "pozzolanic activity" includes two parameters, namely the maximum amount of lime that a pozzolana can combine and the rate at which such combination occurs. Therefore, the pozzolanic activity of calcined sludges was investigated by measuring the consumption of calcium hydroxide with time on pastes composed of 50 % calcium hydroxide and 50 % calcined sludge, by weight. Pastes were prepared at equivalent consistencies according to the French standard NFP 15-402 and cast into cylindrical molds ($\varnothing$ = 20 mm; h = 40 mm) using a shaking table.

The samples were demolded after 3 days of hydration and immersed in lime-saturated water for 4 or 25 days. Therefore, it was possible to evaluate the calcium hydroxide consumption after 3, 7, and 28 days of hydration. This measurement was done by differential thermal analysis (DTA) on 600 mg of powder less than 100 μm. The surface area of the residual calcium hydroxide peak was measured and compared to that of a paste containing 50 % calcium hydroxide and 50 % ground silica. The temperature rate was 10°C/min. The different calcined sludges were rated according to their reactivity and compared to a commercial metakaolin supplied by Engelhard (U.S.A.).

Development of building materials containing calcined paper sludge

The use of calcined paper sludge as a mineral admixture in blended cements was not investigated, due to the specifications of European standard ENV 197-1. This standard only recognizes as admixtures the following supplementary materials: ground granulated blast-furnace slag, coal fly ash, limestone filler, siliceous filler, and silica fume. Other materials can only be added in the proportion of 5 %. Therefore, calcined sludge was directly introduced in different concretes and compared to other pozzolans.

RESULTS AND DISCUSSION

Analysis of calcined sludges

XRD and FT-IR spectrometry data showed that kaolinite was completely dehydroxylated during calcination and transformed into metakaolinite. Calcite was somewhat decarbonated, specially in sludges containing the lowest proportions of kaolinite (i.e, GMD and GE). The presence of a IR band at 3645 cm^{-1} revealed the formation of $Ca(OH)_2$ after calcination and cooling. The intensity of the band allowed the determination of the quantity of calcium hydroxide present in the final product. The results obtained are shown in Table 3. Table 3 also presents the particle size distribution and the BET specific surface area of calcined sludges.

From Table 3, it appears that important agglomerations occurred during the calcination of BEG, ESP, BSE, and DES sludges, in which 14 to 27 % of the particles were found to be larger than 100 μm. The average diameter of the particle size distribution was in the range of 7 μm (I) to 29 μm (BEG). The ultra-fine particle content (< 5 μm) varied from 18 % (DK) to 40 % (I) and the BET surface area ranged from 5.7 m^2/g (DES) to 15.9 m^2/g (BEG).

Table 3. Quantity of $Ca(OH)_2$ formed during the calcination of sludges at 700°C for 5 hours (w_t %) and physical properties of calcined sludges

Sludge	Ca(OH)$_2$ (%)	Particle size distribution (%) < 5 μm	< 10 μm	< 100 μm	D$_{50}$ (μm)	BET surface area (m^2/g)
BEG	0	20	32	73	29	15.9
BEN	0	30	62	98	8	8.4
GE	36	20	40	91	17	10.6
I	12	40	68	100	7	16.5
HD	15	28	54	94	10	10.2
GMD	27	15	30	96	20	9.2
GME	11	18	34	95	17	9.4
DK	7	22	48	95	11	7.0
BSE	9	20	32	74	19	16.0
DES	5	28	56	86	9	5.7
BL	0	30	48	90	11	12.6
KNP	7	25	45	97	14	14.6
HAK	10	28	48	95	11	10.2
GEN	2	26	50	97	10	7.0
PM	6	30	58	95	9	10.7
ESP	2	22	42	73	16	9.6

Lime consumption of calcined sludges

The quantity of Ca(OH)$_2$ combined by calcined sludges in pastes containing 50 % calcium hydroxide and 50 % calcined sludges is shown in Table 4. After 3 days of hydration, the best pozzolans were calcined sludges I, BL, BEN, BEG, and PM; the less reactive were GE, GMD, GME and DK. This rating was confirmed after 7 and 28 days, except for sludge GME which consumed a lot of lime between 7 and 28 days. After 28 days of hydration, all the lime was consumed by calcined sludges BL, BEG, and KNP which presented a kaolinite content varying from 43.3 % to 61.4 % in the inorganic fraction. The kaolinite content itself seems to be insufficient to explain the reactivity of the calcined sludge. This result was confirmed by the comparison made between the performances of calcined sludges and those of a commercial metakaolin (MK). The metakaolinite content of MK was 97 % and its average particle size distribution was 1 µm. The quantities of Ca(OH)$_2$ consumed by MK were 24 % after 3 days of hydration, 45 % after 7 days, and 94 % after 28 days. Ten calcined sludges were more reactive at early ages than MK, despite a lower metakaolinite content. This parameter only influences the 28 d-pozzolanic activity: only 3 sludges were more efficient than MK (BEG, BL, and KNP).

Table 4. Ca(OH)$_2$ combined by calcined sludges (%)

Sludge	Time of hydration (days)		
	3 d.	7 d.	28 d.
BEG	50	92	100
BEN	55	73	75
GE	5	24	26
I	62	70	75
HD	22	31	43
GMD	3	9	10
GME	20	34	61
DK	18	24	41
BSE	35	63	73
DES	40	58	61
BL	57	90	100
KNP	17	56	100
HAK	32	56	68
GEN	27	56	66
PM	50	68	75
ESP	45	73	80

Therefore, a multiple linear regression analysis was undertaken to explain the pozzolanic activity of calcined sludge at the different hydration times (3, 7, and 28 days). The parameters investigated were:
- the lime consumption, C,
- the quantity of calcium hydroxide formed during the calcination, D,
- the kaolinite content of the inorganic fraction, K,
- the BET specific surface area, SBET,
- the quantity of particles lower than 10 μm, G10,
- the average diameter of the particle size distribution, D50.

The software used for this task was SPSS for MS WINDOWS. The main findings of this investigation are as follows:

1. after 3 days of hydration, the pozzolanic activity of the calcined sludge can be explained by 4 parameters (D, K, G10, and D50):

$C_3 = -0.77 D + 0.61 K + 1.67 G10 + 2.30 D50 - 89$
$r^2 = 0.86$

2. after 7 days of hydration, the main parameters are K, and D:

$C_7 = 1.07 K - 1.17 D + 24.5$
$r^2 = 0.76$

3. after 28 days of hydration, the main parameters are also K, and D:

$C_{28} = 1.18 K - 1.42 D + 34$
$r^2 = 0.84$

A general conclusion may therefore be drawn for practical use. The main parameters influencing the pozzolanic activity of calcined sludge are:
- the quantity of kaolinite in the inorganic fraction,
- the quantity of calcium hydroxide produced during the calcination, and which has to be minimized,
- the quantity of particles smaller than 10 μm, which has to be maintained at a certain level to get a high activity at early ages.

A previous study [15] has also shown that superficial defects occurred during the calcination and enhanced the reactivity of the calcined sludges at early ages. The high $Ca(OH)_2$ combination by calcined sludges was also explained by the reaction which took place between the remaining calcium carbonate ($C\overline{C}$) in the sludge, calcium hydroxide (CH) and metakaolin (MK) according to the following equation proposed by J. Ambroise [12]:

$$AS_2 + CH + C\overline{C} + 8H \longrightarrow C_4A\overline{C}H_{11} + C\text{-}S\text{-}H$$

The presence of monocarboaluminate $C_4A\overline{C}H_{11}$ was detected both by DTA (endothermic peak at 220°C) and XRD (rays at $2\theta = 11.7°$ and $2\theta = 23.5°$).

Properties of concretes

Industrial production of calcined sludge was made by means of a multi-hearth furnace, at a temperature ranging from 630° to 700°C. BEG sludge was used for these tests.

In a first series of experiments, calcined sludge was compared to other pozzolans at a cement replacement of 15 %: commercial metakaolin (MK), fly ash (FA) and silica fume (SF). the cement was a CEM I CPA 52.5 according to European standards. The mixture proportion of the control concrete was:
- cement: 350 kg/m^3
- sand (0/4 mm): 785 kg/m^3
- gravel (14/20 mm): 1,070 kg/m^3
- water: 177 l/m^3
- superplasticizer (SP): 9.8 l/m^3.

Its flow was 160 mm. In other concretes, 15 % cement was replaced by each admixture, and the water and superplasticizer were adjusted to get about the same workability. The compositions and main properties of the concretes are shown in Table 5.

Table 5. Comparison of different mineral admixtures

Admixture	Water (l/m^3)	SP (l/m^3)	W/(C+A)	Slump (mm)	Compressive strength (MPa)			
					2 d.	7 d.	28 d.	90 d.
Control	177	9.8	0.51	160	33	77	74	74
BEG	178	13.3	0.51	150	27	61	76	80
MK	177	13.3	0.51	150	25	61	68	78
FA	167	8.8	0.48	150	25	60	70	76
SF	200	10.9	0.57	165	30	65	71	76

Table 5 points out that all these admixtures behaved very well and led to high strength at 28 days. Calcined sludge developed the same performance as commercial metakaolin.

In a second series of tests, calcined sludge (BEG) was compared to cristobalite (C), silica fume (SF), finely ground blast-furnace slag (GBFS) and metakaolin (MK), in the development of very high strength concrete (> 100 MPa). The results obtained are summarized in Table 6.

Table 6. Design of very-high strength concrete

	Cristobalite Silica fume		Blast-furnace slag	Calcined sludge	Metakaolin
Composition (kg/m³)					
Cement	525	525	525	525	525
Sand (0/4 mm)	670	670	670	670	670
Gravel (6/10 mm)	1,010	1,010	1,010	1,010	1,010
Ultra-fine	52.5	52.5	52.5	52.5	52.5
Superplasticizer	20.2	22.7	17.2	32.5	25.0
Water	140	141.5	138	147.8	143.0
Properties					
Slump (mm)	530	370	590	550	550
28 d. Strength (MPa)	127	120	125	110	130

From Table 6, it can be concluded that calcined sludge is suitable for very high-strength concrete but its use needs more superplasticizer and water than other ultra-fine particles.

The last series of tests consisted of introducing calcined sludge instead of metakaolin in coloured concrete blocks. The presence of metakaolin is useful to avoid efflorescences due to the carbonation of the remaining calcium hydroxide in normal portland cement. These industrial tests were done without any change in the production process management. The compositions and properties of these blocks are reported in Table 7.

According to French standards, two conditions are required for blocks: 14 d. compressive strength higher than 4 MPa, and coefficient of capillarity lower than 5 %. Table 7 shows that all the mixtures fulfilled these requirements and the replacement of metakaolin by calcined sludge decreased the water absorption of blocks by 33 %. A microstructural investigation carried out using DTA and FT-IR spectrometry pointed out that there was no unreacted calcium hydroxide left in mixtures 2 to 4, while some free $Ca(OH)_2$ remained in mixture 1 cast with metakaolin. Calcined sludge should be more efficient than metakaolin to prevent coloured concrete efflorescences.

Table 7. Design of concrete blocks

Concrete reference	1	2	3	4
Composition (kg)				
Cement	75	75	75	75
Aggregate (0/10 mm)	675	675	675	675
Metakaolin	32	-	-	-
Calcined sludge	-	32	42	22
Mineral admixture	32	32	22	42
Plasticizer	2.1	2.1	2.1	2.1
Colouring pigment	2.5	2.5	2.5	2.5
Water	48.8	50.5	52.1	39.1
Properties				
Compressive strength at 14 days (MPa)	7.9	7.5	6.3	7.2
Capillarity				
- absorbed water (g)	60	40	60	20
- coefficient of capillarity (standard NFP 14-102)	2.04	1.36	2.04	0.68

CONCLUSION

The transformation of paper sludges into a high-value added material represents a reliable and long-term issue for the paper industry. Despite a lower kaolinite content, some calcined sludges are able to develop higher pozzolanic activity than metakaolin obtained by calcination of clays. The main parameters playing a role in this activity are the kaolinite content in the organic fraction and the quantity of calcium hydroxide produced by calcite decarbonation during the calcination. These two parameters are closely linked and, when considered separately, insufficient to describe the reactivity of calcined sludges. At early ages, the proportion of particles lower than 10 µm also influences the reactivity. The average diameter of the particle size distribution and the BET specific surface area do not play a main role in this activity.

The tests carried out on concretes show that calcined sludge is a suitable admixture for high strength concrete and coloured concrete, in which it quickly reduces the amount of free calcium hydroxide and therefore, decreases the risk of efflorescences.

ACKNOWLEDGEMENT

The authors gratefully acknowledge support from the European Community, which funded the project under the Brite/Euram Programme (BRE2 CT 94/0626, Project No. PL-8174).

REFERENCES

1. D. Clark Keynote address. PPI Recycling'96 conference. London (U.K.), 13-14 November 1996. (1996).
2. M. Anon. Recovered paper statistical highlights. American Forest & Paper Association, Washington. (1993).
3. M. Anon. The papermaker. N° 7, p. 14. (1994).
4. L. Webb. A host of options available for sludge. Pulp and Paper International, Vol. 38, N° 11, pp. 44-48. (1996).
5. M.A. Douglas, J. Latva-Somppi, and V.V. Razbin. Tappi Journal, Vol. 77, N° 5, p. 109. (1994).
6. J.A. Sohara. Recycling mineral fillers from de-inking sludges. PPI Paper Recycling'96 Conference. London (U.K.), 13-14 November 1996. (1996).
7. J. Wendell. Water retardant covering material for solid waste landfills. U.S. Patent 5, 054, 406. (1991).
8. L. Ernstbrunner. Reconstruction of a cement plant for fiber residue processing. Ciments, Bétons, Plâtres, Chaux, N° 815, pp. 244.248. (1995).
9. W. M.J. Frederick, K Lisa, J.R. Lundy et al. Energy and materials recovery from recycled paper sludge. Tappi Journal, Vol. 79, N° 6, pp. 123-131. (1996).
10. R.C. Mielenz. Mineral admixtures: History and background. Concrete International, August 1983, pp. 34-42. (1983).
11. M. Murat, J. Ambroise, and J. Péra. Hydration reaction and hardening of calcined clays and related minerals. Cement and concrete Research, Vol. 15, pp. 261-268. (1985).
12. J. Ambroise. Elaboration de liants pouzzolaniques à moyenne température et étude de leurs propriétés physico-chimiques et mécaniques. Thesis (in French), Lyon, 165 p. (1984).
13. J. Ambroise, S. Martin-Calle, and J. Péra. Pozzolanic behaviour of thermally activated kaolin. In: Proceedings of the Fourth CANMET/ACI International Conference on "Fly Ash, silica Fume, Slag and Natural Pozzolans in Concrete". ACI Special Publication SP-132, Vol. 1, pp. 731-748, (Editor: V.M. Malhotra). (1992).
14. S. Husson, B. Guilhot, and J. Péra. Influence of different fillers on the hydration of C_3S. In: Proceedings of the 9th International Congress on the Chemistry of Cement. New-Delhi (India), 23-28 Novembre 1992, Vol. 3,

pp. 232-239, (Published by: National Council for Cement and Building Materials, India). (1992).

15. J. Péra, and A. Amrouz. Development of highly reactive metakaolin from paper sludge. Journal of Advanced Cement Based Materials, Vol. 7, N° 2, pp. 49-56. (1998).

FINE - GRAINED CEMENTLESS CONCRETE CONTANING SLAG FROM FOUNDRY

S.I. Pavlenko
Siberian State University of Industry
42, Kirov Street
Novokuznetsk, RUSSIA, 654007
V.I. Malyshkin
KHAKASENERGO
74, Pushkin Street
Abakan, RUSSIA, 664019

ABSTRACT

The Department of Civil Engineering of the Pavlodarsky Tractor Plant (PTP) has developed the composition and technology of fine-grained concrete with ground basic slag as a binding material and acidic slag sand with a particle size of 0 to 5 mm as an aggregate. Both basic and acidic slag are waste products of three foundry departments of the PTP with produces 30 000 tons acidic slag annually. These slag are dumped polluting the environment and occupying land. Concrete developed included 400 to 700 kg/m^3 (depending on the compressive strength) basic slag, 1000 to 1200 acidic slag sand, 230 to 280kg/m3 by water and 0,3% by weight of binding materials of plasticizing admixture (technical grade lignosulfonate). The slump of a mixture was 4 to 6 cm and the average density was 1800 to 1900 kg/m^3 .The technology of processing acidic slag into sand with a particle size of 0 to 5mm and basic slag into powder has also been developed. Schematic diagrams of this technological process are presented in the paper. The studies showed the possibility and reliability of producing concrete from by-products and their use in the construction of single, two-story houses as well as in the production of small blocks and unburnt brick.

INTRODUCTION

The Pavlodarsky Tractor Plant in Kasakhstan has faced the following problems: the problem of utilizing waste products from steelmaking and iron foundry processes and shortage of housing construction for workers due to lack of cement and aggregates for lightweight and heavy concretes.

The objective of the present study was to solve the above problems by investigating slags in order to develop on their basis concrete for load-bearing and non-load-bearing structures for use in the construction of cast in-situ houses.

Composition of fine-grained slag concrete and technology of processing slags into sand with a particle size of 0 to 5mm and into powder with a fineness of 4000 to 4500 cm^2/g have been developed by the Siberian State Academy of Mining & Metallurgy. The concrete does not contain any natural or artificial porous aggregates.

MATERIALS

The materials used in the investigation were acid slags from the iron foundry department and the steelmaking department No I (SD-1), basic slag from the steel-making department No 2 (SD-2). Physical properties and chemical analysis of the materials are given in Tables 1 and 2, respectively.

Table 1. Physical Properties of Slags from PTP

Characteristics of Slag	Slag from SD-1	Slag from SD-2	Cupola Slag
Bulk Density, kg/m^3	1240	1250	1250
Absolute Density, kg/m^3	2450	2750	2800
Crushabilily, %	30	19	13
Cylinder Crushing Strength, MPa	2.44	2.50	2.78
Specific Efficiency, pc/g	0.4	3.1	7.2
Colour	green	yellow-grey	grey

Slag from the Steelmaking Department No 1

Slag from SD-1 is an acid slag and from its granulametric composition it is referred to as coarse sand. It is of a green colour, consisting of particles of various shapes including thin-walled spirals. It has an unstable structure and after being boiled thrice, it loses up to 50% weight. It contains up to 20% iron oxides and metallic inclusions. Therefore, prior to utilization of slag in concrete it was ground to sand, with a particle size distribution of 0 to 5 mm, the metallic inclusions and iron oxides being removed by magnetic separation.

Slag from the Steelmaking Department No 2

Slag from SD-2 is basic. It is yellow-grey in colour and consists of particles 70 mm and even larger in size. It loses 12% of its weight after being boiled three times. The process of slag structure failure is attributable to the polyamorphic transformation of dicalcium silicate from β- to γ-modification. Free lime (up to 6%) which is present in slag in various degrees of overburning is hydrated and the resulting increase in the volume also assists the failure of slag. A decision was made to grind this slag into a powder having a specific surface of 4000 to 4500 cm^2/g, which helped to eliminate the negative properties of slag and utilization of the positive ones (i.e. release a free lime to react with silica and water).

Table 2. Chemical Composition of Slags

Oxides, %	Slag from SD-1	Slag from SD-2	Cupola Slag
SiO	30.95	30.06	57.76
Al_2O_3	6.88	2.65	4.40
Fe_2O_3	11.97	1.99	13.77
FeO	8.82	0.71	1.93
CaO (total)	8.95	20.00	10.09
CaO (free)	1.76	4.58	2.12
MgO	11.79	22.34	1.62
Na_2O	1.32	2.34	0.95
K_2O	0.10	0.24	0.58
TiO_2	1.01	0.50	0.85
SO_3	0.43	0.76	0.37
MnO	17.67	18.24	9.59
P_2O_5	0.07	0.06	0.10
Loss on ignition	-	0.13	1.94

Cupola Slag

Cupola slag is acid and by its fineness modulus (3,71) is referred to as coarse sand.

Technical grade lignosulfonate (TGL)

TGL is a waste product of pulp and paper industry. It was used as a plasticizer for concrete mixture in order to reduce the amount of mixing water, while the detergent called "Progress", a secondary sodium alkyl sulphate, was used in concrete as an air-entraining admixture.

OBJECTIVE OF STUDY

The objectives of this work were to: 1) create the optimum composition of cementless fine-grained concrete based on a slag waste from a tractor plant which does not contain any natural or artificial porous aggregates; 2) create a technology of processing of a basic slag into a powder to be used as a cementitious material and of acidic slag into a sand to be used as an aggregate.

EXPERIMENTAL

It is well know (1,2,3) that high-calcium ashes and slag have an increased content of free calcium oxide (over 5%) which may result in an irregular volume change and destruction of concrete. To eliminate the irregularity of the volume change (hydration of free CaO after concrete hardening), the following methods were used: slag grinding (breakage of melted strutures and covers of CaO), previous hydration of the material, autoclave treatment, chemical binding.

Therefore, slag from SD-2 was ground in a laboratory ball mill into a powder with a specific surface of 4000 to 4500 cm^2/g which was used as a cementitious material and a fine-dispersed aggregate for the fine-grained slag concrete.

Slag sand from SD-1 and iron foundry department with a particle size distribution of 0 to 5 mm produced by grinding in a laboratory roller crusher was used as fine aggregate.

In selecting optimum compositions of concrete mixtures, the method of rational planing of experiment (4,5) was used. The values of primary factors (contents of slag sand, water, plasticizing and air-entraining admixtures) were changed in a given range and their effect on secondary factors (average density, compressive strength of fine-grained slag concrete) was studied. The primary factors range is given in Table3.

Concrete was mixed in a laboratory fixed-drum concrete mixer. Test cubes, 100x100x100 mm in size, were cured in a laboratory steam-curing chamber at 900°C using a 3+10+3 hour cycle. By changing the values of the primary factors in accordance with Table 3 and by calculating the average resulting values the optimum compositions of fine-grained slag concrete mixtures for various strength grades were obtained. The proportioning of the concrete mixtures is summarized in Tables 4, 5.

Table3. Primary Factors Values for the Ingredients of Slag Concrete Mixture

Materials	Quantities, kg/m³		
	1	2	3
Acid Slag Sand	700	1100	1500
Basic Fine-Grained Slag	300	500	700
Water	250	350	450
Plasticizing Admixture TGL	0	0.3	0.9
Air-Entraining Admixture	0	3	9

Table 4. Optimum Compositions of Fine-Grained Concrete on the Basis of Ground Slag and Slag Sand for Non-Load-Bearing Structures

Characteristics	Concrete Brand			
	50	50	75	75
Acid Slag Sand SD-1, kg/m³	690	-	765	-
Acid Slag Sand from Foundry Department, kg/m³	-	680	-	750
Basic Fine-Ground Slag SD-2, kg/m³	500	520	500	515
Water, kg/m³	290	297	295	298
TGL, %	0,3	0,3	0,3	0,3
Air-Entraining Admixture, %	2	2	1	1
Mixture Slump, cm	4-6	4-6	4-6	4-6
Average Density of Mixture, kg/m³	1480	1497	1560	1563
Compressive Strength at 28 days, MPa	5,6	5,3	8,1	7,4
Compressive Strength at 180 days, MPa	6,0	5,6	8,6	7,8

It can be seen that the fine-grained slag concrete has the compressive strengths of 5 to 8 MPa and 10 to 25 MPa for non-load-bearing and load-bearing structures, respectively. The best results were obtained by utilizing as an aggregate acid slag from a steelmaking department.

Table 5. Optimum Compositions of Fine-Grained Concrete on the Basis of Ground Slag and Slag Sand for Load-Bearing Structures

Characteristics	Concrete Brand					
	100	100	150	150	200	200
Acid Slag Sand SD-1, kg/m^3	1077	-	1110	-	1142	-
Acid Slag Sand from Foundry Department, kg/m^3	-	1067	-	1103	-	1140
Basic Fine-Ground Slag SD-2, kg/m^3	500	510	500	510	500	508
Water, kg/m^3	278	283	280	285	268	273
TGL, %	0,3	0,3	0,3	0,3	0,3	0,3
Air-Entraining Admixture, %	-	-	-	-	-	-
Mixture Slump, cm	4-6	4-6	4-6	4-6	4-6	4-6
Average Density of Mixture, kg/m^3	1855	1860	1890	1898	1910	1921
Compressive Strength at 28 days, MPa	10,8	10,1	16,2	15,3	21,0	19,7
Compressive Strength at 180 days, MPa	12,0	11,2	20,2	17,4	25,8	22,3

The data on physic-mechanical, deformation properties, frost resistance and heat conductivity of concrete are given in Table 6, and durability studies of this concrete are given in our work (6).

The results obtained meet the requirements of the Building Code (7) for classes 50 to 200 fine-grained concretes used for the construction of low-rise buildings.

Table 6. Physic-Mechanical, Deformation and Special Properties of Fine-Grained Slag Concrete

Properties	Concrete Strength Group				
	For Non-Load-Bearing Structures		For Load-Bearing Structures		
	50	75	100	150	200
Axial Tensile Strength, MPa	0.67-0.79	0.88-1.12	0.98-1.23	1.54-1.58	1.94-2.13
Flexural Strength, MPa	0.94-1.11	1.06-1.20	1.39-1.73	2.17-2.33	2.84-3.12
Prism Strength, MPa	3.69-3.83	6.15-6.30	7.07-8.10	11.26-21.72	15.30-15.80
Shrinkage, mm/m	0.74-0.77	0.68-0.72	0.29-0.32	0.31-0.38	0.32-0.37
Initial Modulus of Elasticity, MPa	6.7 -7.3	7.9 -8.4	18.3 – 19.6	19.1 – 20.6	21.8 -23.4
Creep minus Shrinkage, mm/m	0.36-0.38	0.31-0.34	0.20-0.22	0.22-0.24	0.24-0.26
Compressibility, mm/m	1.24-1.39	1.02-1.26	0.82-0.91	0.85-0.87	0.97-1.12
Tensibility, mm/m	0.24-0.27	0.18-0.22	0.09-0.11	0.08-0.09	0.11-0.12
Frost Resistance, cycle	55-75	80-100	—	—	—
Heat Conductivity (in a dry state), Kcal/m°C	0.33-0.35	0.36-0.39'	—	—	—

PROCESSING OF ACID AND BASIC SLAGS INTO SAND AND POWDER

The practical results of this work were the development of technological schemes and regulations. At present, the project for processing acid and basic slag's into sand with a particle size of 0 to 5 mm and basic slag sand into powder with a specific surface of 4000 to 4500 cm2/g are being developed. The annual output of slag's from three departments of the Pavlodarsky tractor plant is 80000 tons including 30000 tons of basic slag.

Processing of Slag's into Sand

As mentioned above, the basic slag from SD-2 has a shape of a rubble consisting of lumps of 70 mm in size and larger. To process slag into sand, its four-

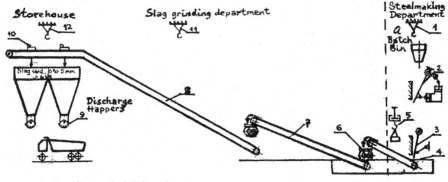

stage grinding is needed (Fig. 1).

1 - Tackle	7 - Belt Conveyer No 2
2,3 - Alligators	8 - Belt Conveyer No 3
4 - Belt Conveyer	9 - Sector Gate
5 - Iron Separator	10 - Electric Dumper
6 - Roller Crusher	11,12 - Tackles

Fig. 1. Schematic Diagram of Processing Slag's into Sand

The first two stages are performed at SD-2 which cuts expenses for a construction of a grinding department and its maintenance. The third and the fourth stages of grinding are made in the adjacent accommodation.

The first stage of grinding by a hydraulically driven jam crusher at SD-2 (2), raw materials to the batch bin being delivered from a pouring platform by an electric tackle having a lifting capacity of 5 tons (1). From here slag is dumped.

A SMD-116 jaw crusher with a complex motion of a movable jaw is used in the second stage for medium and fine grinding (3). Twin-roller crushers are used in the third and fourth stages of grinding. They are simple in design, easy to maintain and not large. In grinding hard rock, twin-roller crushers can provide a degree of grinding up to 6 (8) with the degree of grinding being 5 to 1.65 when grinding slags. Providing degree of grinding of 4 to 5 to 3.5 to 4 in the third and fourth stages, respectively, it is possible to produce slag sand with a fineness modulus of 2.5 to 3 mm complying with the sand standards for concrete.

In order to diminish dust formation, conveyers (4,7,8) handling slag are equipped throughout their width and length with chutes from chain conveyers for supplying water over them in the amount required. This also assists partial hydration of free lime in Ca(OH)$_2$.

The technological scheme provides for processing acid slags from SD-1 and iron foundry department into sand. Depending on their structure and grain composition they may be subjected to one or two stages of grinding in roller crushers.

Processing of Slag Sand into Powder

In order to produce powder (4000 to 4500 cm^2/g), the basic slag sand is ground in a multi-chamber tube ball mill (Fig. 2, position 2).

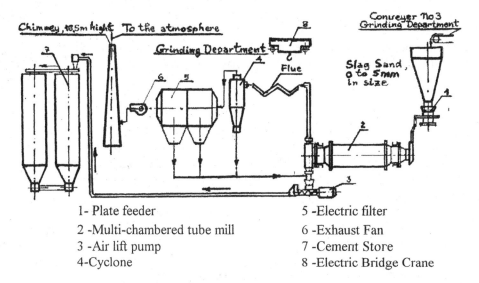

1- Plate feeder

2 -Multi-chambered tube mill

3 -Air lift pump

4-Cyclone

5 -Electric filter

6 -Exhaust Fan

7 -Cement Store

8 -Electric Bridge Crane

Fig. 2. Schematic. Diagram of Processing of Basic Slag Sand into Powder

To eliminate dust and heat during dry grinding, the tube mill is provided with a set or aspiration equipment consisting of an exhaust fan (6), an electric filter (5), a cyclone (4) and a flue (9). Dust is stored in cement silo banks (6). For assembly, dismantling and repair work, a bridge electric crane with a lifting capacity of 20 tons will be provided for in the grinding department (8).

The above facilities are being designed. Slag sand and powder will be delivered by a motor transport to the ready-mixed concrete plant and then by truck mixers to the construction of dwellings.

By the present work, the plant would solve the following problems: 1)utilization of by-products and 2) providing housing construction with a cheap material. The plant has already got orders for partial sale of slag and powder to building companies.

CONCLUSIONS

1. Using the waste products from the foundry department of the PTP (basic and acid slag's) it is possible to organize the production of fine-grained cementless concrete of 5 to 20 MPa strength grades. Fine-ground basic slag may be used as a cementitious material and acidic slag processed into sand as an aggregate.
2. Five strength grades (50, 75, 100, 150 and 200) of the concrete developed for non-load-bearing and load-bearing structures are in accordance with the requirements of Building Code and State Standard.
3. Technological schemes of processing acid slag into sand with a particle size distribution of 0 to 5 mm and basic slag into a cementitious material with a specific surface of about 4500 cm^2/g have been suggested.

REFERENCES

[1] S.I. Pavlenko "Lightweight cementless concrete on the basis of high-calcium fly ash and slag sand from TPP"; pp 95-106 in *Blended Cements in Construction*. Edited by R.N. Swamy. Elsevier Applied Science, London and New York. 1991.

[2] M.A.Savinkina and A.T. Logvinenko "Ashes Produced from the Kansko-Achinsky Brown Coal"; p. 46, Nauka, Novosibirsk, 1979.

[3] J. Papayianni, "Concrete with high-calcium fly ash". pp. 261-284 in *CANMET/ACI International Conference on Advances in Concrete Technology*, Athens, Greece 1992.

[4] M.M. Protodyakonov and R.I. Tedder, "Method of rational experiment planning", p 70, Nauka Pablishing House, Moscow, 1970.

[5] V.A. Voznesensky, "Statistical methods of experiments planning in technico-economical investigations", p192, Edition Statistiks, 1974.

[6] S.I. Pavlenko and V.S. Boqusevich, "Durability Studies of Concrete Containing Low Cement Content and Fine Particles of Foundry Slag"; pp.1069-1086 in *ACI SP 145-59*, Editor by V.M. Malhotra, Detroit, MI, 1994.

[7] Building Code 2.03.01-84, "Concrete and reinforced concrete structures", Gosstroy USSR, Moscow, 1985.

[8] A.A. Borshchevsky and A.C. Olgin, "Mechanical equipment for the production of building materials and articles", p. 386, Higher School, Moscow , 1987.

SPECIAL CEMENTS

AND CONCRETES

VOLUME STABILITY OF DENSIFIED CEMENT PASTES

Hua Ai and J. Francis Young
Departments of Materials Science and Engineering
and Civil Engineering
University of Illinois at Urbana-Champaign

ABSTRACT

Low water:cement ratio (w/c) cement pastes containing silica fume provide high strength and low permeability and ionic diffusivity. These effects are well documented in the literature, but much less is known about the volume stability of these materials. This paper presents preliminary results that show the effects of w/c and silica fume content on chemical (autogeneous) shrinkage, drying shrinkage and thermal expansion of cement pastes.

INTRODUCTION

The use of silica fume to make strong, durable concrete is now well-established technology throughout the world[1,2]. In the United States silica fume is primarily used in bridge deck overlays, due to its increased resistance to chloride penetration, or in high strength concrete (~100 MPa compressive strength). Silica fume is also essential to create ultra-high-strength concretes and mortars (> 150 MPa compressive strength)[2]. The ability of dispersed silica fume particles (0.1-0.3 μm) to pack between the larger cement particles (Figure 1) is a key aspect in the enhanced performance of these materials. A high range water-reducing agent (i.e. a superplasticizer), together with high intensity mixing, is required to disperse both silica fume and cement so that the dense packing can occur most efficiently.

Using this technique values of w/(c+s) < 0.20 and compressive strength > 250 MPa can be obtained routinely with castable systems[3-6]. The designations DSP (Densified with Small Particles) or RPC (Reactive Powder Concrete) have

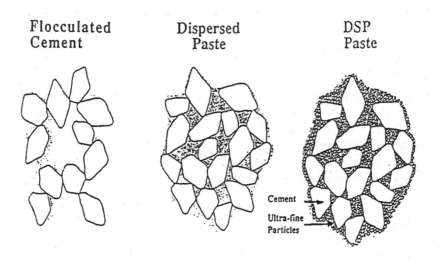

Floccculated Cement   Dispersed Paste   DSP Paste

Cement
Ultra-fine Particles

Figure 1.   Schematic representation of particle packing in a DSP cement.

been used to describe these very dense materials. However there is a smooth transition between ultra-high-strength concretes and DSP mortars.

DSP cement pastes not only have low w/(c+s) ratios, but consequently a large fraction of the cement remains unhydrated. Therefore it would be expected that these materials would show low dimensional change. However, there are reports of transverse cracking in full-depth sections of silica fume concrete in bridge decks[7]. Such cracking could be due to combined effects of chemical shrinkage, drying shrinkage and thermal contraction creating high internal stresses[8]. Low w/c pastes are known to be susceptible to chemical shrinkage and this is enhanced by the addition of silica fume[9-12].

In this paper we present measurements of chemical shrinkage, drying shrinkage and thermal expansion of DSP cement pastes. This is part of a larger study looking at the effects of microstructure on volume stability.

EXPERIMENTAL

Materials and Sample Preparation

The cement used is a white Portland cement (Lehigh Cement Co., Allentown, PA) characterized by a high C_3S content (> 70%), low C_3A content (<5%) and low alkali content (N-equiv. < 0.4%). The silica fume (Elkem Materials Inc., Pittsburgh, PA) is high in silica (> 95%) and with a surface area ~20m^2/g. The superplasticizer is a solid form of sulfonate naphthalene-formaldehyde condensate (Mighty 100) manufactured by the Kao Chemicals, Japan.

Cement and silica fume were blended in an attritor mill (Szegvari Attritor System, Union Process, Akron, OH) in an iso-propanol slurry for 20 min. The alcohol was evaporated, and the dry blend mixed with the appropriate amount of water in which the superplasticizer is dissolved. Mixing was done in a planetary mixer for 5-10 min. The resulting paste was cast into a steel mold to form a rectangular block 110 mm x 100 mm x 20 mm.

Measurements

1.   Chemical Shrinkage

The method adopted is a modification of the approach of Jensen and Hansen [12]. Corrugated plastic tubing (25.4 mm i.d) was cut into ~300 mm length. One end was sealed with tape and the tubing filled from the other end with cement paste. Air bubbles were removed using a combination of rodding and vibration. The open end was also sealed and two steel studs were imbedded at each end of the specimens. The tube was supported horizontally for 14 hours, after this time the paste had set sufficiently that it could be placed vertically in a dial gage comparator. The time between readings was initially 2 hours, extending to several hours and finally every 24 hours toward the end of the measurements.

2.   Drying Shrinkage

Thin slabs were cut from the rectangular block at the end of the desired wet curing period (8 weeks) using a diamond saw (see Figure 2). The dimensions of the slabs were 100 mm x 18 mm x 2.5 mm. The slabs were fitted with steel pins, then placed in a closed container conditioned to fixed relative humidity (rh) using saturated salt solutions (KNO$_2$ for 50% rh; LiCl for 11% rh). Samples were removed periodically to record length changes (dial comparator to 0.0025 mm) and mass change (analytical balance to 0.0001 g).

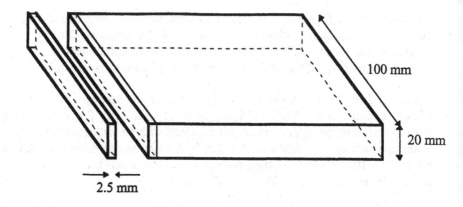

100 mm

20 mm

2.5 mm

**Figure 2   Schematic showing
how slabs are cut from a cement paste block.**

3.   Thermal Expansion/Contraction

Thin slabs were placed in a specially designed glass tube filled with lime water
which was immersed in a stirred water bath with a temperature controller (see
Figure 3). The slabs are held vertically in a centering slot that allows free
movement. A dial comparator contacts the top of the specimen and provides
continuous monitoring of the length of the slab, as the water bath temperature is
raised from 25°C to 60°C. The temperature within the tube is also monitored
continuously with a mercury thermometer. Heating is almost linear with time
(0.5°C/min.), but cooling is slow and non-linear.

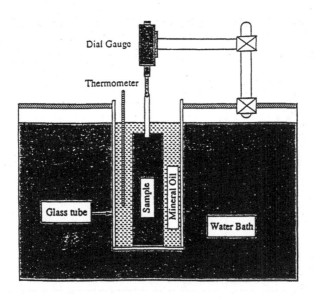

Figure 3.   Schematic diagram of the thermal expansion measurement setup.

RESULTS AND DISCUSSION

1.   <u>Chemical Shrinkage</u>

Chemical shrinkage data are presented in Figures 4 and 5. With the simple set up used measurements could not start before 14 hrs., so that some early shrinkage is not measured, since length changes begin around final set. This does not affect the comparative observations, although some systematic error is introduced. It can be seen that shrinkage increases as the w/c is lowered below 0.3. The addition of silica fume increases the chemical shrinkage to a large degree. These results are in good agreement with published data [10-12], and indicate that chemical shrinkage is of a similar magnitude to drying shrinkage (see below).

The data suggest that chemical shrinkage can be divided into two components: one associated with hydration which levels off within 200 hrs. and one associated with pozzolanic reactions of silica fume[13]. The total length changes are substantial and are in line with recent experimental observations that the internal rh of the paste can fall as low as 75% in the presence of silica fume[10,11]. Therefore subsequent drying shrinkage of such pastes must be influenced by the state of self-dessication.

Figure 4   Chemical shrinkage of pastes at different w/c ratio

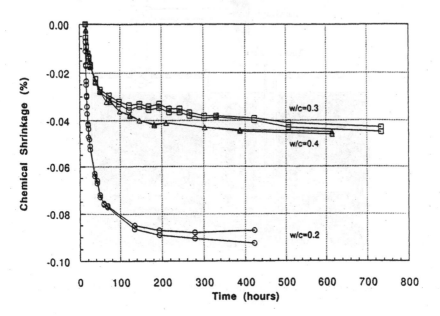

Figure 5

Effect of silica fume addition on chemical shrinkage, w/(c+s)=0.3

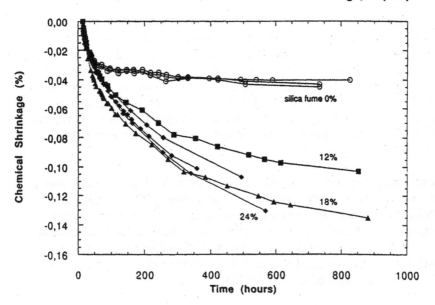

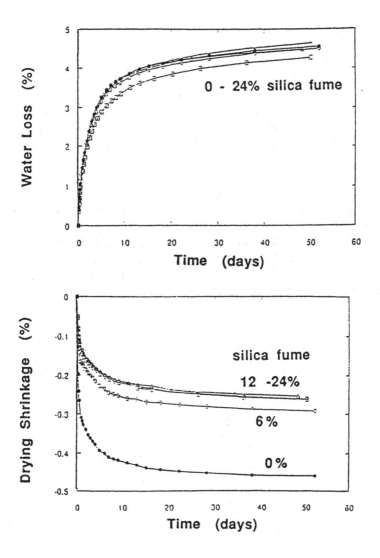

Figure 6.   Shrinkage and water loss as a function of time for w/c = 0.2 cement pastes.

2.   Drying Shrinkage

Figure 6 shows typical shrinkage and weight loss plotted against time. Note that despite the thin cross-section, shrinkage develops very slowly and equilibrium is not reached even after 50 days drying. The rate at which shrinkage develops is much slower than observed for pastes at a higher w/c or for mesoporous Vycor glass[14]. It can be noted that the addition of silica fume greatly reduces the observed shrinkage, but not the weight loss. The presence of silica fume is more important than the actual amount.

Since the addition of silica fume greatly increases the chemical shrinkage (Fig. 5) it is possible that the reduction in drying shrinkage merely reflects self-desiccation of the relatively massive block during curing. To check this hypothesis thin slabs for drying shrinkage measurements were cut from a block one day after casting and allowed to cure separately for 8 weeks. Self-dessication is unlikely to occur under these conditions. The results are shown in Figure 7 as shrinkage-weight loss curves.

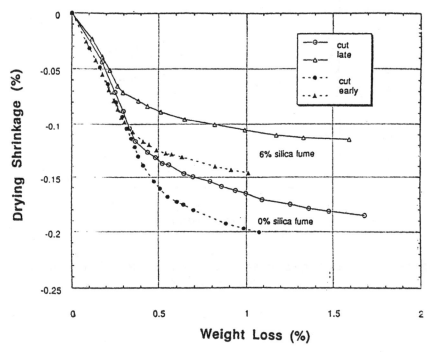

Figure 7.   Shrinkage v. weight loss for cement paste slabs dried at 11% rh. (a) sawn after 24 hours curing (early); (b) sawn after 8 weeks curing (late).

The decrease in weight loss is consistent with greater hydration in the absence of self-dessication and a reduction in porosity. The increase in drying shrinkage is consistent with an absence of chemical shrinkage, a higher volume fraction of C-S-H and a finer pore structure. Therefore, the total shrinkage of a cement paste should be considered as a combination of chemical and drying shrinkage. The two are interdependent, so that measured drying shrinkage depends on the extent of self-dessication during curing. A large proportion of the observed shrinkage is due to slow redistribution of water within the microstructure after the larger capillaries have been emptied.

3.   Thermal Expansion/Contraction

The experimental set up was calibrated using bars fabricated from Invar and 6061 Aluminum alloy. Figure 8 shows typical data for these materials from which the coefficient of thermal expansion could be calculated. Good

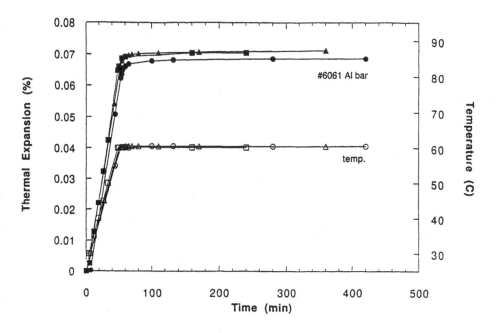

Figure 8   Calibration of thermal expansion cell
with aluminum alloy bars

Figure 9   Effect of temperature cycling on thermal response of cement pastes, w/(c+s)=0.40

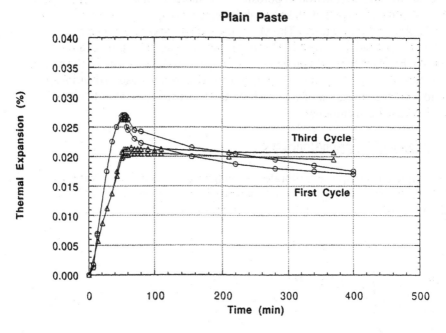

Plain Paste

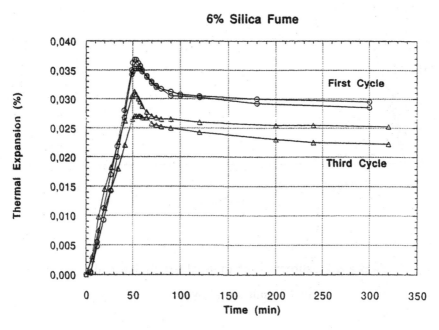

6% Silica Fume

agreement was obtained with literature values. Cement pastes showed anomalous behavior in that after initial expansion a time-dependent contraction is observed, while the temperature is kept constant at the new value (Figure 9). This phenomenon has been observed earlier[15,16]. The time dependent contraction gradually diminished after 3-4 cycles between 25 and 60°C. It also is not observed in pastes (without silica fume) cured at 80°C. When silica fume is added the time-dependent contraction is still observable after several thermal cycles (Figure 9). Increasing proportions of silica fume increases the initial expansion and the subsequent time-dependent contraction; so that the net expansion is independent of silica fume (Figure 10). Whereas the initial expansions are quite reproducible, the contractions are much more variable.

All these observations point towards the redistribution of water within the pore structure as a likely explanation. Redistribution of water in mesoporous silica gels has been studied[17]. At high rates of heating, expansion is observed as water expands within the pores; at lower rates, a contraction may be observed as water moves out of the pore system. Sabri and Illston[16] observed delayed expansion only in pastes near saturation. When pastes were conditioned at lower relative humidities (as low as 11% rh) prior to thermal measurements a delayed contraction occurred. Dry pastes did not show any delayed expansion or contraction. Under our experimental conditions mesoporous Vycor glass showed a contraction (Figure 11) which gradually diminished after three cycles. Thus, we tentatively conclude that the subsequent thermal contraction of cement pastes reflects slow redistribution of water in very fine mesopores.

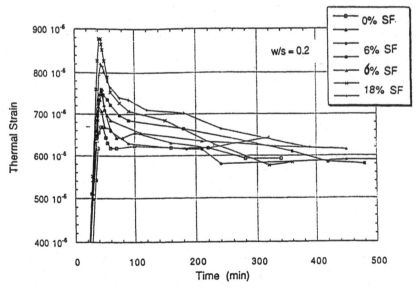

Figure 10. Effect of silica fume on the thermal response of cement pastes.

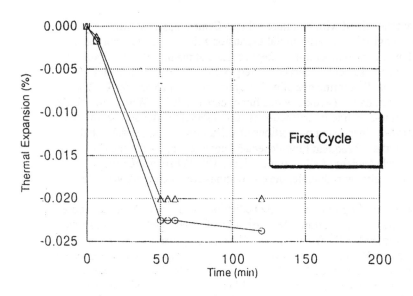

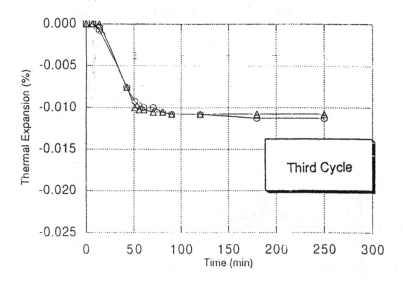

Figure 11.   Thermal expansion of Vycar glass.

CONCLUSIONS

Length changes of low w/c cement pastes with and without silica fume have been measured. The following are the principal experimental results.

1. The addition of silica fume increases the amount of chemical shrinkage observed during sealed curing. Chemical shrinkage of DSP pastes is comparable to drying shrinkage.

2. This reduction of drying shrinkage is reduced by the addition of silica fume is ascribed to the enhanced chemical shrinkage that precedes drying of the hardened paste.

3. The initial thermal expansion of a cement paste is increased by the addition of silica fume. But such additions also increases subsequent iso-thermal contraction, so that the net thermal expansion is independent of silica fume additions.

4. The anomalous thermal behavior is ascribed to the redistribution of water in fine mesopores, but further work is needed to substantiate this hypothesis.

ACKNOWLEDGMENTS

We thank Lehigh Cement Co. and Elkem Materials Inc. for gifts of cement and silica fume respectively. This study was supported by the Federal Aviation Administration through the Center of Excellence at the University of Illinois (Grant No. IDOT 95-C-001).

REFERENCES

1. V. M. Malhotra, V. S. Ramachandran, R. F. Feldman and P. C. Aitcin, Condensed Silica Fume in Concrete, (CRC Press, Boca Raton, FL., 1987) 221 pp.

2. P. C. Aitcin, High Performance Concrete, (E & FN Spon, London, U.K.) 336 pp.

3. J. F. Young, "Engineering Advanced Cement-Based Materials for New Applications" pp. 103-112 in Concrete Technology: New Trends,

Industrial Applications, A. Aguado, R. Gettu and S. P. Shah, Eds. (E & FN Spon, London, U.K., 1995).

4. H-Y Hsieh, "Packing Characteristics and Interfacial Microstructure of DSP Cement Mortars", M.S. Thesis, Univ. Illinois 1992.

5. P. Richard and M. Cheyrezy, "Composition of Reactive Powder Concrete", Cem. Concr. Res., 25 (7) 1501-1511 (1995).

6. Y. Malier and P. Richard, "High Performance Concrete" Custom Designed Concrete - A Review of the French Experience and Prospects for Future Developments" pp. 55-80 in High Strength Concrete: An International Perspective SP-167 (Amer. Concr. Inst., Farmington Hills, MI, 1996).

7. John Bugler, private communication.

8. a) O. Bjontegaard, E. J. Sellevold and T. A. Hammer, "High Performance Concrete at Early Ages: Self-generated Stresses due to Autogeneous Shrinkage and Temperature", pp. 1-7;
   b) H. Hedlund and G. Westman, "Modelling Dimensional Changes in Low Water/Cement Ratio Pastes" pp. 158-173; in Self-Dessication and It's Importance in Concrete Technology, B. Persson and G. Fagerlund, eds (Lund Univ., Lund, Sweden, 1997).

9. RILEM Commission 42-CEA, "Properties of Set Concrete at Early Ages", Mater. Struct. 84 426-   (1981).

10. B. Persson and G. Fagerlund, eds, Self-Dessication and Its Importance in Concrete Technology Rpt. TVBM-3075, 255 pp (Lund Univ., Lund, Sweden, 1997).

11. O. Mejlhede Jensen and P. Freiesleben Hansen: (a) "Autogeneous Relative Humidity Change in Silica Fume-Modified Cement Pastes", Adv. Cement Res., 7 (25) 33-38 (1995); (b) "Autogeneous Deformation and Change of the Relative Humdity in Silica Fume-Modified Cement Paste" ACI Mater. J., 93 539-543 (1996).

12. O. Mejlhede Jensen and P. Freiseleben Hansen, "A Dilatometer for Measuring Autogeneous Deformation in Hardening Portland Cement Pastes", Maer. Struct., 28 406-409 (1995).

13. P. Lu, G. K. Sun and J. F. young, "Phase Composition of DSP Cement Pastes" J. Amer. Ceram. Soc., 76 1003-7 (1993).

14. C. L. Huang, "Drying Shrinkage and Microstructure of Hydrated Cement Pastes", Ph.D. Thesis, Univ. Illinis, 1983.

15. R. A. Helmuth, "Dimensional Changes of Hardened Portland Cement Pastes Caused by Temperature Changes," Highway Res. Bd., Proc. 40 (1960).

16. S. Sabri and J. M. Illston, "Immediate and Delayed Expansion of Hardened Cement Pastes", Cem. Concr. Res., 12 (3) 199-208 (1982).

17. G. W. Scherer, H. Hdach and J. Phalippou, "Thermal Expansin of Gels: A Novel Method for Measuring Permeability", J. Non-Cryst. Solids, 130 157-170 (1991).

PARTICLE PACKING CHARACTERISTICS AND EFFECTS ON HIGH STRENGTH CEMENTS

H. Li, M. R. Silsbee and D. M. Roy
Materials Research Laboratory
The Pennsylvania State University
University Park, Pa 16801

ABSTRACT

A Type-I cement was ground and fractions of the ground materials were mixed with fractions of the as-received material. The mixture was used to manufacture MDF and DSP cements. The particle packing effects on cement strength were observed. Calculated packing density showed that a specific addition of the fine component (it varied according to the fine component particle size distribution, from about 20 to 40% addition) would lead to the maximum packing density. Both MDF and DSP cement strength developments followed the same general trends predicted by calculated packing density, but reached their maximum at a lower fine component addition compared to calculated packing density. The difference was attributed to increased surface forces and the reduction of workability. Analysis shows that the particle-packing effect would show the maximum advantage in lower W/C systems. The optimal particle packing conditions are also discussed in this paper.

INTRODUCTION

The development of high strength cementitious materials has been the subject of intense scrutiny in recent years. The most common approach has been to attempt to eliminate the larger pores in the system, as for instance in DSP cement where ultra-fine silica fume is used, resulting in a denser matrix. The effect of improved particle size distribution (PSD) has been reported based on experimental[1] and theoretical[2] efforts. Particle packing effects in cementitious systems are different from other ceramic bodies. The particle size distribution in a cement-water system is constantly changing due to the continual hydration of the cement, leading to a constant changing density. It has been demonstrated that the addition of a fine component to a coarse packed system can help eliminate larger pores and yield higher strength. However, this effect has not been optimized in reference to either optimum size or amount of the fine component.

In order to explore the above issues we have conducted a study of the effect of PSD in:

1. A macro-defect-free (MDF) cement system. The MDF system was chosen, since the cement components in this system hydrate only very slowly, and thus particle packing effect can been seen clearly; and

2. A DSP cement system. The DSP system will have correspondingly higher levels of hydration when compared to MDF cement; hence, the effect of hydration on the particle packing can be more clearly seen.

The goal of this study was to develop a basic understanding of particle packing and its effect on strength development in cementitious systems.

A finely ground fraction of the same cement as the matrix was used in our studies. The ground cement has the same chemical composition as the matrix, eliminating a variable. If a less-reactive filler[3] such as silica fume is used, the interface bonds are weaker allowing cracks and flaws to propagate more easily, resulting in lower strengths during the early stages of hydration (see Fig.1). The ground cement is a very active component, which provides more chemically bonded inter-locks, yielding higher strengths.

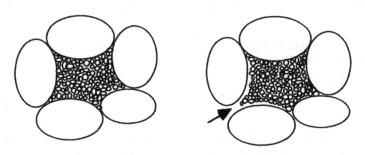

Fig 1   Weaker bonds along the interface

EXPERIMENTAL MATERIALS

The cement used was an ASTM Type-I cement. For MDF formulations polyacrylamide-acrylic-acid-sodium-salt (40% carboxyl, MW>10,000,000) was used as the polymer component with a small amount of glycerin used as a plasticizer. For the DSP, system naphthalene formaldehyde condensate Mighty 150 supplied by Boremco was used as the high-range water reducer. De-ionized water was used in all cases.

EXPERIMENTAL PROCEDURES

Fine cement preparation

The as-received Type-I cement was ground in an attrition mill for 1,2,3 and 4 hours. Alcohol was used as the coolant and dispersion agent to increase the grinding efficiency. Particle size distributions of the as-received and ground materials were measured after grinding and drying using an x-ray sedimentation technique (Sedigraph^{TM}).

Sample preparation

1. MDF cement sample preparation:

As-received Type-I cement with or without additions of ground cement was mixed in a kneader style mixer and extruded into plates. The formulation used is shown in Table 1. The polymer, water and glycerin were premixed by hand.

The cement powder components (as-received and ground materials) were also premixed by hand. The materials were then mixed for 10 minutes and passed through the extruder three times forming a cylinder. Finally, the doughy mixture was extruded as a plate. The sample was then sealed in a plastic bag for 24 hours, then unsealed and allowed to cure for the required time. The MDF samples were all cured at room temperature (25°C) and ambient humidity. Prior to strength testing the samples were cut into specimens with nominal dimensions of 3.6mm x 2.8mm x 20.3mm using a low-speed diamond saw. The resulting bars were then cured at 80°C and low humidity for 4 hours. Prior to strength testing the sample dimensions were measured.

Table I. Formulations used for MDF materials

Component	Weight percentage
cement	78.5
de-ionized water	13.0
polyacrylamide	8.0
glycerin	0.5

2. DSP cement sample preparation:

The dry as-received Type-I cement was premixed with the required amounts of the ground cements. After premixing the materials were mixed in a high-speed blender for 3 minutes with a W/C = 0.20 and 0.20 wt.% of Mighty150 by weight of cement. The DSP paste was then cast into small disc shaped specimens and vibrated for 3 minutes. The samples were cured at 23°C for 24 hours, then demolded and cured at 38°C until testing.

Prior to strength testing the DSP samples were cut into specimens with dimensions of 3.6mm x 2.8mm x 20.3mm using a low-speed diamond saw. The sample dimensions were measured.

RESULTS AND DISCUSSION

Particle Packing Characteristics

1.Particle Size Distribution: The particle size distribution of the ground cement was measured using a x-ray sedimentation technique. The recorded particle size ranged from 100 μm to 0.36 μm. The PSD's of ground cements are shown in Figure 2.

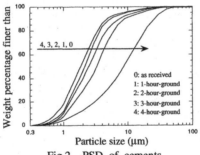

Fig.2   PSD of cements

The particle size distribution found on milling of powder can often be characterized using a Rosin-Rammler distribution:

$$R(x) = 100\exp[-(x/x')^n]$$

where $R(x)$ is the residue or percentage of particles with size larger than x, x' is the size for which 36.8% (weight) of the particles are larger, this has also been referred to as the characteristic particle size of a powder system and n is the size dispersion index. In this study x' is reduced from 11.5μm in the as-received cement to 2.2 μm after 4 hours of grinding, about 1/5 of the original value. The mean particle size X (50% particle size larger than) is also a useful parameter for characterizing a particle size distribution.

The characteristic size x' and X of the ground cements are shown in Figure 3. It is very clear that both x' and X are reduced dramatically at grinding times of up to 1 hour and then only slowly further reduced at longer grinding times.

The same tendencies are seen in the development of BET surface areas. Figure 4 shows that the surface area is increased from 1.08 m^2/gm for the as-received material to 5.57 m^2/gm for cements ground for 4 hours, about 4- fold increase. The development of x', X and specific surface area can be explained based on grinding theory[4].

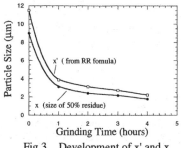
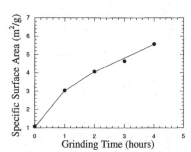

Fig 3   Development of x' and x   Fig.4 Development of Surface area

2.Packing Density in Cementitious Systems: The theoretical self-packing densities of the ground cements were calculated and are shown in figure 5. Referring to Figure 2 it can be seen that as grinding continues it is at the expense of the large size fraction. That is the range of particle sizes present is becoming smaller. In other words, the ground fractions are progressing towards a mono-sized distribution. This results in a lessening of packing efficiency.

For additions of ground materials at 5, 10, 15, 20, 25 and 30 wt.% respectively, the theoretical packing density was calculated using an approach described elsewhere[5] (Figure 6). In a 2-component system as fine material is added the packing density first increases and then decreases. This occurs as the fine particles first fill the pores with dimensions larger than the particles then begin to self-pack increasing the porosity.

For the 2-component mixtures used in this study the maximum packing density is about 81.5% of the theoretical density. In order to maintain dense packing only enough water to fill the pores can be added. Therefore the corresponding critical W/C is:

$$W/C = (0.185 \times 1)/(0.815 \times 3.1) = 0.07$$

where 1 and 3.1 are the specific gravity of water and cement respectively. If the W/C is greater than the critical amount, the excess water will separate the particles from each other and the system will expand. For a 1-component system (packing density=0.725) the critical W/C=0.12. Based on these observations it is obvious that if dense packing is desired, the W/C should be much lower than normal. However, in many cases as was found, a higher W/C is required to allow mixing.

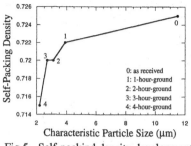

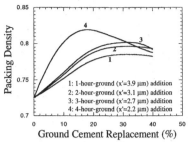

Fig.5 Self-packind density development

Fig.6 Packing density development

Theoretically, for a mono-sized close packed system, there are two kinds of pores that can be filled by smaller particles. The first type of pore is octahedrally coordinated. The number of octahedral sites is equal to that of the particles in the system. The second type of site is tetrahedrally coordinated. There are twice as many tetrahedrally coordinated pores as there are particles. When fine particles are added the maximum size particle that can be accommodated without forcing the original particles apart is 0.414R for the octahedral sites, where R is the radius of the original particles[6]. The tetrahedral sites can accommodate particles with a radius of 0.224R [6]. For example, in a mono-size system, if particles of 10 μm are close packed, only other particles smaller than about 4 μm can fit into the larger pores, and only particles smaller than about 2 μm can fit into the smaller pores. This means that particles larger than 4 μm can not fit into pores created by the 10 μm particles. There will always be a finite limit and volume fraction of fine particles available to fill the pores between larger particles. In particular as the distribution becomes more mono-sized the packing efficiency will decrease. This is the reason for the changes in packing shown in Figure 5.

The above discussion implies that in a 2-component system there will be a finite upper limit above which the addition of a second fine component will not further increase the packing density. However, if the addition of increased levels of fine particles results in a finer overall pore structure, the strength may still increase.

Effect of Packing on Bending Strength

In cementitious systems, the pores and other voids act as strength limiting flaws. It is also well known that bending strength is sensitive to flaws and cracks. Therefore, in this study 3-point bending strength was used to characterize strength development.

1. MDF cement: Strength development in MDF samples is shown in Figure 7. In all cases, the addition of a fine cement component of up to 10% resulted in a strength increase compared to as-received material. When the cement had been ground for only 1 hour (X=3.1μm, x'=3.9 μm) the strength continued to increase with up to 15 wt.% addition. Further additions of the fine component began to result in a strength decrease. However, in all cases examined here the addition of a fine component resulted in a strength increase compared to the as-received materials. Up to some level, the change in strength in the MDF system followed the same general trends as the calculated packing densities. However, as the percentage of fine component continued to increase the strength behavior began to deviate from the trends indicated by the

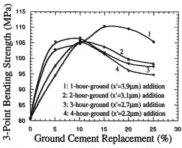

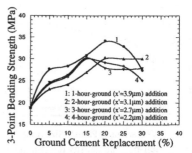

Fig. 7 MDF cement 7-day strength

Fig.8 DSP cement 7-day strength

calculated packing densities. This is likely due to the increased effect of surface forces on mixing and increased reaction rates in the system resulting in higher porosity than otherwise might be expected.

2. DSP cement : Figure 8 and Figure 9 show the strength development in the DSP systems at 7 and 28 Days. Although the maximum strengths were observed to occur at higher levels of replacement (15-20%) than in the MDF system (10%), otherwise the strength development observed in the DSP samples followed the same general trends as that observed in the MDF system.

One potential explanation for the reduced strengths at higher levels of replacement is that as the replacement increases the chemically bound water content is increased as shown in Figure 10. With limited pore space, the increase content of hydrated materials may result in crack formation[7,8].

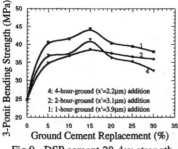

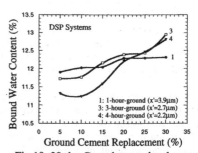

Fig.9 DSP cement 28-day strength

Fig.10 28-day Bound water development

CONCLUSIONS
1. Theoretical analysis indicated that particle packing effect would have a maximum advantage in lower W/C systems, for 1- and 2- component systems the critical W/C is about 0.12 and 0.07

respectively. In order to realize the full potential of optimum particle packing in cementitious systems, the W/C must also be optimized.

2. In a system containing a range of particle sizes, the dispersion of the sizes is important. The ratio of coarse to fine particles must be controlled to develop maximum packing. For mono-size (radius = R) distribution, the particle size of a fine component added larger than about 0.4R will not fit in the pores.

3. In a 2-component system, as the content of fine component increases, 3-point bending strength first increases then decreases. The strength improvement follows the same general trends as predicted by the calculated packing densities. However, the maximum observed strengths occur at a lower level of replacement than would be inferred by packing calculations. The deviation may be at least partially attributed to several phenomena among which are the increased roles of surface forces, loss of workability and cracking due to increased hydration.

REFERENCES

1. M.Yudenfreund, I. Odler and S. Brunauer, "Hardened Portland Cement Pastes of Low Porosity," Cem. Concr. Res. Vol.2 (3), 313 (1972).

2. W. Aiqin, Z. Chengzhi and Z. Ningsheng, "Study of The Influence Of The Particle Size Distribution on The Properties of Cement," Cem. Concr. Res. Vol.27 (5), 685 (1997).

3. S. Chatterji, N. Thaulow and P. Christensen, "Pozzolanic Activity of Byproduct Silica-Fume From Ferro-Silicon Production," Cem. Concr. Res. Vol.12 (6), 781 (1982).

4. G. C. Lowrison, Crushing and Grinding - The Size Reduction of Solid Materials, CRC Press Inc.(1974).

5. D. M. Roy, B. E. Scheetz, R. I. A. Malek, D. Shi, V. Johansen and P. J. Andersen, Concrete Components Packing Handbook SHRP-C-624, Strategic Highway Research Program, Federal Highway Administration, Washington, D. C., (1993) 161pp.

6. R. C. Evans, Introduction to Crystal Chemistry, The Cambridge University Press, (1964).

7. I. Odler, M. Yudenfreund, J. Skalny and S. Brunauer, "Hardened Portland Cement Pastes of Low Porosity," Cem. Concr. Res. Vol. 2 (4), 463 (1972).

8. N. Barowski, P. Abadzhiev and M. Markova, "Study of Pore Structure in Cement Stone with Microsilica Powder (silica fume)," Cements Research Progress, p.226, Am. Cer. Soc. (1988).

INVESTIGATION OF THE MICROSTRUCTURE OF FLY ASH-FLUORGYPSUM BINDER

Yan Peiyu[a]   **Lin** Xian[b]   **Yang** Wenyan[c]   **You** Yi[a]

a: Department of Civil Engineering, Tsinghua University; Beijing 100084, China
b: Research Institute of Engineering Technology of C.N.P.C., Tianjin 300451, China
c: State Key-Lab of Tribology, Tsinghua University; Beijing 100084, China

ABSTRACT

The influences of the curing conditions and the activator on the hydration process and mechanical properties of a binder consisting of fly ash, fluorgypsum and Portland cement were studied. After curing the pastes at 20°C in air, the dominant hydrates were gypsum, CSH gel and ettringite. They intermix well resulting in a dense paste structure. Thus high strength and good water resistance were obtained at later ages of hydration. Initial steam curing impeded the transformation of anhydrite into gypsum and accelerated the pozzolanic reaction of the fly ash. This enhanced the early strength, but did not improve the later strength increasing of the binder. Additional activator shortened the setting time and decreased the compressive strength of the binder, due to a poorer paste structure composed mainly of crystalline hydrates.

INTRODUCTION

Fly ash and chemical gypsum are two major industrial by-products. They are generated annually at quantities of over one hundred million tons in China alone. It is well known that fly ash is a pozzolanic material. It can demonstrate satisfactory hydraulic performance when intermixed with activators such as Portland cement, lime, gypsum, anhydrite, etc[1,2]. Fly ash has been successfully used as a blending materials in cement or a mineral admixture in concrete[3,4]. Gypsum has been utilized as a set regulator in the production of Portland cement or as a raw materials in the fabrication of wall elements[5,6]. However less than 50% of the fly ash and even a lower percentage of gypsum have been utilized till now in China. The remainder occupies considerable land and contaminates air and water sources, creating serious environmental problems. Therefore, there is sustained interest in establishing ways in which they may be used to manufactory valuable building materials[7].

Some composite binders and prefabricated building elements based on gypsum and fly ash were studied[8-10]. Amongst them, an excellent cementitious binder consisting of fly ash and fluorgypsum (discharged during the production of fluorite) was developed[11]. It exhibits high strength, good volume stability and superior water resistance.

In the present paper the influences of the curing conditions and activator upon the hydration process, the microstructure of the hardened paste and the mechanical properties of the binder were studied to determine the optimum conditions for using this binder.

EXPERIMENTS

The raw materials used in this study were same as those previously described[11]. According to this earlier work[11], superior hydraulic properties can be obtained when the composition of the binder has the proportions of fly ash(FA): fluorgypsum(FG): Portland cement(PC) = 55:31:16. In the present study, most of the samples were prepared in these proportions except for two in which 5% extra alum (commercial product) was added to the mixture as an activator. The designation of the samples and their composition, water:binder ratios and curing conditions are shown in Table I.

Fluorgypsum, fly ash, Portland cement and alum were blended in the proportions shown in Table I. The mixture was ground in a ball mill to a fineness similar to that of Portland cement. The pastes were mixed by machine for 3 minutes, cast into 40×40×160 mm moulds and then vibrated for 2 minutes. The samples were kept in air at 20±2°C for 24 hours. After demoulding, the samples were cured at the different conditions shown in Table I until the date of testing.

After strength determination, the sample residue was washed with alcohol and dried at 80°C for 5 hours to stop hydration. The dried sample powder was used for determination of hydrate phases by X-ray diffraction (XRD) and differential thermal analysis (DTA). The morphologies of the pastes were investigated on their fresh fracture surface using scanning electron microscopy (SEM).

Table I
The Composition, Water Binder Ratio and Curing Conditions of Studied Samples

Sample	FA:FG:PC (%)	Activator (%)	Water Binder Ratio	Curing Conditions
G	55:31:16	-	0.22	at 20°C in air
GW	55:31:16	-	0.22	at 20°C under water
GH	55:31:16	-	0.22	initially steam curing at 60°C for 6 hour, then at 20°C in air
GHW	55:31:16	-	0.22	initially steam curing at 60°C for 6 hour, then at 20°C under water
GA	55:31:16	5	0.24	at 20°C in air
GAW	55:31:16	5	0.24	at 20°C under water

RESULTS AND DISCUSSION

Identification of Hydration Products

The XRD patterns of samples G and GW are shown in Fig. 1. After adding water into the binder, the Portland cement that possesses only a small part of the binder hydrates quickly due to the dilution effect of the fly ash, whose pozzolanic reaction is not activated in the early hydration period. $Ca(OH)_2$ crystallizes and CSH gel forms in the paste. A high alkaline pore solution soon forms. This enhances the transformation of anhydrite to gypsum in large quantities. The characteristic peaks of $Ca(OH)_2$ appear in the XRD pattern of sample G hydrated for 3 days. Strong peaks of gypsum and remaining anhydrite exist, but those of clinker are very weak. With progressive hydration, more anhydrite transforms into gypsum, and some $Ca(OH)_2$ and gypsum are consumed to activate the pozzolanic reaction of the fly ash. At 28 days, the peak intensity of anhydrite and $Ca(OH)_2$ decreases, and that of gypsum increases slightly in the XRD pattern of sample G.

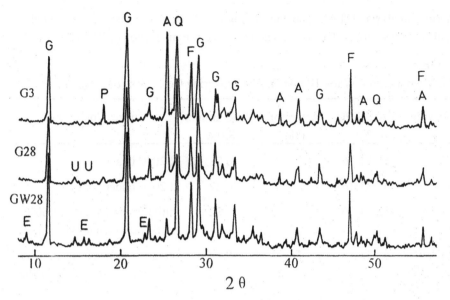

Fig. 1 XRD patterns (Cu Kα) for sample G cured at 20°C in air or under water

G: Gypsum; A: Anhydrite; P: Portlandite; E: Ettringite; Q: Quartz; F: CaF$_2$; U: Unknown phase

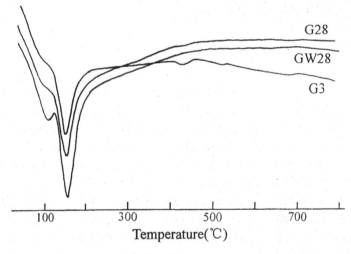

Fig. 2 DTA curves of sample G cured at 20°C in air or under water

When curing the samples under water, there is enough water supply to enhance the formation of gypsum and the pozzolanic reaction of the fly ash, which consume $Ca(OH)_2$ and gypsum to form CSH gel and ettringite. Thus in the XRD pattern of sample GW hydrated for 28 days, the $Ca(OH)_2$ peaks disappear from the XRD pattern; those of anhydrite weaken greatly and those of ettringite appear. The quartz in the fly ash and the CaF_2 in fluorgypsum remain constant. There is an unknown hydrate (U) whose characteristic peaks are situated at d=0.606nm and 0.544nm in both samples G and GW at 28 days.

The DTA curves of samples G and GW are shown in Fig.2. A quantity of gypsum, whose endothermic effect is located at about 150°C, and some $Ca(OH)_2$, whose endothermic effect is located at about 430°C, can be detected in the early hydration period. The overlapping endothermic effect of CSH gel and ettringite in the region between 100° and 110°C is very weak. At 28 days of hydration, the endothermic effects of gypsum, CSH gel and ettringite increase slightly and that of $Ca(OH)_2$ weakens in sample G; however, in sample GW the endothermic effects of CSH gel and ettringite increase significantly, that of $Ca(OH)_2$ disappears, and that of gypsum does not change. The results are similar to those of the XRD investigations.

The XRD patterns and DTA curves of the pastes initially cured in steam at 60°C for 6 hours, then in air or under water at 20°C are shown in Fig.3 and Fig.4. After initial steam curing, only a small amount of anhydrite is transformed into gypsum. Anhydrite remains in large quantities. Though the transformation of anhydrite is impeded, the hydration of Portland cement and the pozzolanic reaction of fly ash are accelerated at elevated temperature. There are no XRD characteristic peaks and DTA endotherm of $Ca(OH)_2$ at 3 days. All of the $Ca(OH)_2$ produced during the hydration of cement takes part in the pozzolanic reaction of fly ash and is consumed. The endotherm of CSH gel is more intense at 3 days in sample GH than in sample G. The endotherms of gypsum, CSH gel and ettringite, which increase with of longer times of hydration, are stronger at 28 days in sample GHW than in sample GH. There is a weak endotherm at about 350°C on the DTA curve of paste GHW at 28 day. This temperature region is considered as one of dehydroxylation of calcium aluminate hydrate[12]. The unknown hydrate shown in the XRD patterns may be a kind of calcium aluminate hydrate represented by this endotherm.

The additional alum accelerates the transformation of anhydrite and the

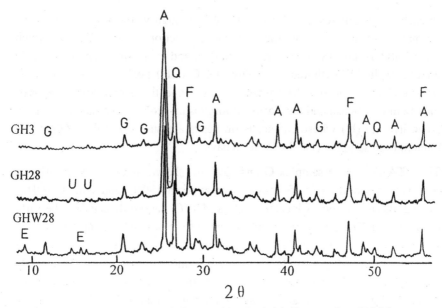

Fig.3 XRD patterns (Cu Kα) for sample GH initially cured at 60°C in steam for 6 hours, then at 20°C in air or under water

G: Gypsum; A: Anhydrite; E: Ettringite; Q: Quartz; F: CaF$_2$; U: Unknown phase

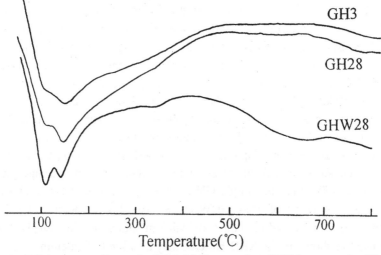

Fig. 4 DTA curves of sample GH initially cured at 60°C in steam for 6 hours, then at 20°C in air or under water

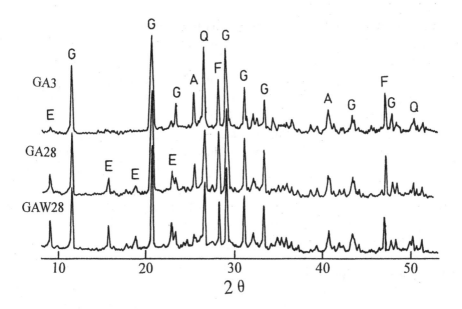

Fig. 5 XRD patterns (Cu Kα) for sample GA cured at 20°C in air or under water

G: Gypsum; A: Anhydrite; E: Ettringite; Q: Quartz; F: CaF_2;

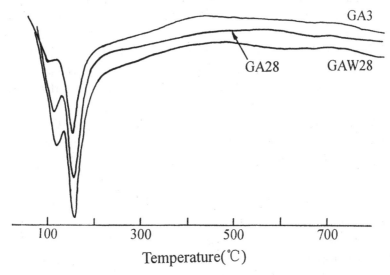

Fig. 6 DTA curves of sample GA cured at 20°C in air or under water

pozzolanic reaction of fly ash. The initial and final setting time is shortened from 8.5 hours and 13 hours of paste G to 1.25 hours and 2.6 hours of paste GA. In the XRD pattern (Fig.5) of paste GA hydrated for 3 days, the peak intensity of gypsum is much stronger than that of anhydrite. The characteristic peaks of ettringite appear, which can be detected in pastes G and GH only at later stages of hydration. The characteristic peaks of $Ca(OH)_2$ normally produced during the hydration of Portland cement do not appear. The $Ca(OH)_2$ is depleted to activate the pozzolanic reaction of fly ash yielding CSH gel and ettringite. Their endotherms in the DTA curves are also stronger than those of pastes G and GH at the same age of hydration (Fig.2,4,6). With the prolongation of hydration, the quantities of CSH gel and ettringite increase, that of anhydrite decreases, and that of gypsum changes little. There are no obvious quantitative differences of hydrates between pastes GA cured in air and under water.

Morphologies of hardened pastes

The microstructure of hardened pastes of different samples alternates. At 3 days, there is mainly amorphous hydrate, CSH gel, in the sample G (Fig.7). It is intermixed with gypsum microcrystals and anhydrite residue and precipitated amongst fly ash particles to solidify the paste. Typical large lathe-like gypsum and cubic anhydrite crystals are seldom seen in the paste. They are coagulated firmly by CSH gel. Elemental analysis with the energy dispersion spectrum indicates that the gypsum that is formed contains more than 10% of Si and Al to form a hydraulic solid solution. These guarantee that the binder has good water resistance. In sample GH (Fig.8), a thick hydrate layer has formed on the surface of the fly ash particles due to the accelerating effect of steam curing. A few anhydrite crystals are dispersed in the amorphous hydrate matrix. In paste GA (Fig.9), the dominant hydrates are lathe-like gypsum and rod-like ettringite crystals. There is much less amorphous phase in paste GA than in pastes G and GH. This results in a porous paste structure, and therefore poor mechanical properties.

At 28 days, substantial amorphous CSH gel covers the fly ash particles and bonds them firmly in paste G (Fig. 10). Small needle-like ettringite crystals are uniformly distributed in the paste. Some plate- and rod-like gypsum crystals can be found. The hardened paste is very dense. Thus it possesses excellent strength and water resistance. Paste GH exhibits a similar morphology to paste G (Fig.11), but higher degree of reaction of the fly ash and more ettringite crystals than paste

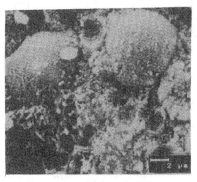

Fig. 7 Paste G cured for 3 days

Fig. 8 Paste GH cured for 3 days

Fig. 9 Crystalline hydrates in paste GA cured for 3 days

Fig. 10 Paste G cured for 28 days

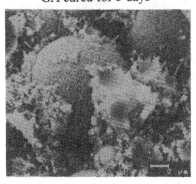

Fig. 11 Paste GH cured for 28 days

Fig. 12 Overlapped crystalline hydrates in paste GA cured for 28 days

G. On the contrary, besides rare amorphous hydrates, the major hydrates are crystalline gypsum and ettringite in the paste GA (Fig.12). These lathe- and plate-like crystals overlap each other to form a stable paste structure that possesses satisfactory strength and water resistance.

Mechanical properties of binders

Table II　　　　　　　　The strength of binders (MPa)

	Flexural Strength				Compressive Strength			
	3 d	7 d	28 d	91 d	3 d	7 d	28 d	91 d
G	3.32	3.61	7.32	7.75	17.8	25.2	52.7	79.4
GW			3.99	7.11			59.3	77.3
GH	4.08	5.07	6.47	7.28	24.3	31.9	49.2	68.9
GHW			7.21	7.35			56.3	68.1
GA	2.47	4.45	7.00	9.98	9.6	26.4	38.7	57.0
GAW			5.12	5.75			41.3	58.3

Table II shows the strengths of the binders at different ages. The early strengths of the binders are not very high because of the low content of Portland cement and the slowness of pozzolanic reaction of fly ash. The strengths of all the samples increase significantly as the hydration proceeds due to the continuous pozzolanic reaction of the fly ash. Initial steam curing increases the early strength of the binders. Sample GH, cured initially in steam and then in air, has a lower rate of strength increase than sample G cured only in air. The strength of sample G has surpassed that of sample GH at 28 days. At 91 days, the strength of sample G is 15% higher than that of sample GH. The pastes cured under water have higher compressive strengths at 28 days than those cured in air. At 91 days, the pastes cured both under water and in air have nearly the same compressive strengths. Thus this binder has excellent water resistance. Additional activator decreases the strengths of the binders at all hydration ages due to the formation of a poor paste structure, though it accelerates the hydration reaction of the binders at early ages. At 91 days, sample G had the highest compressive strength, and sample GA had the lowest. Thus 5% alum as an activator may be too high to improve the paste structure and the properties of the binders. Further studies are needed to determine the optimum type and content of the activator.

CONCLUSION

Curing conditions influence the hydration process and the mechanical properties of fly ash-fluorgypsum binders to a great degree. The major hydrates are gypsum, CSH gel and ettringite for the samples with or without activator cured at 20°C in air. Curing the ordinary binder paste at 20°C in air results in a dense paste structure, and therefore the highest later strength. Initial steam curing impedes the transformation of anhydrite into gypsum and enhances the pozzolanic reaction of fly ash. It can increase the early strength, but can't affect the later strength of the binder. Five percent of additional alum as an activator shortens the setting time and decreases the strength of the binder due to the formation of a poor paste structure. Uniform intermix of amorphous CSH gel with hydraulic gypsum solid solution and ettringite microcrystals insures that the binder has excellent water resistance.

REFERENCES

1. A. Xu and S. L. Sarkar, Microstructural Study of Gypsum Activated Fly Ash Hydration in Cement Paste, *Cem. Concr. Res.*, 21, 1137-1147(1991).
2. W. Ma and P.W. Brown, Hydrothermal Reactions of Fly Ash with $Ca(OH)_2$ and $CaSO_4 2H_2O$, *Cem. Concr. Res.*, 27(8), 1237--1248(1997).
3. V.M. Malhotra and R.T. Hemmings, Blended Cements in North America—A Review, *Cem. Concr. Compo.*, 17, 23-35 (1995).
4. S. Nagataki, Mineral Admixtures in Concrete: State of the Art and Trends, *ACI SP144*, 447-482(1994).
5. J. Lou, Fabrication of White Portland Cement with Fluorgypsum as Setting Regulator, *Cement* (Chinese), 1997(7), 12-13.
6. H. Wang and Z. Zhang, The Study on the Composite Binder Elements Consisting of Gypsum-Cement-Pozzolanic Materials, *China Cem. Concr. Prod.*, 1996(1), 57-59.
7. M. Singh and M. Garg, Durability of Cementitious Binder Derived from Industial Wastes, *Mater. Struc.*, 30(12), 607-612(1997).
8. L. Coppola, G. Belz, G. Dinelli and M. Collepardi, Prefabricated Building

Elements Based on FGD Gypsum and Ashes from Coal-fired Electric Generating Plants, *Mater. Struc.*, 29, 305-311(1996).

9. M. Singh and M. Garg, Relationship between Mechanical Properties and Porosity of Water-Resistant Gypsum Binder, *Cem. Concr. Res.*, 26(3), 449-456(1996).

10. K. Ikada and T. Tomisaka, Fundamental Studies on the Preparation and Strength of Steam-cured Porous Materials Made from Mixture of Fly Ash, Gypsum and Lime, *J. Ceram. Jap. Soc.*, 97,468-474(1989)

11. P. Yan and Y.You, Studies on the Binder of Fly Ash-Fluorgypsum-Cement, *Cem. Concr. Res.*,28(1), 135-140 (1998).

12. J.I.Bhatty, A Review of the Application of Thermal Analysis to Cement Admixture System, *Thermochimica Acta*, 189, 313(1991).

INDEX

Glassy silica, 71
Gomes, João Castro, 109
Goñi, S., 93, 455
Grattan-Bellew, P.E., 295
Guerrero, A., 93
Gypsum, 81, 409

Hanehara, Shunsuke, 25
Hansen, Will, 191
Hara, Naomichi, 71
Hazrati, K., 143
Heat curing, 25
Heavy metals, 309
Heidemann, D., 395
Hidalgo, A., 285
Homogenous paste expansion, 379
Hooton, R.D., 435
Hübert, C., 395
Hydration, 93, 109

Igusa, Takeru, 45
Impact loading, 163
Impact strength, 217
Incineration, 455
Inoue, Norihiro, 71
Interfacial transition zone, 3
Interfacial zone, 357
Ion interaction, 143
  influence of Ca^{++}, Na^+, and K^+ ions, 395

Jalali, ﬠ id, 109
Jennite, 71
Jensen, Elin, 191
Ji, Xihuang, 231
Jin, Weihua, 367
Johansen, V., 143

Kaolinite, 467
Kawamura, Mitsunori, 261, 277, 357
Kurdowski, Wieslaw, 81
Kuroi, Tokio, 179

Lactic acid, 419
Lee, Richard J., 379
Li, Guanshu, 339
Li, H., 501
Lin, Xian, 509
Luxán, M.P., 93

Macías, A., 93, 455
Maltais, Y., 143
Malyshkin, V.I., 481
Marchand, J., 143
Metakaolin, 409, 447
  high-reactivity, 447
Meyer, Christian, 367
Micro-silica, 435
Microstructure, 45, 339, 419, 509
  of hardened cement, 25
Migration, 143
Mindess, S., 163
Mineral admixture, 419
Mix design criteria, 45
Miyazawa, Shingo, 179
Modeling, 143, 367
Mohamed, Ashraf R., 191
Moranville, Micheline, 339
Mortars, 309
  expansion, 295
  high-strength, 179

Neutron diffraction, 127
Noma, Hiroaki, 71
Olek, Jan, 447
Oven curing, 25

Packing density, 246
Particle packing density, 501
Particle size distribution, 501
Pavement performance, 191
Pavlenko, S.I., 481
Pavlik, Vladimír, 309
Perꞔ, Jean, 419, 467
Permeability, 45
Pessimum, 367
Petersson, Ö., 247
Plastic fibers, 179
Pore size distribution, 3, 25
Pore solutions, 3, 285
  chemistry, 447
Pore system, 3
Porosity, 3, 81
Potentiometry, 285
Pozzolan, 467
Pozzolanic activity, 455
Pozzolanic reaction, 25

Printed and bound by CPI Group (UK) Ltd, Croydon, CR0 4YY